applied
numerical
analysis

curtis f. gerald, california state polytechnic college, san luis obispo

applied
numerical
analysis

addison–wesley publishing company
reading, massachusetts
menlo park, california · london · amsterdam · don mills, ontario · sydney

This book is in the

ADDISON–WESLEY SERIES IN MATHEMATICS

Consulting Editor :
Richard S. Varga

Second printing, November 1973

ISBN 0-201-02337-7
DEFGHIJK-MA-79876

preface

Numerical techniques of solving scientific and engineering problems are of growing importance, and the subject has become an essential part of the training of applied mathematicians, engineers, and scientists. One reason for this is that numerical methods can give the solution when ordinary analytical methods fail, as for example in finding the roots of transcendental equations, or in solving nonlinear differential equations. Numerical methods have almost unlimited breadth of application.

A second and perhaps more important reason for the lively interest in numerical procedures is their interrelation with digital computers. The price one pays for the general applicability of the numerical scheme is arithmetic complexity. Since digital computers provide a nearly effortless way (from the point of view of the user) to perform the simple but long and tedious computations involved in problem-solving by the numerical method, the advent of these marvelous servants has revived interest in what would otherwise be only a specialized field of applied mathematics. Furthermore, since computers operate essentially by repetitive arithmetic operations, the way that computers are programmed to solve scientific problems is by the use of procedures which we know as *numerical methods*. The interaction is two-fold—numerical methods require the computer to do the calculations in all but the simplest cases; computers derive their program logic from the body of knowledge covered in numerical analysis.

This text has been written for the undergraduate student of mathematics, engineering, and science. A knowledge of the calculus is assumed, and while an introductory course in differential equations is advantageous, it is not essential.

Several more important topics from the calculus course which are required in the analysis of errors and in the derivation of numerical methods, such as Taylor series expansions and the mean value theorems, are reviewed in the appendix.

The emphasis throughout the book is on applications of numerical procedures. Many of the examples and exercises solve mathematical models corresponding to real physical situations. These serve to motivate the student, but, even more important, they provide him with examples of that frequently neglected phase of training of students, the relationship between the corresponding physical and mathematical problem. Simplified physical problems are chosen for the most part because the average undergraduate student is not knowledgeable in widely different fields of technology.

One cannot properly apply numerical methods without some knowledge of their limitations. For that reason, this text is entitled *Applied Numerical Analysis* in order to imply that it goes beyond a mere recital of numerical procedures. The error analysis of a numerical method is much more difficult than its derivation or even its application, but without some treatment of errors and of the related topics of convergence and stability, a book on numerical methods provides only recipes and not understanding. This makes for a considerable dilemma in an introductory undergraduate text. An attempt has been made to bring error analysis to the level of the sophomore or junior student by avoiding formality in the statement of theorems and by less than rigorous demonstrations in many cases. In a number of instances a proof is given for a special case only, with a discussion covering the more general situation. The tone is informal throughout. The author is convinced that formal proofs repel many students, and the recitation of a set of conditions on a function that are needed to support the proof, but are given before the need for them arises, seems artificial to him. None of the proofs are given in this rigid and formalistic style but are presented as a discussion between the reader and the author. Nonetheless, the important items in theory are not neglected, although no pretense of rigor is claimed.

Because of the great interweaving of numerical analysis with computer programming, many recent authors have written books suggesting that the two topics must then be studied simultaneously, but the experience of the classroom shows this is fallacious. Normally, the student has more trouble in writing and debugging his program than he does in incorporating the numerical method. Hence, if he combines his first exposure to numerical methods with program writing, he essentially ignores the numerical analysis aspect. This author avoids this difficulty, while retaining the interrelationship, by treating these concurrently but independently. Each chapter is first presented with attention focused on the numerical methods; after these have been developed, consideration is given to the programming of the methods. Students are expected first to solve exercises with slide rules and desk calculators to establish an intimate feel for the procedure. Following this, they may be required to program them for the computer. Instructors have some options in this regard—some teachers prefer to postpone programming until the second semester entirely, returning to selected earlier topics for computer application after the student has developed some maturity in the art of numerical analysis. The author has used this plan with success. Since FORTRAN is the most widely used language for programming scientific problems, it is used in the program examples. FORTRAN II is employed because it is thought to be accessible to more students than higher levels.

There is sufficient material in this book for a full year's course, but by selecting topics, a shorter course may be taught. To provide for such flexibility, some sections are provided which will be review material for students covering the entire text, but which will introduce the fundamental ideas of certain chapters that may have been omitted. Many of these cover items in a different way so that there is little idle repetition. For example, a section covering finite difference approximations to derivatives permits study of the chapters on boundary value problems and partial differential equations independent of interpolating polynomials. A minimum treatment of numerical analysis would include material on transcendental and algebraic equations, sets of linear equations, ordinary initial value and boundary value problems, and partial differential equations. A judicious selection of topics and less than normal attention paid to error analysis may permit such a minimum treatment in three or four quarter units.

San Luis Obispo   C. F. G.
January 1970

contents

1
solution of
nonlinear
equations

In applied mathematics we frequently need to find the roots of an equation, and much of algebra is devoted to the "solution of equations." In simple situations, this consists of a rearrangement to exhibit the value of the unknown variable as a simple arithmetic combination of the constants of the equation. For second-degree polynomials, this can be expressed by the familiar *quadratic formula*. For third- and fourth-degree polynomials formulas exist but are so complex as to be rarely used; for higher-degree equations it has been proved that finding the solution through a formula is impossible. Most transcendental equations (involving trignometric or exponential functions) are likewise intractable.

Even though it is difficult if not impossible to exhibit the solution of such equations in explicit form, numerical analysis provides methods whereby a solution may be found, or at least approximated as closely as desired. Many of these numerical procedures follow a scheme which may be thought of as providing a series of successive approximations, each more precise than the previous one, so that enough repetitions of the procedure gives eventually an approximation which differs from the true value by less than some arbitrary error tolerance. Numerical procedures are thus seen to resemble the limit concept of mathematical analysis.

1. METHOD OF HALVING THE INTERVAL

The first numerical procedure that we shall study is that of *interval-halving*.* Consider the cubic

$$f(x) = x^3 + x^2 - 3x - 3 = 0.$$

At $x = 1$, f has the value -4. At $x = 2$, f has the value $+3$. Since the function is continuous, it is obvious that the change in sign of the function between $x = 1$ and $x = 2$ guarantees at least one root on the interval $(1,2)$. See Fig. 1.1.

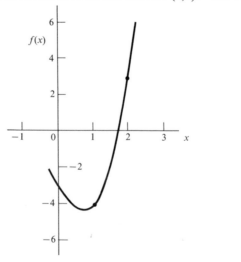

Fig. 1.1

---

* The method, also known as the *Bolzano method*, is of ancient origin.

Suppose we now evaluate the function at $x = 1.5$ and compare to the function values at $x = 1$ and $x = 2$, as in Table 1.1. Since the function changes sign between $x = 1.5$ and $x = 2$, the root lies between these values. We can obviously continue this interval-halving to determine a smaller and smaller interval within which the root must lie. For this example, continuing the process leads eventually to an approximation to the root at $x = \sqrt{3} = 1.7320508075 \ldots$.

Table 1.1. Method of Halving the Interval

x	$f(x) = x^3 + x^2 - 3x - 3$
$x_1 = 1$	-4.0
$x_2 = 2$	3.0
$x_3 = 1.5$	-1.875
$x_4 = 1.75$	$0.17187 \cdot \cdot \cdot$
$x_5 = 1.625$	$-0.94335 \cdot \cdot \cdot$
$x_6 = 1.6875$	$-0.40942 \cdot \cdot \cdot$
$x_7 = 1.71875$	$-0.12478 \cdot \cdot \cdot$
$x_8 = 1.73437 \cdot \cdot \cdot$	$0.02198 \cdot \cdot \cdot$
$\cdot$	$\cdot$
$\cdot$	$\cdot$
$\cdot$	$\cdot$
$1.73205 \cdot \cdot \cdot$	$-0.00000 \cdot \cdot \cdot$

The entries in Table 1.1 illustrate the necessity to represent values of the argument, x, as well as for the function, $f(x)$, only approximately by carrying a limited number of decimal figures. Even when one retains a certain number of significant digits, as in floating point operations on digital computers, there is some inaccuracy in our work. Note that this is true in all computations, not just in numerical methods. We shall give attention to such "round-off errors" later. The distinction between *numerical methods* and *numerical analysis* is that the latter term implies the consideration of errors in the procedure used. Certainly the blind use of any calculation method without concern for its accuracy is foolish.

Whether one rounds to the nearest fractional value or whether one chops off the extra digits will make a difference in the effect of the round-off errors. In Table 1.1, the figures have been chopped after five places, which is similar to the action of most digital computers.

Beyond the limitation on accuracy due to retaining only a limited number of figures in our work, there is an obvious limitation if we terminate the procedure itself too soon. One important advantage of the interval-halving method, beyond its simplicity, is the knowledge as to the accuracy of the current approximation to the root. Since a root must lie between the x-values where the function changes sign, the error in the last approximation can be no more than one-half the last interval of which it is the midpoint. For other methods, the accuracy determination is much more difficult.

The student will frequently have only a slide rule available as a computational tool. This limits one to about three significant figures, which is probably adequate while learning the method. Some problems should be done on a desk calculator if only to become familiar with such devices, and one's experience with numerical analysis will be incomplete unless some computer programs are written and run.

2. METHOD OF INTERPOLATION

While the interval-halving method is easy and has simple error analysis, it is not very efficient. For most functions, we can improve the rate at which we converge to the root. One such method is the method of interpolation.* Suppose we assume the function is linear over the interval (x_1,x_2), where $f(x_1)$ and $f(x_2)$ are of opposite sign. From the obvious similar triangles in Fig. 2.1 we can write†

$$\frac{x_2 - x_3}{x_2 - x_1} = \frac{f(x_2)}{f(x_2) - f(x_1)},$$

$$x_3 = x_2 - \frac{f(x_2)}{f(x_2) - f(x_1)}(x_2 - x_1).$$

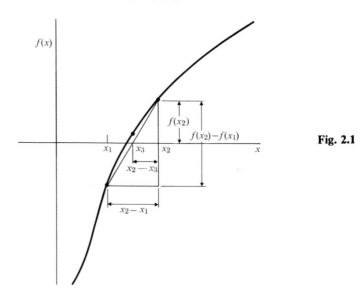

Fig. 2.1

---

*   This is also known as the method of false position, and by the Latinized version *regula falsi*. It is also a very old method.

†   Notice that since $(f(x_2) - f(x_1))/(x_2 - x_1)$ is the slope of the secant line, which approximates the slope of the function in the neighborhood of the root, the equation can be considered to be $x_3 = x_2 - f(x_2)/(\text{slope of function})$. Compare to the Newton method, in the next section.

Table 2.1. Method of Linear Interpolation

x	$f(x) = x^3 + x^2 - 3x - 3$
$x_1 = 1.0$	-4
$x_2 = 2.0$	3
$x_3 = 1.57142$	-1.36449
$x_4 = 1.70540$	-0.24784
$x_5 = 1.72788$	-0.03936
$x_6 = 1.73140$	-0.00615
$x_7 = 1.73194$	

$$x_3 = 2 - \frac{3}{7}(2 - 1) = 1.57142$$

$$x_4 = 2 - \frac{3}{4.36449}(2 - 1.57142) = 1.70540$$

$$x_5 = 2 - \frac{3}{3.24784}(2 - 1.70540)$$

$$x_6 = 2 - \frac{3}{3.03936}(2 - 1.72788)$$

We then compute $f(x_3)$ and interpolate linearly between the values at which the function changes sign, giving x_4. Repetition of this will give improving estimates of the root. Table 2.1 shows the results of this method for the same polynomial discussed in Section 1. The method appears to be somewhat faster than the method of halving the interval, giving about the same accuracy after six steps as was obtained in eight. It is intuitively obvious that the speed with which the successive approximations approach the zero of the function will depend on the degree to which the function departs from a straight line in the interval of consideration. In other words, the rate of convergence will be related to the rate of change of the slope of the curve, which is measured by the magnitude of the second derivative. We shall not pursue this further for this method.

There is one obvious way that we can further improve the method of linear interpolation. Instead of requiring that the function have opposite signs at the two values used for interpolation, we can choose the two values nearest the root (as indicated by the magnitude of the function at the various points) and interpolate or extrapolate from these. Usually the nearest values to the root will be the last two values calculated. This makes the interval under consideration shorter and hence improves the assumption that the function can be represented by the line through the two points.

Table 2.2 shows the calculations according to this scheme, which is known as the *secant method*.* Here again, the example illustrates a more rapid convergence: x_6 is more accurate than was x_7 by linear interpolation.

---

* So called because the line through two points on a curve is the secant line.

Table 2.2. Secant Method

x	$f(x) = x^3 + x^2 - 3x - 3$
$x_1 = 1.0$	-4
$x_2 = 2.0$	3
$x_3 = 1.57142$	-1.36449
$x_4 = 1.70540$	-0.24784
$x_5 = 1.73513$	0.02920
$x_6 = 1.73199$	

$$x_3 = 2 - \frac{3}{7}(2 - 1)$$

$$x_4 = 2 - \frac{3}{4.36449}(2 - 1.57142)$$

$$x_5 = 1.70540 - \frac{-0.24784}{1.11665}(1.70540 - 1.57142)$$

$$x_6 = 1.73513 - \frac{0.02920}{-0.27704}(1.73513 - 1.70540)$$

It is important not to extrapolate to a root from two points whose functional values are of the same sign when knowledge is lacking that a real root is nearby. Figure 2.2 illustrates such a fallacy of searching for a root which is not there. This is especially important in a computer program, since the successive calculated values are usually not apparent as soon as they are computed, as they are in a hand computation.

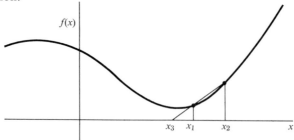

Fig. 2.2

The methods we have discussed are not limited to polynomials, of course. Table 2.3 shows the computations to find the root of the equation

$$3x + \sin x - e^x = 0.$$

The trignometric term is, of course, evaluated with the x-value in radians. The first two functional evaluations isolate a root between $x = 0$ and $x = 1$. In hand

Table 2.3. Finding Zero of Transcendental Function by Secant Method

x	$f(x) = 3x + \sin x - e^x$
$x_1 = 0$	-1.000
$x_2 = 1$	1.1232
$x_3 = 0.47$	0.3629
$x_4 = 0.35$	-0.0262
$x_5 = 0.358$	-0.0061
$x_6 = 0.3604$	

$$x_3 = 1 - \frac{1.1232}{2.1232}(1 - 0) = 0.4710, \text{ say } 0.47$$

$$x_4 = 0.47 - \frac{0.3629}{1.3629}(0.47 - 0) = 0.3449, \text{ say } 0.35$$

$$x_5 = 0.47 - \frac{0.3629}{0.3891}(0.47 - 0.35) = 0.3581, \text{ use } 0.358$$

$$x_6 = 0.358 - \frac{-0.0061}{0.0201}(0.358 - 0.35) = 0.3604$$

computation it saves effort if fewer decimal places are carried in the initial computations, adding decimal places only as the accuracy improves.

Note that all the methods we have been using require an initial estimate of the root we are computing. This is true for nearly all numerical schemes of solving equations. One normally finds such starting values by initial trial and error computations, or by making a rough graph of the function. We later discuss some methods that are self-starting for polynomials.

3. NEWTON'S METHOD

One of the most widely used methods of solving equations is *Newton's method.**
Figure 3.1 gives a graphical description. Starting from an initial estimate which is not too far from a root, x_1, we extrapolate along the tangent to its intersection with the x-axis, and take that as the next approximation. This is continued until either the successive x-values are sufficiently close, or until the value of the function is sufficiently near zero.†

---

* Newton did not publish any extensive discussion of this method, but solved a cubic polynomial in *Principia* (1686). The version given here is considerably improved over his original example.
† Which criterion should be used often depends on the particular physical problem for which the equation applies. Customarily, agreement of successive x-values to a specified tolerance is required.

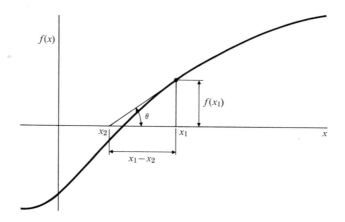

Fig. 3.1

The calculation scheme follows immediately from the right triangle shown in Fig. 3.1, which has the angle of inclination of the tangent line to the curve at $x = x_1$ as one of its acute angles:

$$\tan \theta = f'(x_1) = \frac{f(x_1)}{x_1 - x_2}, \qquad x_2 = x_1 - \frac{f(x_1)}{f'(x_1)}.$$

We continue the calculation scheme by computing

$$x_3 = x_2 - \frac{f(x_2)}{f'(x_2)},$$

or, in more general terms,

$$x_{n+1} = x_n - \frac{f(x_n)}{f'(x_n)}, \qquad n = 1, 2, 3, \ldots$$

Such a scheme which specifies the method of performing a computation is termed an *algorithm*. When we use the algorithm to compute successive approximations by continuing application of the scheme, we say we are *iterating*. The successive x-values are frequently called *iterates*.

Newton's algorithm is widely used because, at least in the near neighborhood of a root, it is more rapidly convergent than any of the methods so far discussed. We show in a later section that the method is quadratically convergent, by which we mean that the error of each step approaches proportionality to the square of the error of the previous step. The net result of this is that the number of decimal places of accuracy nearly doubles at each iteration.

In applying Newton's method to polynomials, it is most efficient to evaluate $f(x_i)$ and $f'(x_i)$ by use of synthetic division. We illustrate this by the same cubic polynomial as has been used before, $x^3 + x^2 - 3x - 3 = 0$, which has a root at $x = \sqrt{3}$. We begin with the value $x = 2$. We utilize the remainder theorem to evaluate $f(2)$, and evaluate $f'(2)$ as the remainder when the reduced polynomial (of degree two here) is divided by $(x - 2)$:

$$
\begin{array}{c|cccc}
x_1 = 2 & 1 & 1 & -3 & -3 \\
 & & 2 & 6 & 6 \\
\hline
 & 1 & 3 & 3 & 3 \leftarrow \text{remainder} = f(2) \\
 & & 2 & 10 & \\
\hline
 & 1 & 5 & 13 \leftarrow & \text{second remainder} = f'(2)
\end{array}
$$

$$x_2 = 2 - \frac{3}{13} = 1.76923 \ldots$$

$$
\begin{array}{c|cccc}
x_2 = 1.76923 & 1 & 1 & -3 & -3 \\
 & & 1.76923 & 4.89940 & 3.36048 \\
\hline
 & 1 & 2.76923 & 1.89940 & 0.36048 \\
 & & 1.76923 & 8.02957 & \\
\hline
 & 1 & 4.53846 & 9.92897 &
\end{array}
$$

$$x_3 = 1.76923 - \frac{0.36048}{9.92897} = 1.73292.$$

Similarly

$$x_4 = 1.73292 - \frac{0.00823}{9.47487} = 1.73205.$$

The value of x_4 is correct to five decimals. To observe the improvement in accuracy, consider the successive errors:

	Error	Number of correct figures
$x_1 = 2$	0.26895	1
$x_2 = 1.76923$	0.03718	2
$x_3 = 1.73292$	0.00087	4
$x_4 = 1.73205$	0.00000	6

In order to compute with five decimal places, as in this example, a desk calculator was used. (If the student has access to an electronic calculator with storage for two or more values, he will find it especially well adapted to this method. Only the underlined values need to be written down, in addition to the original coefficients.)

The initial value at which Newton's method is begun can make a considerable difference. For example, if this problem is started with $x_1 = 1$, the following values result:

x	$f(x)$	$f'(x)$
$x_1 = 1$	-4	2
$x_2 = 3$	24	30
$x_3 = 2.2$	5.888	15.92
$x_4 = 1.83015$		

From here on, the convergence is rapid, for we are using iterates similarly near the root as in the previous example.

After a first root is found, one normally proceeds to determine additional roots from the reduced polynomial (whose coefficients are in the third row of the synthetic division tableau). This makes the computations somewhat shorter. In the example, the reduced equation is a quadratic, so the quadratic formula would be used, but if a higher degree polynomial were being solved, Newton's method employing synthetic division would be employed to improve an initial estimate of a second root. The process is then repeated until the reduced equation is of second degree.

It should be observed that this procedure can be subject to unexpected errors. Since the first root is determined only approximately, the coefficients of the reduced equation are themselves not exact and the succeeding roots are subject to not only round-off errors and the errors that occur when iterations are terminated too soon, but also to inherited errors residing in the nonexact coefficients. Some polynomials are extremely sensitive in that small changes in the value of the coefficients cause large differences in the roots. One can avoid this difficulty by returning to the original polynomial, perhaps after using the above scheme to locate approximate values for all the roots. Removing roots in order of increasing magnitude is said to minimize the difficulty, and the use of double precision arithmetic will further help preserve accuracy.

It is of interest to develop the synthetic division algorithm and to establish the remainder theorems. The scheme is also the most efficient way to evaluate polynomials and their derivatives in a computer program.

Write the nth degree polynomial* as

$$P_n(x) = a_1 x^n + a_2 x^{n-1} + \ldots + a_n x + a_{n+1}.$$

We wish to divide this by the factor $(x - x_1)$, giving a reduced polynomial $Q_{n-1}(x)$ of degree $n - 1$, and a remainder, b_{n+1}, which is a constant:

$$\frac{P_n(x)}{x - x_1} = Q_{n-1}(x) - \frac{b_{n+1}}{x - x_1}.$$

Rearranging yields

$$P_n(x) = (x - x_1)Q_{n-1}(x) + b_{n+1}.$$

Note that at $x = x_1$,

$$P_n(x_1) = (0)\big(Q_{n-1}(x)\big) + b_{n+1}$$

which is the remainder theorem: The remainder on division by $(x - x_1)$ is the value of the polynomial at $x = x_1$, $P_n(x_1)$.

---

* It is well to avoid zero subscripts for the coefficients so that when a computer program is written we do not have to resubscript to avoid the forbidden zero subscript in FORTRAN.

If we differentiate the last equation, we get

$$P'_n(x) = (x - x_1)Q'_{n-1}(x) + (1)Q_{n-1}(x) + 0.$$

Letting $x = x_1$, we have

$$P'_n(x_1) = Q_{n-1}(x_1).$$

We evaluate the Q-polynomial at x_1 by a second division whose remainder equals $Q_{n-1}(x_1)$.

We now develop the synthetic division algorithm, writing $Q_{n-1}(x)$ in form similar to $P_n(x)$:

$$P_n(x) = a_1x^n + a_2x^{n-1} + \ldots + a_nx + a_{n+1}$$
$$= (x - x_1)Q_{n-1}(x) + b_{n+1}$$
$$= (x - x_1)(b_1x^{n-1} + b_2x^{n-2} + \ldots + b_{n-1}x + b_n) + b_{n+1}.$$

Multiplying out and equating coefficients of like terms in x, we get

Coef of x^n: $a_1 = b_1$ $b_1 = a_1$

      x^{n-1}: $a_2 = b_2 - b_1x_1$ $b_2 = a_2 + b_1x_1$

      x^{n-2}: $a_3 = b_3 - b_2x_1$ or $b_3 = a_3 + b_2x_1$

        •

        •

        •

      x: $a_n = b_n - b_{n-1}x_1$ $b_n = a_n + b_{n-1}x_1$

Const: $a_{n+1} = b_{n+1} - b_nx_1$ $b_{n+1} = a_{n+1} + b_nx_1$.

The general form is $b_i = a_i + b_{i-1}x_1$ by which all the b's except b_1 may be calculated. If this is compared to the synthetic divisions above, it is seen to be identical except that we now have a vertical array. The horizontal layout is easier for hand computation. For evaluation of the derivative, a set of c values is computed from the b's in the same way the b's are computed from the a's.

Synthetic division is also known as the *nested multiplication* method of evaluating polynomials. Consider the fifth-degree polynomial, evaluated at $x = x_1$:

$$a_1x_1^5 + a_2x_1^4 + a_3x_1^3 + a_4x_1^2 + a_5x_1 + a_6.$$

We can rewrite this as

$$\left(\left(\left(\left(a_1x_1 + a_2\right)x_1 + a_3\right)x_1 + a_4\right)x_1 + a_5\right)x_1 + a_6.$$

In the original form, $5 + 4 + 3 + 2 + 1 = 15$ multiplications are required, plus five additions. In the nested form, only five multiplications are required, plus five additions; it is obviously the more efficient method.

Comparing with the equations $b_2 = a_1x_1 + a_2$ and $b_i = a_i + b_{i-1}x_1$ for synthetic division, we see that the successive terms are formed in exactly the same way, so that synthetic division and nested multiplication are two names for the same thing.

Newton's method applies to nonalgebraic equations equally well. Evaluation of the function and its derivative by synthetic division is, of course, not possible and one must normally determine the derivative function analytically. It should be stressed that the equation must be put in the form $f(x) = 0$ before taking the derivative. We illustrate with the example $e^x = 3x + \sin x$, with $x_1 = 0$:

$$f(x) = 3x + \sin x \div e^x = 0,$$
$$f'(x) = 3 + \cos x - e^x,$$

$$f(0) = -1, \qquad f'(0) = 3, \qquad x_2 = 0 - \frac{-1}{3} = 0.3333,$$

$$f(0.3333) = -0.0684, \quad f'(0.3333) = 2.5493, \quad x_3 = 0.3333 - \frac{-0.0684}{2.5493} = 0.3601,$$

$$f(0.3601) = -0.000773, \quad f'(0.3601) = 2.5024, \quad x_4 = 0.3601 - \frac{-0.000773}{2.5024} = 0.3604.$$

In some cases Newton's method will not converge. Figure 3.2 illustrates this situation. Starting with x_1, one never reaches the root r. We shall develop the analytical condition for this, as well as show that the method is quadratically convergent in a later section.

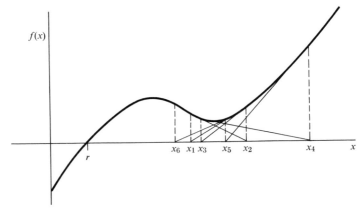

Fig. 3.2

4. METHOD OF ITERATION

We now discuss another method that is of general applicability, and which also lets us develop some necessary theory. We begin with the equation $f(x) = 0$, and rearrange it into an equivalent expression of the form

$$x = g(x), \qquad \text{such that if } f(r) = 0, \qquad r = g(r).$$

Under suitable conditions, which we develop below, the algorithm

$$x_{n+1} = g(x_n), \qquad n = 1, 2, 3, \ldots$$

will converge to a zero of $f(x)$. Consider a simple example:

$$f(x) = x^2 - 2x - 3 = 0,$$

which has obvious roots at $x = 3$, $x = -1$.

Rearranging yields:

$x = \sqrt{2x + 3}$, so $g(x) = \sqrt{2x + 3}$. Starting with $x_1 = 4$, we get

$$x_2 = \sqrt{11} = 3.316,$$
$$x_3 = \sqrt{9.632} = 3.104,$$
$$x_4 = \sqrt{9.208} = 3.034,$$
$$x_5 = \sqrt{9.068} = 3.011,$$
$$x_6 = \sqrt{9.022} = 3.004.$$

The various iterates appear to converge to $x_\infty = 3$.

The equation $f(x) = x^2 - 2x - 3 = 0$ can be rearranged in other ways also. For example, $x = \dfrac{3}{(x-2)}$ is an alternative rearrangement of form $x = g(x)$. If $x_1 = 4$,

$$x_2 = 1.5,$$
$$x_3 = -6,$$
$$x_4 = -0.375,$$
$$x_5 = -1.263,$$
$$x_6 = -0.919,$$
$$x_7 = -1.028,$$
$$x_8 = -0.991,$$
$$x_9 = -1.003.$$

Note that this converges, but to the root at $x = -1$, and that the iterates oscillate rather than converge monotonically.

Consider a third rearrangement:

$$x = \frac{x^2 - 3}{2}.$$

From $x_1 = 4$ we get

$$x_2 = 6.5,$$
$$x_3 = 19.635,$$
$$x_4 = 191.0,$$

which obviously is diverging.

We now study what conditions are needed for convergence. Our algorithm is

$$x_{n+1} = g(x_n).$$

Let $x = r$ be a solution to $f(x) = 0$, so $f(r) = 0$ and $r = g(r)$. Subtracting, and then multiplying and dividing by $(x_n - r)$, we have

$$x_{n+1} - r = g(x_n) - g(r) = \frac{g(x_n) - g(r)}{(x_n - r)}(x_n - r).$$

If $g(x)$ and $g'(x)$ are continuous on the interval from r to x_n, the mean value theorem* lets us write

$$x_{n+1} - r = g'(\xi_n)(x_n - r),$$

where ξ_n lies between x_n and r.

If we define the error of the ith iterate as $e_i = x_i - r$, and $e_{i+1} = x_{i+1} - r$, we can write

$$e_{i+1} = g'(\xi_i)e_i.$$

Taking absolute values, we get

$$|e_{i+1}| = |g'(\xi_i)| \cdot |e_i|.$$

Now suppose that $|g'(x)| \leq K < 1$ for all values of x in an interval of radius h about r. If x_1 is chosen in this interval, x_2 will also be in the interval and the algorithm will converge, since

$$|e_{n+1}| \leq K|e_n| \leq K^2|e_{n-1}| \leq \cdots \leq K^n|e_1|.$$

In summary, if $g(x)$ and $g'(x)$ are continuous on an interval about a root r of the equation $x = g(x)$, and if $|g'(x)| \leq 1$ for all x in the interval, then $x_{n+1} = g(x_n)$, $n = 1, 2, 3, \ldots$ will converge to the root $x = r$ provided that x_1 is chosen in the interval. Note this is a sufficient condition only, since for some equations convergence is secured even though not all the conditions hold.†

Since the above demonstration shows that $e_{i+1} = g'(\xi_i) \cdot e_i$, it is obvious that the rate of convergence is rapid if $|g'(x)|$ is small in the interval. If the derivative is negative, the errors alternate in sign, giving oscillatory convergence. As the iterates get closer to the root r, the values of $g'(\xi_i)$ approach the constant value $g'(r)$ because the ξ's are squeezed into smaller intervals about r. In the limit, each error becomes proportional to the previous error. For this reason, the method is sometimes called *linear iteration*.

Even though the proportionality between successive errors is true only in the limiting situation, if we assume that the errors are proportional, we can develop an acceleration technique called *Aitken acceleration*, that is often useful:

$$|e_n| = |x_n - r| = K^{n-1}e_1$$

or

$$x_n = r + K^{n-1}e_1.$$

By direct substitution it is found that

$$\frac{x_n x_{n+2} - x^2_{n+1}}{x_{n+2} - 2x_{n+1} + x_n} = \frac{(r + K^{n-1}e_1)(r + K^{n+1}e_1) - (r + K^n e_1)^2}{(r + K^{n+1}e_1) - 2(r + K^n e_1) + (r + K^{n-1}e_1)}$$

$$= \frac{r(K^{n+1} - 2K^n + K^{n-1})e_1}{(K^{n+1} - 2K^n + K^{n-1})e_1} = r.$$

---

* The Appendix reviews certain calculus principles, including this theorem.
† The analytical test that $|g'(x)| < 1$ is often awkward to apply. A constructive test is to merely observe whether the successive x_i values converge. In a computer program it is worthwhile to determine if $|x_3 - x_2| < |x_2 - x_1|$.

From three successive estimates of the root, x_1, x_2, and x_3, we can extrapolate to an improved estimate. Since the assumption of constant ratio between successive errors is false, our extrapolated value is not generally exact, but it is usually improved. One uses this technique by calculating two values beginning with x_1, extrapolating, calculating two new values, extrapolating again, etc.

For hand computation, a different form is useful to avoid the round-off problem that occurs in subtracting large numbers of nearly the same magnitude. Define*

$$\Delta x_i = x_{i+1} - x_i,$$
$$\Delta^2 x_i = \Delta(\Delta x_i) = \Delta(x_{i+1} - x_i) = x_{i+2} - 2x_{i+1} + x_i.$$

Our acceleration scheme becomes

$$r = x_n - \frac{(\Delta x_n)^2}{\Delta^2 x_n} = \frac{x_n x_{n+2} - x^2_{n+1}}{x_{n+2} - 2x_{n+1} + x_n}.$$

The differences are most readily computed in a table. We illustrate with the iterates from the first example in this section:

$$f(x) = x^2 - 2x - 3 = 0,$$
$$x_{n+1} = \sqrt{2x_n + 3}, \qquad x_1 = 4.$$

x	Δx	$\Delta^2 x$
$x_1 = 4.000$		
	0.684	
$x_2 = 3.316$		0.472
	0.212	
$x_3 = 3.104$		

The accelerated estimate is

$$r = 4.000 - \frac{(0.684)^2}{0.472} = 3.009.$$

We have jumped ahead about two iterations.

5. CONVERGENCE OF NEWTON'S METHOD

We now use the result of the previous section to show a criterion for convergence of Newton's method. The algorithm

$$x_{n+1} = x_n - \frac{f(x_n)}{f'(x_n)}, \qquad n = 1, 2, 3, \ldots$$

is of the form $x_{n+1} = g(x_n)$ for which successive iterations converge if $|g'(x)| < 1$. Since

$$g(x) = x - \frac{f(x)}{f'(x)},$$
$$g'(x) = 1 - \frac{f'(x)f'(x) - f(x)f''(x)}{(f'(x))^2} = \frac{f(x)f''(x)}{(f'(x))^2}.$$

---

* We shall deal quite extensively with such differences in a later chapter.

Hence if

$$\left| \frac{f(x)f''(x)}{(f'(x))^2} \right| < 1$$

on an interval about the root r, the method will converge for any initial value x_1 in the interval. The condition is sufficient only, and requires the usual continuity and existence of $f(x)$ and its derivatives. Note that $f'(x)$ must not be zero.

Now we show that Newton's method is quadratically convergent. Since r is a root of $f(x) = 0$, $r = g(r)$. Since $x_{n+1} = g(x_n)$, we can write

$$x_{n+1} - r = g(x_n) - g(r).$$

Let us expand $g(x_n)$ as a Taylor series* in terms of $(x_n - r)$, with the second derivative term as the remainder:

$$g(x_n) = g(r) + g'(r)(x_n - r) + \frac{g''(\xi)}{2}(x_n - r)^2,$$

where ξ lies in the interval from x_n to r.
Since

$$g'(r) = \frac{f(r)f''(r)}{(f'(r))^2} = 0$$

because $f(r) = 0$ (r is a root), we have

$$g(x_n) = g(r) + \frac{g''(\xi)}{2}(x_n - r)^2.$$

Letting $x_n - r = e_n$, we have

$$e_{n+1} = x_{n+1} - r = g(x_n) - g(r) = \frac{g''(\xi)}{2}e^2_n.$$

Each error is (in the limit) proportional to the square of the previous error, i.e., Newton's method is quadratically convergent.†

6. BAIRSTOW'S METHOD FOR QUADRATIC FACTORS

The methods so far considered are difficult to use to find a complex root of a polynomial. It is true that Newton's method works satisfactorily provided that one begins with an initial estimate which is complex-valued, but in a hand computation, performing the multiplications and divisions of complex numbers is awkward. There is no problem in a computer program if complex arithmetic capabilities exist.

For polynomials, the complex roots occur in conjugate pairs if the coefficients are all real-valued. For this case, if we extract the quadratic factors that are the

---

* See Appendix for a review of Taylor series.
† If $f'(x) = 0$ at $x = r$ (hence a multiple root), the rate of convergence will not be quadratic. For a double root, it can be shown that convergence is linear.

products of the pairs of complex roots, we can avoid complex arithmetic because such quadratic factors have real coefficients. We first develop the algorithm for synthetic division by a trial quadratic, $x^2 - rx - s$, which is hopefully near to the desired factor of the polynomial:

$$P_n(x) = a_1 x^n + a_2 x^{n-1} + \cdots + a_{n+1}$$
$$= (x^2 - rx - s)Q_{n-2}(x) + \text{remainder}$$
$$= (x^2 - rx - s)(b_1 x^{n-2} + b_2 x^{n-3} + \cdots + b_{n-2}x + b_{n-1}) + b_n(x-r) + b_{n+1}.$$

[The remainder is the linear factor, $b_n(x-r) + b_{n+1}$, written in this form to provide later simplicity. If $x^2 - rx - s$ is an exact divisor of $P_n(x)$, b_n and b_{n+1} will both be zero.] The negative signs in the factor also are for later simplifications.

On multiplying out and equating coefficients of like powers of x, we get

$$
\begin{array}{lcl}
a_1 = b_1 & & b_1 = a_1 \\
a_2 = b_2 - rb_1 & & b_2 = a_2 + rb_1 \\
a_3 = b_3 - rb_2 - sb_1 & & b_3 = a_3 + rb_2 + sb_1 \\
a_4 = b_4 - rb_3 - sb_2 & \text{or} & b_4 = a_4 + rb_3 + sb_2 \\
\quad \vdots & & \quad \vdots \\
a_n = b_n - rb_{n-1} - sb_{n-2} & & b_n = a_n + rb_{n-1} + sb_{n-2} \\
a_{n+1} = b_{n+1} - rb_n - sb_{n-1} & & b_{n+1} = a_{n+1} + rb_n + sb_{n-1}.
\end{array}
$$

(6.1)

We would like b_n and b_{n+1} to both be zero, for that would show $x^2 - rx - s$ to be a quadratic factor of the polynomial. This will normally not be so; if we properly change the values of r and s, we can make the remainder zero, or at least make its coefficients smaller. Obviously b_n and b_{n+1} are both functions of the two parameters r and s. Expanding these as a Taylor series for a function of two variables* in terms of $(r^* - r)$ and $(s^* - s)$, where $(r^* - r)$ and $(s^* - s)$ are presumed small so that terms of higher order than the first are negligible, we obtain

$$b_n(r^*,s^*) = b_n(r,s) + \frac{\partial b_n}{\partial r}(r^* - r) + \frac{\partial b_n}{\partial s}(s^* - s) + \cdots,$$

$$b_{n+1}(r^*,s^*) = b_{n+1}(r,s) + \frac{\partial b_{n+1}}{\partial r}(r^* - r) + \frac{\partial b_{n+1}}{\partial s}(s^* - s) + \cdots.$$

Let us take (r^*,s^*) as the point at which the remainder is zero, and call $r^* - r = \Delta r$, $s^* - s = \Delta s$ (Δr and Δs are increments to add to the original r and s to get the new values r^* and s^* for which the remainder is zero). Then

$$b_n(r^*,s^*) = 0 \doteq b_n + \frac{\partial b_n}{\partial r}\Delta r + \frac{\partial b_n}{\partial s}\Delta s,$$

$$b_{n+1}(r^*,s^*) = 0 \doteq b_{n+1} + \frac{\partial b_{n+1}}{\partial r}\Delta r + \frac{\partial b_{n+1}}{\partial s}\Delta s.$$

---

* The Appendix reviews this.

All the terms on the right are to be evaluated at (r,s). We wish to solve these two equations simultaneously for the unknown Δr and Δs, so we need to evaluate the partial derivatives.

Bairstow showed that the required partial derivatives can be obtained from the b's by a second synthetic division by the factor $x^2 - rs - s$ in just the same way that the b's are obtained from the a's. Define a set of c's by the relations shown below at the left, and compare these to the partial derivatives in the right columns:

$$c_1 = b_1$$

$$\frac{\partial b_1}{\partial r} = \frac{\partial a_1}{\partial r} = 0 \qquad\qquad \frac{\partial b_1}{\partial s} = \frac{\partial a_1}{\partial s} = 0$$

$$c_2 = b_2 + rc_1$$

$$\frac{\partial b_2}{\partial r} = r\frac{\partial b_1}{\partial r} + b_1 = b_1 = c_1 \qquad\qquad \frac{\partial b_2}{\partial s} = \frac{\partial a_2}{\partial s} + r\frac{\partial b_1}{\partial s} = 0$$

$$c_3 = b_3 + rc_2 + sc_1$$

$$\frac{\partial b_3}{\partial r} = r\frac{\partial b_2}{\partial r} + b_2 = c_2 \qquad\qquad \frac{\partial b_3}{\partial s} = r\frac{\partial b_2}{\partial s} + s\frac{\partial b_1}{\partial s} + b_1 = b_1 = c_1$$

$$c_4 = b_4 + rc_3 + sc_2$$

$$\frac{\partial b_4}{\partial r} = r\frac{\partial b_3}{\partial r} + b_3 + s\frac{\partial b_2}{\partial r} \qquad\qquad \frac{\partial b_4}{\partial s} = r\frac{\partial b_3}{\partial s} + s\frac{\partial b_2}{\partial s} + b_2$$

$$= b_3 + rc_2 + sc_1 = c_3 \qquad\qquad = b_2 + rc_1 = c_2$$

$$\vdots \qquad\qquad \vdots \qquad\qquad \vdots$$

$$c_n = b_n + rc_{n-1} + sc_{n-2}$$

$$\frac{\partial b_n}{\partial r} = r\frac{\partial b_{n-1}}{\partial r} + b_{n-1} + s\frac{\partial b_{n-2}}{\partial r} \qquad\qquad \frac{\partial b_n}{\partial s} = r\frac{\partial b_{n-1}}{\partial s} + s\frac{\partial b_{n-2}}{\partial s} + b_{n-2}$$

$$= b_{n-1} + rc_{n-2} + sc_{n-3} = c_{n-1} \qquad\qquad = b_{n-2} + rc_{n-3} + sc_{n-4} = c_{n-2}$$

Hence the partial derivatives that we need are equal to the properly corresponding c's. Our simultaneous equations become

$$-b_n = c_{n-1}\Delta r + c_{n-2}\Delta s,$$
$$-b_{n+1} = c_n\Delta r + c_{n-1}\Delta s.$$

By Cramer's rule

$$\Delta r = \frac{\begin{vmatrix} -b_n & c_{n-2} \\ -b_{n+1} & c_{n-1} \end{vmatrix}}{\begin{vmatrix} c_{n-1} & c_{n-2} \\ c_n & c_{n-1} \end{vmatrix}}$$

$$\Delta s = \frac{\begin{vmatrix} c_{n-1} & -b_n \\ c_n & -b_{n+1} \end{vmatrix}}{\begin{vmatrix} c_{n-1} & c_{n-2} \\ c_n & c_{n-1} \end{vmatrix}}.$$

Example. Find the quadratic factors of

$$x^4 - 1.1x^3 + 2.3x^2 + 0.5x + 3.3 = 0.$$

Use $x^2 + x + 1$ as starting factor $(r = -1, s = -1)$ (Frequently $r = s = 0$ are used as starting values if no information as to an approximate factor is known.)

Equations (6.1) lead to a double synthetic division scheme as follows:

	a_1	a_2	a_3	a_4	a_5
$r = -1$	1	-1.1	2.3	0.5	3.3
		-1.0	2.1	-3.4	0.8
$s = -1$		—	-1.0	2.1	3.4
	1	-2.1	-3.4	-0.8	0.7
		-1.0	3.1	-5.5	
		—	-1.0	3.1	b_{n+1}
	1	-3.1	5.5	-3.2	b_n
		c_{n-2}	c_{n-1}	c_n	

Note that the equations for b_2 and c_2 have no term involving s.   The dashes in the above tableau represent these missing factors.   Then

$$\Delta r = \frac{\begin{vmatrix} 0.8 & -3.1 \\ -0.7 & 5.5 \end{vmatrix}}{\begin{vmatrix} 5.5 & -3.1 \\ -3.2 & 5.5 \end{vmatrix}} = \frac{2.23}{20.33} = 0.11, \qquad r^* = -1 + 0.11 = -0.89,$$

$$\Delta s = \frac{\begin{vmatrix} 5.5 & 0.8 \\ -3.2 & -0.7 \end{vmatrix}}{20.33} = \frac{-1.29}{20.33} = -0.06, \qquad s^* = -1 - 0.06 = -1.06.$$

The second trial yields

	1	-1.1	2.3	0.5	3.3
-0.89		-0.89	1.77	-2.68	0.06
-1.06		—	-1.06	2.11	-3.17
	1	-1.99	3.01	-0.07	0.17
		-0.89	2.56	-4.01	
		—	-1.06	3.05	
	1	-2.88	4.51	-1.03	

$$\Delta r = \frac{\begin{vmatrix} 0.07 & -2.88 \\ -0.17 & 4.51 \end{vmatrix}}{\begin{vmatrix} 4.51 & -2.88 \\ -1.03 & 4.51 \end{vmatrix}} = \frac{-0.175}{17.374} = -0.010, \qquad r^* = -0.89 - 0.010 = -0.900,$$

$$\Delta s = \frac{\begin{vmatrix} 4.51 & 0.07 \\ -1.03 & -0.17 \end{vmatrix}}{17.374} = \frac{-0.694}{17.374} = -0.040, \qquad s^* = -1.06 - 0.040 = -1.100.$$

The exact factors are $(x^2 + 0.9x + 1.1)(x^2 - 2x + 3)$.

7. QUOTIENT-DIFFERENCE ALGORITHM

All the methods so far discussed require a starting value sufficiently near to the root or to the quadratic factor being sought. A relatively efficient method to determine all the roots of a polynomial without starting values is the Q-D or quotient-difference algorithm. We present the method without elaboration.*

For the nth degree polynomial

$$P_n(x) = a_1 x^n + a_2 x^{n-1} + \cdots + a_n x + a_{n+1},$$

we form an array of q and e terms, starting the tableau by calculating a first row of q's and a second row of e's:

$$q^{(1)} = -a_2/a_1, \qquad \text{all other } q\text{'s are zero,}$$
$$e^{(i)} = a_{i+2}/a_{i+1}, \qquad i = 1, 2, \cdots, n-1,$$
$$e^{(0)} = e^{(n)} = 0.$$

The start of the array is

$e^{(0)}$	$q^{(1)}$	$e^{(1)}$	$q^{(2)}$	$e^{(2)}$	$q^{(3)} \ldots e^{(n-1)}$	$q^{(n)}$	$e^{(n)}$
	$\dfrac{-a_2}{a_1}$		0		$0 \ldots$	0	
0		$\dfrac{a_3}{a_2}$		$\dfrac{a_4}{a_3}$	$\ldots \dfrac{a_{n+1}}{a_n}$		0

A new row of q's is computed by the equation

$$\text{new } q^{(i)} = e^{(i)} - e^{(i-1)} + q^{(i)},$$

using terms from the e and q rows just above. Note that this algorithm is "e to right minus e to left plus q above." A new row of e's is now computed by the equation

$$\text{new } e^{(i)} = (q^{(i+1)}/q^{(i)})e^{(i)};$$

"q to right over q to left times e above." The example in Table 7.1, isolates the roots of the quartic

$$P_4(x) = 128x^4 - 256x^3 + 160x^2 - 32x + 1$$

by continuing to compute rows of q's and then e's until all the e-values approach zero. When this occurs, the q-values assume the values of the roots. Since the method is slow to converge, it is generally used only to get approximate values which are then improved by Newton's method.

If the polynomial has a pair of conjugate complex roots, one of the e's will not approach zero but will fluctuate in value. The sum of the two q-values on either side of this e will approach r and the product of the q above and to the left times the q below and to the right approaches $-s$ in the factor $x^2 - rx - s$. Two equal roots behave similarly.

---

* Henrici (1964) discusses the method in some detail.

Table 7.1. Example of Q-D Method

$$P(x) = 128x^4 - 256x^3 + 160x^3 - 32x + 1$$

(handwritten column labels above printed headers: $E(1)$, $Q(1)$, $E(2)$, $Q(2)$, $E(3)$, $Q(3)$, $E(4)$, $Q(4)$, $E(5)$)

$e^{(0)}$	$q^{(1)}$	$e^{(1)}$	$q^{(2)}$	$e^{(2)}$	$q^{(3)}$	$e^{(3)}$	$q^{(4)}$	$e^{(4)}$
	2.000		0		0		0	
0		−0.625		−0.200		−0.031		0
	1.375		0.425		0.169		0.031	
0		−0.193		−0.079		−0.006		0
	1.182		0.539		0.242		0.037	
0		−0.088		−0.036		−0.001		0
	1.094		0.591		0.277		0.038	
0		−0.048		−0.017		−0.000		0
	1.046		0.622		0.294		0.038	
0		−0.028		−0.008		−0.000		0
	1.018		0.642		0.302		0.038	
0		−0.018		−0.004		−0.000		0
	1.000		0.656		0.304		0.038	
0		−0.012		−0.002		−0.000		0
	0.988		0.666		0.306		0.038	
0		−0.008		−0.001		−0.000		0
	0.980		0.673		0.307		0.038	
0		−0.005		−0.001		−0.000		0
	0.975		0.677		0.308		0.038	

True values of roots are 0.96194, 0.69134, 0.30866, and 0.03806.

Table 7.2 shows the result of the method for the polynomial

$$(x - 1)(x - 4)(x^2 - x + 3) = x^4 - 6x^3 + 12x^2 - 19x + 12.$$

Factors are $(x - 4)(x - 1)(x^2 - x + 3)$.

$q^{(1)}$ converging to 4. $q^{(4)}$ converging to 1.

Since $e^{(2)}$ does not approach zero, $q^{(2)}$ and $q^{(3)}$ represent a quadratic factor

$$r \doteq q^{(2)} + q^{(3)} = 1.456 - 0.466 = 0.990.$$
$$s \doteq - (-6.426)(-0.466) = -2.995.$$

Factor is $x^2 - rx - s = x^2 - 0.990x - (-2.995)$.

Note that one cannot compute the first q and e rows if one of the coefficients in the polynomial is zero, for division by zero is undefined.   In such a case, we change the variable to $y = x - 1$.   (Subtracting 1 from the roots of the equation is an arbitrary choice, but this facilitates the reverse change of variable to get the roots of the original equation after the roots of the new equation in y have been found.)

For example, if $f(x) = x^4 - 2x^2 + x - 1 = 0$, we let $y = x - 1$ and use repeated synthetic division to determine the coefficients of $f(y) = 0$.   The successive

Table 7.2. Q-D Method with Complex Roots

$$P(x) = x^4 - 6x^3 + 12x^2 - 19x + 12$$

$e^{(0)}$	$q^{(1)}$	$e^{(1)}$	$q^{(2)}$	$e^{(2)}$	$q^{(3)}$	$e^{(3)}$	$q^{(4)}$	$e^{(4)}$
	6.000		0		0		0	
0		−2.000		−1.583		−0.632		0
	4.000		0.417		0.951		0.632	
0		−0.208		−3.610		−0.420		0
	3.792		−2.985		4.141		1.052	
0		0.164		5.008		−0.107		0
	3.956		1.859		−0.974		1.159	
0		0.077		−2.624		0.127		0
	4.033		−0.842		1.777		1.032	
0		−0.016		5.538		0.074		0
	4.017		4.712		−3.687		0.958	
0		−0.019		−4.333		−0.019		0
	3.998		0.398		0.627		0.977	
0		−0.002		−6.826		−0.030		0
	4.000		−6.426		7.423		1.007	
0		0.003		7.885		−0.004		0
	4.003		1.456		−0.466		1.010	

remainders on dividing by $x - 1$ are the coefficients of $f(y)$:

```
1     0    −2     1    −1    | 1
      1     1    −1     0
─────────────────────────
1     1    −1     0    (−1)
      1     2     1
─────────────────────
1     2     1    (1)
      1     3
───────────────
1     3    (4)
      1
─────────
1    (4)
(1)
```

$$f(y) = y^4 + 4y^3 + 4y^2 + y - 1.$$

We proceed to find the roots of $f(y) = 0$, and then get the roots of $f(x) = 0$ by adding 1.

8. OTHER METHODS

We mention two other methods that are sometimes used. *Graeffe's method* finds values for all the roots of a polynomial directly from its coefficients without requiring starting values. It is based on the fact that if the roots are all different and widely separated, then for the polynomial

$$P_n(x) = a_1x^n + a_2x^{n-1} + \cdots + a_nx + a_{n+1},$$

the roots are given by

$$r_1 \doteq -\frac{a_2}{a_1}, \qquad r_2 \doteq -\frac{a_3}{a_2}, \qquad \cdots, \qquad r_n \doteq -\frac{a_{n+1}}{a_n}. \tag{8.1}*$$

In order to separate the roots of the given polynomial, it is converted to another polynomial whose roots are the negative squares of the original roots. After enough repetitions of the root-squaring operation, the relations of (8.1) give the values of the roots, provided no multiple roots occur.† The method is laborious and is used less often than the Q-D algorithm.

Muller's method is an interpolation method that used quadratic interpolation rather than linear as in Section 4. A second-degree polynomial is made to fit three points near a root, $(x_1, f(x_1))$, $(x_2, f(x_2))$, $(x_3, f(x_3))$, and the proper zero of this quadratic, using the quadratic formula, is used as the improved estimate of the root. The process is then repeated using the set of three points nearest the root being evaluated. The technique is not restricted to polynomial equations. The original reference [Muller (1956)] should be consulted for details.

9. PROGRAMMING OF METHODS TO SOLVE EQUATIONS

Typical of the ways that FORTRAN programs may be written to solve for roots of equations are the accompanying computer printouts, illustrating the use of the half-interval search and Bairstow methods.

In the program for half-interval search (Program 1), an arithmetic statement function is used to define the problem in the form $f(x) = 0$. To use the program for a different problem, a new arithmetic statement function would be substituted for this card. We use the device, in testing if $f(a)$ and $f(b)$ are of opposite sign, that their product must be negative. The example is illustrated with the function $e^{-x} - \sin(\pi x/2) = 0$, and finds the root near $x = 0.5$. $x_1 = 0$ and $x_2 = 1$ are input values. The computation is ended when $f(x_n) < 10^{-5}$.

In this program, as well as all of the others in this text, the input data cards are listed immediately after the END statement.

The Bairstow method program (Program 2) is written to find all the quadratic factors of a polynomial of degree six or less. It does this by continuing to find the factors of the reduced equation until it has been reduced to degree one or two.

---

* We use the notation $r_1 \doteq -\dfrac{a_2}{a_1}$ to indicate approximate equality of the two quantities.

† Scarborough (1950) discusses the method, including its application in cases of multiple and complex roots.

Program 1

```
ZZJOB 5                        CSC  001    GERALD, C. F.
ZZFORX5
*LIST PRINTER
C  SOLUTION TO F(X) = 0 BY INTERVAL HALVING METHOD.
C  CHANGE FUNCTION DEFINITION CARD TO SOLVE NEW EQUATION.
       F(X) = EXPF(-X) - SINF(3.1415926*X/2.)
C  READ IN X1, X2 VALUES, TOL.
       READ 100, X1, X2, TOL
  100 FORMAT (2F10.0, E10.0)
C  COMPUTE F(X1), F(X2) AND TEST IF OPPOSITE SIGNS OR ZERO.
       FX1 = F(X1)
       FX2 = F(X2)
       IF (FX1*FX2) 1, 2, 3
    3 TYPE 200
  200 FORMAT (80H FUNCTION VALUES WITH INITIAL X VALUES ARE NOT OF OPPOS
     1ITE SIGNS. CHANGE INPUTS. )
       CALL EXIT
C  WHEN PRODUCT OF FX1*FX2 IS ZERO, X1 OR X2 IS A ROOT.
    2 IF (FX1) 4, 5, 4
    5 PRINT 201, X1
  201 FORMAT (22H THE INPUT VALUE, X = ,F6.3, 27H IS A ROOT OF THE EQUAT
     1ION. )
    4 IF (FX2) 6, 7, 6
    7 PRINT 201, X2
       CALL EXIT
    6 PRINT 202
  202 FORMAT (55H ERROR. FX1 AND FX2 NOT ZERO THOUGH PRODUCT TESTS ZERO.
     1 )
       CALL EXIT
C  BEGIN INTERVAL HALVING PROCESS. LIMIT TO 25 TIMES.
    1 DO 30 I = 1, 25
       X = (X1 + X2)/2.
       FX = F(X)
       PRINT 206, X, FX
  206 FORMAT (8H AT X = , F10.7, 8H F(X) = ,F10.6)
       IF (ABSF(FX) - TOL) 8, 8, 9
C  TEST WHETHER F(X) AND F(X1) ARE OF OPPOSITE SIGNS.
C     RESET VALUES ACCORDINGLY.
    9 IF (FX*FX1) 10, 11, 12
   10 X2 = X
       GO TO 30
   12 X1 = X
       FX1 = FX
   30 CONTINUE
       PRINT 204
  204 FORMAT (33H NON-CONVERGENT IN 25 ITERATIONS. )
       CALL EXIT
   11 PRINT 205
  205 FORMAT (7H ERROR. )
       CALL EXIT
    8 PRINT 203, X, I
  203 FORMAT (16H THE VALUE, X = ,F10.7, 12H IS A ROOT. , I3, 26H ITERAT
     1IONS WERE REQUIRED. )
       CALL EXIT
       END
0.        1.                  1.E-5
ZZZZ
AT X =    .5000000 F(X) =    -.100576
AT X =    .2500000 F(X) =     .396117
AT X =    .3750000 F(X) =     .131719
AT X =    .4375000 F(X) =     .011255
AT X =    .4687500 F(X) =    -.045774
AT X =    .4531250 F(X) =    -.017534
AT X =    .4453125 F(X) =    -.003207
AT X =    .4414062 F(X) =     .004006
AT X =    .4433593 F(X) =     .000395
AT X =    .4443359 F(X) =    -.001407
AT X =    .4438476 F(X) =    -.000506
AT X =    .4436035 F(X) =    -.000055
AT X =    .4434814 F(X) =     .000170
AT X =    .4435424 F(X) =     .000057
AT X =    .4435729 F(X) =     .000001
THE VALUE, X =    .4435729 IS A ROOT.  15 ITERATIONS WERE REQUIRED.
```

Program 2

```
ZZJOB 5                        CSC  001      GERALD, C. F.
ZZFORX5
*LIST PRINTER
C   QUADRATIC FACTORS OF POLYNOMIALS TO SIXTH DEGREE, BAIRSTOW METHOD.
C   READ IN COEFFICIENTS ON FIRST CARD, COEFFICIENT OF X**6 FIRST.
C   READ IN INITIAL VALUES OF QUADRATIC FACTOR COEFFICIENTS, TEST VALUE, MAXIMUM
C      NUMBER OF ITERATIONS AND DEGREE OF POLYNOMIAL ON SECOND CARD.
        DIMENSION A(9), B(9), C(9)
        PRINT 1000
 1000 FORMAT (1H1,29H DEMONSTRATION PROGRAM OUTPUT)
        READ 1, (A(I), I = 3,9)
      1 FORMAT (7F10.0)
        READ 2, R1, S1, TEST, LIM, N
      2 FORMAT (2F10.0, E10.0, 2I3)
C   PUNCH POLYNOMIAL AND HEADINGS.
        PRINT 3
      3 FORMAT (//25H THE ORIGINAL POLYNOMIAL- /)
        PRINT 4
      4 FORMAT (5X, 10HPOWER OF X , 5X, 11HCOEFFICIENT /)
        J = 9 - N
        DO 5 I = J,9
        M = 9 - I
      5 PRINT 6, M, A(I)
      6 FORMAT (10X, I2, 8X, F10.5)
        PRINT 7
      7 FORMAT (//27H THE QUADRATIC FACTORS ARE- /)
C   COMPUTE B AND C ARRAYS.
        B(1) = 0.
        B(2) = 0.
        C(1) = 0.
        C(2) = 0.
        R = R1
        S = S1
     11 KOUNT = 1
     16 DO 8 J = 3,9
        B(J) = A(J) + R*B(J - 1) + S*B(J - 2)
      8 C(J) = B(J) + R*C(J - 1) + S*C(J - 2)
C   COMPUTE DENOMINATOR AND CHECK IF ZERO.
        DENOM = C(7)*C(7) - C(8)*C(6)
        IF ( DENOM) 9, 10, 9
     10 R1 = R1 + 1.
        S1 = S1 + 1.
        GO TO 11
C   COMPUTE DELTA R, DELTA S AND UPDATE R AND S.
      9 DELR = (-B(8)*C(7) + C(6)*B(9))/DENOM
        DELS = (-C(7)*B(9) + B(8)*C(8))/DENOM
        R = DELR + R
        S = DELS + S
C   CHECK IF COMPUTATION COMPLETE BY MEETING TOLERANCE.
        IF ( ABSF(DELR) + ABSF(DELS) - TEST) 12, 12, 13
     13 IF (KOUNT - LIM) 14, 15, 15
     15 PRINT 24, LIM
     24 FORMAT (25H DOES NOT CONVERGE AFTER , I4, 11H ITERATIONS   )
        CALL EXIT
     14 KOUNT = KOUNT + 1
        GO TO 16
C   PRINT OUT FACTOR JUST CALCULATED.
     12 PRINT 17, R, S
     17 FORMAT ( 8H X**2 + , F10.5, 5H X + , F10.5)
C   SET COEFFICIENTS OF REDUCED EQUATION INTO A ARRAY AND REPEAT BAIRSTOW IF
C      DEGREE IS GREATER THAN TWO.
        N = N - 2
        IF (N - 2) 18, 19, 20
     20 DO 21 K = 3,9
     21 A(K) = B(K - 2)
        GO TO 11
C   OUTPUT FINAL FACTOR.
     19 PRINT 22, B(5), B(6), B(7)
     22 FORMAT (1H ,F10.5, 7HX**2 + , F10.5, 4HX + , F10.5)
        CALL EXIT
     18 PRINT 23, B(6), B(7)
     23 FORMAT (1H ,F10.5, 4HX + , F10.5)
        CALL EXIT
        END
```

(cont.)

```
0.          1.          -17.8       99.41       -261.218  352.611    -134.106
0.          0.                      1.E-5  20   5
ZZZZ
```

THE ORIGINAL POLYNOMIAL-

POWER OF X	COEFFICIENT
5	1.00000
4	-17.80000
3	99.41000
2	-261.21800
1	352.61100
0	-134.10600

THE QUADRATIC FACTORS ARE-

```
X**2 +     4.19999 X +    -2.09999
X**2 +     3.29999 X +    -6.20000
    1.00000X +   -10.30000
```

Program 2, cont.

Subscripts are advanced by 2 so first and second b and c values can be computed by the same equation as the rest of them by defining the first two as zero. Original values of r and s may be input to the program; after the first factor is found, the r and s coefficients of that factor are used as starting values for the next factor. If a zero denominator should occur, the r and s values are arbitrarily incremented by one to avoid this. The number of repetitions is limited to not more than LIM times by testing a counter. The iterations are terminated when the sum of the changes to r and s is less or equal to the value of TEST.

The operation of the program is illustrated with the fifth-degree polynomial $x^5 - 17.8x^4 + 99.41x^3 - 261.218x^2 + 352.611x - 134.106 = 0$, which has the exact factors

$$x^2 - 4.2x + 2.1,$$
$$x^2 - 3.3x + 6.2,$$
$$x - 10.3.$$

PROBLEMS

Section 1

1. The equation $x^2 - 2 = 0$ has the obvious roots $\pm\sqrt{2} = \pm 1.414214$. Use the method of halving the interval to evaluate the positive root beginning with the interval [1,2]. How many iterations are required to evaluate the root correct to four decimal places? After this many iterations, what is the bound to the error as measured by one-half the last interval?

2. The equation $x^3 - x + 1 = 0$ has only one real root. Find the real root by interval-halving.

3. The quadratic $(x - 0.4)(x - 0.6) = x^2 - x + 0.24$ has zeros at $x = 0.4$ and $x = 0.6$, of course. Observe that the end points of the interval [0,1] are not satisfactory to begin the interval-halving method. Graph the function, and from this deduce the boundaries of intervals which will converge to each of the zeros. If the end

points of the interval [0.5, 1.0] are used to begin the search, what is a bound to the error after five iterations? What is the actual error after five repetitions of interval-halving?

4. Interval-halving applies to any continuous function, not just to polynomials. Find where the graphs of $y = 3x$ and $y = e^x$ intersect by first finding the root of $e^x - 3x = 0$ correct to four decimals.

5. Use interval-halving to find the smallest positive root of:

(a) $\tan x - x - 1 = 0$ (b) $x^3 - x^2 - 2x + 1 = 0$
(c) $2e^{-x} - \sin x = 0$ (d) $3x^3 + 4x^2 - 8x - 1 = 0$

Section 2

6. The polynomial $x^3 + x^2 - 3x - 3 = 0$, used as an example in Sections 1 and 2 where the root at $x = \sqrt{3}$ was approximated, has its other roots at $x = -1$ and $x = -\sqrt{3}$. Beginning with two suitable values that bracket the value $-\sqrt{3}$, show that the method of linear interpolation converges to that root.

7. In Problem 6, if one tried as starting values $x = -1.5$ and $x = -1.7$, the function does not change sign, and, hence, they do not qualify for beginning the method of linear interpolation. However, the secant method can begin with these values. Use them to begin the secant method. How many iterations are needed to estimate the root correct to four decimals?

8. Find where the cubic $y = x^3 - x + 1$ intersects the parabola $y = 2x^2$. Make a sketch of the two curves to locate the intersections, and then use linear interpolation and/or the secant method to evaluate the x-values of the points of intersection.

9. Use the method of linear interpolation to solve the equations in Problem 5.

10. Use the secant method to find the root near $x = -0.5$ of $e^x - 3x^2 = 0$.

Section 3

11. Solve Problem 10 using Newton's method.

12. The equation $e^x - 3x^2 = 0$ not only has a root near $x = -0.5$, but also near $x = 4.0$. Find the positive root by Newton's method.

13. Use Newton's method to solve the equations in Problem 5. Use synthetic division for evaluating the polynomials and their derivatives.

14. Use Newton's method on the equation $x^2 = N$ to derive the algorithm for the square root of N:

$$x_{i+1} = \frac{1}{2}\left(x_i + \frac{N}{x_i}\right),$$

where x_0 is an initial approximation to $\sqrt{N}$.

15. (a) If the algorithm of Problem 14 is applied twice, show that

$$\sqrt{N} \doteq \frac{A + B}{4} + \frac{N}{A + B}, \qquad \text{where} \quad N = AB.$$

(b) Show also that the relative error (error/true value) in (a) is approximately

$$\frac{1}{8}\left(\frac{A-B}{A+B}\right)^4.$$

16. The reciprocal of a number N can be computed without dividing, by the algorithm

$$x_{i+1} = x_i(2 - Nx_i).$$

(a) Derive this relation by applying Newton's method to $f(x) = \dfrac{1}{x} - N = 0$.

(b) Beginning with $x_0 = 0.2$ compute the reciprocal of 4 by this algorithm. Compute to six decimals or more, and tabulate the error at each step to observe that the number of places of accuracy approximately doubles each time.

17. Use Newton's method to find the roots of the following polynomials:
   (a) $x^3 - 2x + 1 = 0$ (b) $x^4 + x^3 - 4x^2 - 3x + 3 = 0$
   (c) $x^4 - 4.4x^3 + 9.43x^2 - 14.86x + 7.15 = 0$

18. $(x - 1)^3(x - 2) = x^4 - 5x^3 + 9x^2 - 7x + 2 = 0$ obviously has a root at $x = 2$, and a triple root at $x = 1$. Beginning with $x = 2.1$, use Newton's method once, and observe the degree of improvement. Then start with $x = 0.9$, and note the much slower convergence to the triple root even though the initial error is only 0.1 in each case. Use the secant method beginning with $f(0.9)$ and $f(1.1)$, and observe that just one application brings one quite close to the root in contrast to Newton's method. Explain.

Section 4

19. $f(x) = e^x - 3x^2 = 0$ has three roots. An obvious rearrangement is

$$x = \pm\sqrt{e^x/3}.$$

Show, beginning with $x_0 = 0$, that this will converge to a root near -0.5 if the negative value is used, and that it converges to a root near 1.0 if the positive value is used. Show, however, that this form does not converge to the third root near 4.0 even when a nearly exact starting value is used. Find another form which will converge to the root near 4.0.

20. One root of the quadratic $x^2 + x - 1 = 0 = x(x + 1) - 1$ is at $x = 0.6180$. The equivalent form $x = 1/(x + 1)$ converges to this root beginning at $x_0 = 1$. Carrying four or five decimals, how many steps are required to reach the root (correct to four decimals) by linear iteration? If Aitken acceleration is used after three approximations are available, how many iterations are required?

21. The form $x = 1/(x + 1)$ of Problem 20 will converge to a root of the quadratic for many starting values in addition to $x_0 = 1$. For what starting values will it not converge to a root? [For this problem do not stop with a division by zero, i.e., for $x_0 = -1$, $x_1 = 1/0$. Use $x_2 = \lim_{x_1 \to \infty} 1/(x_1 + 1) = 0$.]

22. The cubic $2x^3 + 4x^2 - 2x - 5 = 0$ has a root near $x = 1$. Find at least three rearrangements that will converge to this root beginning with $x_0 = 1.0$.

23. In Problem 22, the form

$$x = \sqrt{\frac{2x + 5}{2x + 4}}$$

converges particularly rapidly beginning with $x_0 = 1.0$. Compare the method of linear iteration to Newton's method in finding a root of

$$2x^3 + 4x^2 - 2x - 5 = 0.$$

24. Use linear iteration to solve the equations of Problem 5.

Section 5

25. Show that if $P_n(x)$ has a double root at $x = r$, then $P_n'(r) = 0$.

26. Show that if $P_n(x)$ has a root of multiplicity m at $x = r$, then

$$P_n^{(i)}(r) = 0, \qquad i = 1, 2, \ldots, m - 1.$$

27. If $P_n(x) = 0$ has a double root at $x = r$, so that $P_n'(r) = 0$, then the condition

$$\left| \frac{f(x)f''(x)}{(f'(x))^2} \right| < 1$$

does not hold for any interval including r. According to Section 5, we hence have no assurance of convergence to the root at r. The simple equation $(x - 2)^2 = 0 = x^2 - 4x + 4$ has such a double root at $x = 2$, and $f'(x) = 2x - 4$ is zero at $x = 2$. Still, beginning at any finite value of x_0, Newton's method will converge! Reconcile this fact with the convergence criterion of Section 5.

28. For the quadratic $x^2 - 4x + 4 = 0$, which has a double root at $x = 2$, begin with $x_0 = 1$ and compute successive approximations to the root by Newton's method. Tabulate the errors at each step and compare each with the next. Is Newton's method quadratically or only linearly convergent in this case? How could one accelerate the convergence?

29. If $f(x), f'(x), f''(x)$ are continuous and bounded on a certain interval containing $x = r$, and if both $f(r) = 0$ and $f'(r) = 0$, but $f''(r) \neq 0$, show that the form

$$x_{n+1} = x_n - 2\frac{f(x_n)}{f'(x_n)}$$

will converge quadratically if x_n is in the interval. [*Hint:* The algorithm is of the form $x_{n+1} = g(x_n)$. Show that $g'(r) = 0$, using L'Hospital's rule.]

30. The method suggested by Problem 29 extends to a root of multiplicity m:

$$x_{n+1} = x_n - m\frac{f(x_n)}{f'(x_n)}.$$

Show, with suitable restrictions on $f(x)$, that this is quadratically convergent.

31. Two of the four roots of the quartic $x^4 + 2x^3 - 7x^2 + 3 = 0$ are positive. Find these by Newton's method correct to seven decimal places using a desk calculator, and then determine the other roots from the reduced equation. Use synthetic division.

Section 6

32. Beginning with the trial factor $x^2 - 4x + 5$, improve by successive applications of Bairstow's method to find the quadratic factors of

$$x^4 - 3.1x^3 + 2.1x^2 + 1.1x + 5.2.$$

What are the four zeros of this polynomial?

33. Solve Problem 31 by Bairstow's method to get quadratic factors.

34. None of the roots of this fourth-degree polynomial are real. Find these complex roots by resolving into quadratic factors:

$$x^4 + 4x^3 + 21x^2 + 4x + 20 = 0.$$

35. When the modulus of one pair of complex roots is the same as for another pair of complex roots, the Bairstow method is slow to converge. Try your patience on

$$x^4 - x^3 + x^2 - x + 1 = 0,$$

which has as factors $(x^2 + 0.618034x + 1)(x^2 - 1.618034x + 1)$. Start with $x^2 + 0.6 + 1$. What is the modulus of the roots?

Section 7

36. Use the Q-D algorithm to approximate the roots of:
   (a) $x^3 - x^2 - 2x + 1 = 0$ (b) $3x^3 + 4x^2 - 8x - 1 = 0$
   (c) $x^4 - 5x^3 + 9x^2 - 7x + 2 = 0$ (d) $x^4 - 3.1x^3 + 2.1x^2 + 1.1x + 5.2 = 0$

37. The Q-D algorithm is also not very efficient on Problem 35. Solve that problem by Q-D.

38. Solve $x^3 - 2x + 1 = 0$ by using the Q-D algorithm.

Section 9

39. Use the half-interval search program of the text to solve the equations of Problem 5. Determine suitable starting values $(X1, X2)$ by preliminary calculations or sketches. Select a suitable tolerance to end the iterations (TOL).

40. Use the Bairstow method program to find the factors of the polynomials of Problems 32 through 35.

41. Revise the Bairstow program of the text so it will handle polynomials up to degree 20.

42. Write a FORTRAN program to use the secant method to evaluate a root of $f(x) = 0$. Use an arithmetic statement function to define $f(x)$. Input the starting values X1 and X2, a tolerance to end iterations, and a limit to the number of iteration to stop nonconverging loops. Test by solving Problem 8.

43. Write a FORTRAN program to evaluate the zeros of a function by Newton's method. Make it adaptable to transcendental functions by defining both $f(x)$ and $f'(x)$ by arithmetic statement functions. Input X0, TOL, LIM analogously to Problem 42. Test by solving Problem 12.

44. Write a FORTRAN program to solve equations by linear iteration. Define the function $g(x)$ by an arithmetic statement, where $x = g(x)$ is the rearrangement of $f(x) = 0$. Test by solving Problem 23.

45. Write a FORTRAN program to execute the Q-D algorithm. The program will need to test for zero coefficients and to increase the roots to avoid them. Provide for up to sixth-degree polynomials. Testing the e-values to end iterations will not work since some e's may not go to zero. While it is possible to test a combination of q-values to terminate, it is certainly simpler just to program for a fixed number of repetitions of the e- and q-row computations.

MISCELLANEOUS PROBLEMS

46. A "parlor game" question is to ask for solutions to the set of equations

$$x - \sqrt{y} = 7,$$
$$y + \sqrt{x} = 7.$$

After discovering the obvious solution (9,4), a less trivial question is to ask if there are any other real solutions. (The best way to answer this is to consider the two graphs, but it is instructive to find solutions by eliminating x, getting $(7 - y)^2 - \sqrt{y} = 7$, and then squaring to eliminate the radical, which gives a fourth-order equation.)

47. A sphere of density d and radius r weighs $\frac{4}{3}\pi r^3 d$. The volume of a spherical segment is $\frac{1}{3}\pi(3rh^2 - h^3)$. Find the depth to which a sphere of density 0.6 sinks in water as a fraction of its radius.

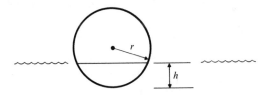

48. The differential equation that expresses the solution to a certain damped vibration problem is

$$y(t) = e^{0.8t}\,(0.33 \cos 8t - 0.22 \sin 8t),$$

where y is the displacement at time t. Find the first three values of t ($t > 0$) for which the displacement is zero.

49. Given

$$x'' + x + 2y' + y = f(t),$$
$$x'' - x + y = g(t), \qquad x(0) = x'(0) = y(0) = 0.$$

In solving this pair of simultaneous second-order differential equations by the Laplace transform method, it becomes necessary to factor the expression

$$(S^2 + 1)(S) - (2S + 1)(S^2 - 1) = -S^3 - S^2 + 3S + 1,$$

so that partial fractions can be used in getting the inverse transform. What are the factors?

50. The solution of boundary value problems by the analytical (Fourier series) method often involves finding the roots of transcendental equations to evaluate the coefficients. For example,

$$y'' + \lambda y = 0, \qquad \dot{y}(0) = 0, \qquad y(1) = y'(1),$$

involves solving $\tan z = z$. Find three values of z other than $z = 0$.

51. Another example of the situation discussed in Problem 50 is

$$y^{IV} - w^2 y = 0, \qquad y(0) = y'(0) = y(1) = y'(1) = 0.$$

The equation has nontrivial solutions only if

$$\cos\sqrt{w} = \frac{1}{\cosh\sqrt{w}}.$$

Find a few values of w for which the differential equation has interesting solutions.

52. In Chapter 3, a particularly efficient method for numerical integration of a function, called Gaussian quadrature, is discussed. In the development of formulas for this method it is necessary to evaluate the zeros of Legendre polynomials. Find the zeros of the Legendre polynomial of sixth order:

$$P_6(x) = \frac{1}{48}(693x^6 - 945x^4 + 315x^2 - 15).$$

[*Note:* All the zeros of the Legendre polynomials are less than one in magnitude, and, for polynomials of even order, are symmetrical about the origin.]

53. The Legendre polynomials of Problem 52 are one set of a class of polynomials known as *orthogonal* polynomials. Another set are the *Laguerre* polynomials. Find the zeros of the following:

(a) $L_3(x) = x^3 - 9x^2 + 18x - 6$ (b) $L_4(x) = x^4 - 16x^3 + 72x^2 - 96x + 24$

54. Still another set of orthogonal polynomials are the *Chebyshev* polynomials. (We will use these in Chapter 13.) Find the roots of

$$T_6(x) = 32x^6 - 48x^4 + 18x^2 - 1 = 0.$$

[Note the symmetry of this function. All the roots of Chebyshev polynomials are also less than one in magnitude.]

2
interpolating
polynomials

We have just examined the question: "Given an explicit function of the independent variable x, what is the value of x corresponding to a certain value of the function?" We now want to consider a question somewhat the reverse of this: "Given values of an unknown function corresponding to certain values of x, what is the behavior of the function?" We would like to answer the question "What is the function", but this is always impossible to determine with a limited amount of data.

Our purpose in determining the behavior of the function, as evidenced by the sample of data pairs $(x, f(x))$, is several-fold. We will wish to approximate other values of the function at values of x not tabulated (interpolation or extrapolation) and to estimate the integral of $f(x)$, and its derivative. The latter objectives will lead us into ways of solving ordinary and partial differential equations.

1. DIFFERENCE TABLES

The problem at hand is considerably simplified if the values of the function are given at evenly spaced intervals of the independent variable, so we first consider this case. It is convenient to arrange the data in a table with x-values in ascending order. We shall conventionally let the letter h stand for the uniform differences of x, $h = \Delta x$. In addition to columns for x and for $f(x)$, we shall tabulate differences of the functional values. Using subscripts to represent the order of the x and $f(x)$ values, we define the first differences of the function as

$$\Delta f_1 = f_2 - f_1, \qquad \Delta f_2 = f_3 - f_2, \qquad \cdots, \qquad \Delta f_i = f_{i+1} - f_i.$$

The second and higher order differences are similarly defined:

$$\Delta^2 f_1 = \Delta(\Delta f_1) = \Delta(f_2 - f_1) = \Delta f_2 - \Delta f_1 = (f_3 - f_2) - (f_2 - f_1)$$
$$= f_3 - 2f_2 + f_1,$$
$$\Delta^2 f_i = f_{i+2} - 2f_{i+1} + f_i,$$
$$\Delta^3 f_1 = \Delta(\Delta^2 f_1) = f_4 - 3f_3 + 3f_2 - f_1, \tag{1.1}$$
$$\Delta^3 f_i = f_{i+3} - 3f_{i+2} + 3f_{i+1} - f_i,$$
$$\cdot$$
$$\cdot$$
$$\cdot$$

$$\Delta^n f_i = f_{i+n} - nf_{i+n-1} + \frac{n(n-1)}{2!}f_{i+n-2} - \frac{n(n-1)(n-2)}{3!}f_{i+n-3} + \cdots.$$

The pattern of the coefficients is the familiar array of coefficients in the binomial expansion. This fact we can prove most readily by symbolic methods, which we postpone until a later section. The second and higher order differences are generally obtained by differencing the previous differences, but Eq. (1.1) shows how any difference can be calculated directly from the functional values.

Table 1.1 illustrates the formation of a difference table. While it might be natural to refer to the first x in the table as x_1, we shall frequently wish to refer to only a portion of the data pairs of the table, so the "first x" loses its significance. The subscripts are useful as an indexing variable to make it easy to refer to a particular entry.   We can then arbitrarily choose the origin for the subscripting variable, which we shall call s, because by the use of negative and zero values we can refer to x-entries that precede x_1.

Table 1.1. Difference Table

s	x	$f(x)$	Δf	$\Delta^2 f$	$\Delta^3 f$	$\Delta^4 f$
-2	x_{-2}	f_{-2}				
			Δf_{-2}			
-1	x_{-1}	f_{-1}		$\Delta^2 f_{-2}$		
			Δf_{-1}		$\Delta^3 f_{-2}$	
0	x_0	f_0		$\Delta^2 f_{-1}$		$\Delta^4 f_{-2}$
			Δf_0		$\Delta^3 f_{-1}$	
1	x_1	f_1		$\Delta^2 f_0$		$\Delta^4 f_{-1}$
			Δf_1		$\Delta^3 f_0$	
2	x_2	f_2		$\Delta^2 f_1$		$\Delta^4 f_0$
			Δf_2		$\Delta^3 f_1$	
3	x_3	f_3		$\Delta^2 f_2$		
			Δf_3			
4	x_4	f_4				

Table 1.2 shows numerically how a difference table is made.   It is customary to stagger the differences between the two values which are subtracted to produce it, so that a finite amount of data gives a table that tapers diagonally at the beginning and the end.   One can form differences of order up to one less than the number of $(x, f(x))$ data pairs.

Table 1.2

x	$f(x)$	Δf	$\Delta^2 f$	$\Delta^3 f$	$\Delta^4 f$
0.0	0.000				
		0.203			
0.2	0.203		0.017		
		0.220		0.024	
0.4	0.423		0.041		0.020
		0.261		0.044	
0.6	0.684		0.085		0.052
		0.346		0.096	
0.8	1.030		0.181		0.211
		0.527		0.307	
1.0	1.557		0.488		
		1.015			
1.2	2.572				

Great care should be exercised to avoid arithmetic errors in the subtractions—the fact that we subtract the upper entry from the lower adds a real source of confusion. One of the best ways to check for mistakes is to add the sum of the numbers in each column to the top entry in the column to its left. This sum should equal the bottom entry in the column to the left.

Since each entry in the difference table is the difference of a pair of numbers in the column to its left, one could recompute one of this pair if it should be erased. As a consequence of this, the entire table could be reproduced, given only one value in each column, if the table is extended to the highest possible order of differences.

When $f(x)$ behaves like a polynomial for the set of data given, the difference table has special properties. In Table 1.3 a function is tabulated over the domain $x = 1$ to $x = 6$, and $f(x)$ obviously behaves the same as x^3. [Note carefully that this does *not* imply that $f(x) = x^3$; the value of $f(x)$ at $x = 7$ might well be 17 instead of $7^3 = 343$. We *only* know the values of $f(x)$ as given in the table.]

Table 1.3. Difference Table for a Function Behaving Like x^3

x	$f(x)$	Δf	$\Delta^2 f$	$\Delta^3 f$	$\Delta^4 f$
0	0				
		1			
1	1		6		
		7		6	
2	8		12		0
		19		6	
3	27		18		0
		37		6	
4	64		24		0
		61		6	
5	125		30		
		91			
6	216				

We observe that the third differences are constant. Consequently, the fourth and all higher differences will be zero. The fact that the nth order differences of any nth degree polynomial are constant is readily proven. We first examine the differences of ax^n:

$$\Delta(ax^n) = a(x + h)^n - ax^n$$
$$= ax^n + anx^{n-1}h + \cdots + ah^n - ax^n$$
$$= (anh)x^{n-1} + \text{terms of lower degree in } x,$$
$$\Delta(anh\, x^{n-1}) = an(n - 1)h^2\, x^{n-2} + \text{terms of lower degree.}$$

Noting that every time a difference is taken the leading term has a power of x one less than originally, we have for a polynomial

$$\Delta P_n(x) = \Delta(a_1 x^n + a_2 x^{n-1} + \cdots + a_n x + a_{n+1})$$
$$= a_1 nh\, x^{n-1} + \text{terms of lower degree,}$$
$$\Delta^2 P_n(x) = a_1 n(n-1)h^2\, x^{n-2} + \text{terms of lower degree,}$$

$$\vdots$$

$$\Delta^n P_n(x) = a_1 n(n-1)(n-2) \cdots (1)\, h^n x^{n-n} = a_1 n!\, h^n. \qquad (1.2)$$

This shows that not only is the nth difference a constant, but that its value is $a_1 n! h^n$. For $P_3(x) = x^3$, $\Delta^3 P_3(x) = (1)(3!)(1)^3 = 6$ when $h = 1$. This is exactly what was found in Table 1.3.

Table 2.1. Effect of Errors on a Difference Table

x	$f(x)$	Δf	$\Delta^2 f$	$\Delta^3 f$	$\Delta^4 f$
0	0				
		1			
1	1		6		
		7		~~6~~ 5	
2	8		~~12~~ 11		~~0~~ 4
		~~19~~ 18		~~6~~ 9	
3	~~27~~ 26		~~18~~ 20		~~0~~ −6
		~~37~~ 38		~~6~~ 3	
4	64		~~24~~ 23		~~0~~ 4
		61		~~6~~ 7	
5	125		30		
		91			
6	216				

2. EFFECT OF ERRORS IN TABLE

Suppose there is an error in one of the $f(x)$ values in the table. Table 2.1 shows how the differences are affected. The data are the same as in Table 1.2, except the value of $f(3)$ has been changed by 1. We show Table 2.1 as a perturbation of Table 1.2 to more clearly show the effect. Observe, in the second difference column, that errors of -1, 2, -1 occur. In the third difference column, they are -1, 3, -3, 1. The fourth differences are not extensive enough to show all of the expected pattern of -1, 4, -6, 4, -1.

If the table is long enough and the effect of errors does not overlap, one can use this principle of error propagation to detect and correct errors. Today, with so much computation being done on a computer, correcting of tables is not as important as it once was. The major consequence to us is that a single error has an effect on many entries in the difference columns as it fans out to the right, and these increase in magnitude. We have shown a gross error in Table 2.1. What one must live with in all tables that are expressed decimally are the little errors due to round-off of our entries—they too fan out and affect the entries with increasing magnitude.

3. INTERPOLATING POLYNOMIALS

When the function which is tabulated behaves like a polynomial (and this we can tell by observing that its nth order differences are constant or nearly so), we can approximate it by the polynomial which it resembles. Our problem is to find the simplest means of writing the nth degree polynomial that passes through $n + 1$ pairs of points, (x_i, f_i), $i = 1, 2, \ldots, n + 1$. Note that such a polynomial is unique —there is only one polynomial of degree n passing through $n + 1$ points. This seems intuitively true, since there are $n + 1$ constants in the polynomial and the $n + 1$ data pairs are exactly enough to determine them. More formally we can reason thus, the proof being by contradiction.

Suppose there are two different polynomials of degree n which are alike at the $n + 1$ points. Call these $P_n(x)$ and $Q_n(x)$ and write their difference:

$$D(x) = P_n(x) - Q_n(x),$$

where $D(x)$ is a polynomial of at most degree n. But since P and Q match at the $n + 1$ pairs of points, their difference, $D(x)$, is equal to zero for all $n + 1$ of these x-values, i.e., it is a polynomial of degree n at most but has $n + 1$ distinct zeros. This is impossible unless $D(x)$ is identically zero. Hence $P_n(x)$ and $Q_n(x)$ are not different—they must be the same polynomial.

Perhaps the easiest way to write a polynomial that passes through a group of equispaced points is the Newton-Gregory forward polynomial:

$$P_n(x_s) = f_0 + s\Delta f_0 + \frac{s(s-1)}{2!}\Delta^2 f_0 + \frac{s(s-1)(s-2)}{3!}\Delta^3 f_0 + \cdots$$

$$= f_0 + \binom{s}{1}\Delta f_0 + \binom{s}{2}\Delta^2 f_0 + \binom{s}{3}\Delta^3 f_0 + \binom{s}{4}\Delta^4 f_0 + \cdots . \tag{3.1}$$

In this equation we have used the notation $\binom{s}{n}$, the number of combination of s things taken n at a time, which is the same as the factorial ratios also shown. Referring to Table 1.1, we now observe that $P_n(x)$ does match the table at all the data pairs (x_i, f_i), $i = 1, 2, \ldots, (n + 1)$. When $s = 0$, $P_n(x_0) = f_0$. If $s = 1$, $P_n(x_1) = f_0 + \Delta f_0 = f_0 + f_1 - f_0 = f_1$. If $s = 2$, $P_n(x_2) = f_0 + 2\Delta f_0 + \Delta^2 f_0 = f_2$. Similarly we can demonstrate that $P_n(x)$ formed according to Eq. (3.1) matches at $n + 1$ points.*

If, over the domain from x_0 to x_n, $P_n(x)$ and $f(x)$ have the same values at tabulated values of x, it is perhaps reasonable to assume they will be nearly the

---

* This demonstration is not a proof, of course. The section on symbolic methods gives perhaps the neatest proof.

same at intermediate x-values.   This assumption is the basis for use of $P_n(x)$ as an interpolating polynomial.   We again emphasize that $f(x)$ and $P_n(x)$ will, in general, not be the same function.   Hence there is some error to be expected in the estimate from such interpolation.   We use the polynomial in Eq. (3.1) as an interpolation polynomial by letting s take on nonintegral values.   Observe that, for any value of x,

$$s = \frac{x - x_0}{h}.$$

Example. Write a Newton-Gregory forward polynomial of degree three that fits Table 1.2 for the four points at $x = 0.4$ to $x = 1.0$.   Use it to interpolate for $f(0.73)$.

To make the polynomial fit as specified, we must index the x's so that $x_0 = 0.4$. It follows then that $f_0 = 0.423$, $\Delta f_0 = 0.261$, $\Delta^2 f_0 = 0.085$, and $\Delta^3 f_0 = 0.096$. We compute s:

$$s = \frac{x - x_0}{h} = \frac{0.73 - 0.4}{0.2} = 1.65.$$

Applying these to Eq. (3.1) with terms through $\Delta^3 f_0$ to give a cubic, we have

$$f(0.73) = 0.423 + (1.65)(0.261) + \frac{(1.65)(0.65)}{2}(0.085) + \frac{(1.65)(0.65)(-0.35)}{6}(0.086)$$

$$= 0.423 + 0.4306 + 0.0456 - 0.0060$$
$$= 0.893. \tag{3.2}$$

The function tabulated in Table 1.2 is tan x, for which the true value is 0.895 at $x = 0.73$.   We hence see that there is an error in the third decimal place.   We should anticipate some error, because, since the third differences are far from constant, our cubic polynomial is a poor representation of the function.   Even so, our interpolating polynomial gave a fair estimate, certainly better than linear interpolation.

Even though the fourth differences are also far from constant, we would hope for some improvement if we approximated $f(x)$ by a fourth-degree polynomial. We can just add one more term to Eq. (3.2) to do this:

$$\binom{s}{4}\Delta^4 f_0 = \frac{(1.65)(0.65)(-0.35)(-1.35)}{4!}(0.211) = 0.0044,$$

$$f(0.73) = 0.893 + 0.0044 = 0.898.$$

Normally the higher degree polynomial is better, but in this instance, adding another term does not improve our estimate; it in fact worsens it slightly.   We shall study the errors of interpolating polynomials in a later section.

The domains of our polynomials are most readily found by working backward from the last difference that is included, drawing imaginary diagonals to the left between the entries.   The x-values included between this fan of diagonals is the domain of the interpolating polynomial.   It will always be found to include one more x-entry than the degree of the polynomial.

4. OTHER INTERPOLATING POLYNOMIALS

It is sometimes convenient to write the interpolating polynomial in other forms. The Newton-Gregory backward polynomial is

$$P_n(x) = f_0 + \binom{s}{1}\Delta f_{-1} + \binom{s+1}{2}\Delta^2 f_{-2} + \binom{s+2}{3}\Delta^3 f_{-3} + \binom{s+3}{4}\Delta^4 f_{-4} + \cdots. \qquad (4.1)$$

It is seen that the differences used here form a diagonal row going upward and to the right,* in contrast to the downward sloping diagonal row of differences used in the Newton-Gregory forward formula.   Trial with various negative integer values of s demonstrates that Eq. (4.1) also matches with data pairs in the table from $x = x_0$ to $x = x_{-n}$.

If the subscripts are suitably chosen, the points where $P_n(x)$ matches the table will be the same as for the Newton-Gregory forward formula, however.   When this is done, the two polynomials are really identical though of a different form.   We illustrate by reworking the same problem as in Section 3.

Choosing $x_0 = 1.0$, so

$$s = \frac{0.73 - 1.0}{0.2} = -1.35,$$

gives the cubic that fits Table 1.2 between $x = 0.4$ to $x = 1.0$; hence

$$f(0.73) = 1.557 + (-1.35)(0.527) + \frac{(-0.35)(-1.35)}{2}(0.181)$$

$$+ \frac{(0.65)(-0.35)(-1.35)}{6}(0.096)$$

$$= 1.557 - 0.7114 + 0.0428 + 0.0049$$

$$= 0.893.$$

One observes that the identical result is obtained as before.

If we again add one more term to make our interpolation correspond to a fourth-degree polynomial, we have

$$f(0.73) = 0.893 + \frac{(1.65)(0.65)(-0.35)(-1.35)}{24}(0.052) = 0.894.$$

In this instance we do improve the estimate, coming closer to the true value of 0.895.   Why did this fourth-degree polynomial not match the fourth-degree one of Section 3?   In the present case, the domain is from $x = 0.2$ to $x = 1.0$, as is found by going back diagonally from the last difference, 0.052, in contrast to the domain of $x = 0.4$ to $x = 1.2$.   In this case the polynomials are not identical.

There is much nonsense in many books about the application of the Newton-Gregory forward polynomial only at the beginning of the table, and the backward formula only at the end.   As our examples clearly show, we may use either formula anywhere in the table by suitably subscripting the x's.   Furthermore,

---

* This ascending diagonal row of differences starting at f_i is called the backward differences of f_i; those in the downward sloping row are called forward differences.

identically the same results are given by *any* interpolating polynomial that ends on the *same* difference entry.

There is a rich variety of interpolation formulas beyond the two we have so far discussed. They differ in the paths taken through the difference table. For example, the Gauss forward goes through the table in a zig-zag path, the first step being a forward one. Stirling's and Bessel's formulas proceed horizontally, using averages of differences, one starting with f_0 and the other halfway between f_0 and f_1. In the next section we present a handy device that permits one to write any of these, knowing only the path that is taken through the difference table.

5. LOZENGE DIAGRAM FOR INTERPOLATION

The identity of all polynomials that fit a table at the same points must mean an interrelation between the coefficients to adjust for the different f-differences that are employed. Figure 5.1 presents in a lozenge diagram (so-called because of the diamond pattern) the various coefficients of all interpolation polynomials. Any of the various forms of interpolating polynomial can be written from this. One may even invent his own polynomial, different in form from any of the named ones.

Certain rules must be followed:

1. One begins at the column of f's, usually at f_0, but not necessarily so. The point of beginning dictates the subscripting, however, and hence the value of x_0 must be consistent with the subscripts of the f's.

2. One proceeds to the right, either diagonally upward or downward to the next difference column, or alternatively horizontally. A term is added for every column crossed.

3. The added term is the entry in the lozenge which is traversed, multiplied by the entry above if the last step was diagonally downward, by the entry below if the last step was diagonally upward, and by the average of entries above and below if it was horizontal. The coefficient of the f-term is always unity, as the interspersed figures indicate.

Kunz (1957) discusses the lozenge in detail, giving the rule if the path progresses from right to left (the term is subtracted) but such variations seem only of academic importance.

Our familiar Newton-Gregory forward polynomial results from a diagonally downward path, shown in Fig. 5.1 by a heavy line. Using the rules given, we have, identically with Eq. (3.1):

$$P_n(x_s) = f_0 + \binom{s}{1}\Delta f_0 + \binom{s}{2}\Delta^2 f_0 + \binom{s}{3}\Delta^3 f_0 + \cdots.$$

The Newton-Gregory backward polynomial results from a diagonally upward path, indicated by a double line:

$$P_n(x_s) = f_0 + \binom{s}{1}\Delta f_{-1} + \binom{s+1}{2}\Delta^2 f_{-2} + \binom{s+2}{3}\Delta^3 f_{-3} + \cdots.$$

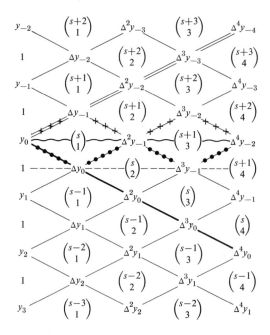

Coefficients for each term depend on the path:

Use term ( )below — Use term above — $\frac{1}{2}$( )Use average of above and $\frac{1}{2}$( )below

Paths for certain polynomials:
Newton-Gregory forward — diagonally downward ———
Newton-Gregory backward — diagonally upward ════
Gauss forward — zig-zag, first step downward ● ● ● ● ●
Gauss backward — zig-zag, first step upward + + + +
Stirling — horizontally, beginning at y_0 ～～～～
Bessel — horizontally, beginning between y_0 and y_1 ─────

Fig. 5.1

Other named polynomials are Gauss forward (dots):

$$P_n(x_s) = f_0 + \binom{s}{1}\Delta f_0 + \binom{s}{2}\Delta^2 f_{-1} + \binom{s+1}{3}\Delta^3 f_{-1} + \binom{s+1}{4}\Delta^4 f_{-2} + \cdots,$$

Gauss backward (crosses):

$$P_n(x_s) = f_0 + \binom{s}{1}\Delta f_{-1} + \binom{s+1}{2}\Delta^2 f_{-1} + \binom{s+1}{3}\Delta^3 f_{-2} + \binom{s+2}{4}\Delta^4 f_{-2} + \cdots,$$

Stirling (wavy line):

$$P_n(x_s) = f_0 + \binom{s}{1}\frac{\Delta f_0 + \Delta f_{-1}}{2} + \frac{\binom{s+1}{2} + \binom{s}{2}}{2}\Delta^2 f_{-1} + \binom{s+1}{3}\frac{\Delta^3 f_{-1} + \Delta^3 f_{-2}}{2} + \cdots,$$

Bessel (dashes): (5.1)

$$P_n(x_s) = \frac{f_0 + f_1}{2} + \frac{\binom{s}{1} + \binom{s-1}{1}}{2}\Delta f_0 + \binom{s}{2}\frac{\Delta^2 f_0 + \Delta^2 f_{-1}}{2} + \frac{\binom{s+1}{3} + \binom{s}{3}}{2}\Delta^3 f_{-1} + \cdots.$$

6. ERROR TERMS AND ERROR OF INTERPOLATION

Since the interpolating polynomial is not, in general, identical with the unknown function $f(x)$ even though they match at certain points, predicting the values of the function at nontabulated points is subject to error. We now develop an expression for the error of $P_n(x)$, an nth degree interpolating polynomial. We write the error function in a form which has the known property that it is zero at the $n+1$ points, from x_0 through x_n, where $P_n(x)$ and $f(x)$ are the same. We call this function $E(x)$:

$$E(x) = f(x) - P_n(x) = (x-x_0)(x-x_1) \cdots (x-x_n)g(x).$$

The $n+1$ linear factors give $E(x)$ the zeros we know it must have and $g(x)$ accounts for its behavior at values other than at $x_0, x_1, \cdots, x_n$. Obviously, $f(x) - P_n(x) - E(x) = 0$, so

$$f(x) - P_n(x) - (x-x_0)(x-x_1) \cdots (x-x_n)g(x) = 0. \qquad (6.1)$$

In order to determine $g(x)$, we now use the interesting mathematical device of constructing an auxiliary function (the reason for its special form becomes apparent as the development proceeds). We call this auxiliary function $W(t)$, and define it as

$$W(t) = f(t) - P_n(t) - (t-x_0)(t-x_1) \cdots (t-x_n)g(x).$$

Note in particular that x has *not* been replaced by t in the $g(x)$ portion. We now examine the zeros of $W(t)$.

Certainly at $t = x_0, x_1, \ldots, x_n$, the W function is zero ($n+1$ times), but it is also zero if $t = x$ by virtue of Eq. (6.1). There are then a total of $n+2$ values of t that make $W(t) = 0$. We now impose the necessary requirements on $W(t)$ for the *law of mean value* to hold. $W(t)$ must be continuous and differentiable. If this is so, there is a zero to its derivative, $W'(t)$, between each of the $n+2$ zeros of $W(t)$, a total of $n+1$ zeros. If $W''(t)$ exists, and we suppose it does, there will be n zeros of $W''(t)$, and likewise $n-1$ zeros of $W'''(t)$, etc., until we reach $W^{(n+1)}(t)$, which must have at least one zero in the interval which has x_0, x_n, or x as end points. Call this value of $t = \xi$. We then have

$$W^{(n+1)}(\xi) = 0 = \frac{d^{n+1}}{dt^{n+1}} \left(f(t) - P_n(t) - (t-x_0) \cdots (t-x_n)(g(x)) \right)_{t=\xi}$$

$$= f^{(n+1)}(\xi) - 0 - (n+1)!g(x). \qquad (6.2)$$

The right-hand side of Eq. (6.2) occurs because of the following arguments. The $(n+1)$st derivative of $f(t)$, evaluated at $t = \xi$, is obvious. The $(n+1)$st derivative of $P_n(t)$ is zero because every time any polynomial is differentiated, its degree is reduced by one, so that the nth derivative is of degree zero (a constant) and its $(n+1)$st derivative is zero. We apply the same argument to the $(n+1)$st degree polynomial in t that occurs in the last term—its $(n+1)$st derivative is a constant and this constant results from the t^{n+1} term and is $(n+1)!$. Of course $g(x)$ is independent of t and goes through the differentiations unchanged. The form of $g(x)$ is now apparent:

$$g(x) = \frac{f^{(n+1)}(\xi)}{(n+1)!}, \qquad \xi \text{ between } (x_0,x_n,x).$$

The conditions on $W(t)$ that are required for this development (continuous and differentiable $n+1$ times) will be met if $f(x)$ has these same properties, because $P_n(x)$ is continuous and differentiable. We now have our error term:

$$E(x) = (x-x_0)(x-x_1) \ldots (x-x_n)\frac{f^{(n+1)}(\xi)}{(n+1)!}. \tag{6.3}$$

We wish to modify this, expressing it in terms of $s = (x-x_0)/h$, to make it more compatible with our interpolating polynomials. Remembering that $x_1 = x_0 + h$, $x_2 = x_0 + 2h$, $\ldots$, so that $(x - x_0) = sh$, $(x - x_1) = sh - h = (s - 1)h$, $(x - x_2) = sh - 2h = (s - 2)h$, $\ldots$, we find that Eq. (6.3) becomes

$$E(x_s) = \frac{(s)(s-1)(s-2) \ldots (s-n)}{(n+1)!} h^{n+1}f^{(n+1)}(\xi) = \binom{s}{n+1}h^{n+1}f^{(n+1)}(\xi),$$

$$\xi \text{ between } (x_0,x_n,x_s). \tag{6.4}$$

Referring again to the lozenge diagram (Fig. 5.1), we observe that the next term after the last one included in an nth degree Newton-Gregory forward interpolating polynomial is $\binom{s}{n+1}\Delta^{n+1}f_0$. One can get the error term of Eq. (6.4) by substituting $h^{n+1} f^{(n+1)}(\xi)$ for the $(n+1)$st difference. This property extends the usefulness of the lozenge diagram to give the error term by this simple change in the next term after the last one included.

Since all polynomials that pass through the same points are identical, this easy rule by which the error term can be written applies to any polynomial developed from the lozenge diagram—just find the next term and change the difference therein to the product of a power of h and the derivative of f of the same order as the difference replaced.

We now can determine which interpolating polynomials give the best estimates of $f(x)$. It makes no difference which form we use—any convenient one is satisfactory, but we should choose the domain where it fits the table to minimize the error term. Changing the points where $P_n(x)$ and $f(x)$ match will cause two changes in the error term; the coefficient involving s will vary, and the value of $f^{(n+1)}(\xi)$ may vary because the interval in which ξ lies changes. Since the function $f(x)$ is, in general, unknown, we certainly do not know its derivatives, and further, the value of ξ is not known except for the intervals that contain it. One then usually chooses the polynomial for which the coefficient involving s has the smallest value. This occurs if the value of x_s at which the polynomial is to be evaluated (the point we are interpolating) lies nearest the midpoint of the interval from x_0 to x_n. Note that this infers that extrapolation will normally be less accurate than interpolation, in accordance with our intuition.

In a case like the example of Sections 3 and 4 where $f(x)$ is known, we can compute the error using Eq. (6.3).   The Newton-Gregory forward cubic interpolating polynomial of Section 3 has an error*

$$\left(\tfrac{8}{4}\right)h^4 f^{iv}(\xi) = \frac{(1.65)(0.65)(-0.35)(-1.35)}{24}(0.2)^4 \left\{ \begin{matrix} 10.1 \\ \\ 395.9 \end{matrix} \right\} = \left\{ \begin{matrix} \text{min. } 0.0003, \\ \\ \text{max. } 0.0134. \end{matrix} \right.$$

The two values within the braces are the maximum and minimum values of the fourth derivative in the interval.   The actual error, 0.002, certainly falls in the range indicated by the error term.   In this example, the computed range for the error was so wide as to be almost useless.   In other cases, the error will be bracketed more satisfactorily.

What if, as is usual, we do not know $f(x)$?   We shall later see (Chapter 5) that the nth derivative and the nth differences are related.   Anticipating some results from that chapter, we have

$$f^{(n+1)}(x) \doteq \frac{\Delta^{n+1}f(x)}{h^{n+1}}.$$

In the absence of knowledge of the function, one may use this relation in the error term computation, provided that the $(n+1)$st differences do not vary too greatly.   This accounts for the rough rule of thumb that the error of interpolation is about the magnitude of the next term beyond that included in the formula.

7. DERIVATION OF FORMULAS BY SYMBOLIC METHODS

In several instances we have presented formulas without proof, although we have demonstrated their validity in specific cases.   Symbolic operator methods are a convenient way to establish these.   We supplement the forward differencing operator Δ with a backward differencing operator ∇ and a *stepping* operator E. These are defined by their actions on a function:

$$\Delta f(x_0) = f(x_0+h) - f(x_0),$$
$$\Delta^2 f(x_0) = \Delta\big(\Delta f(x_0)\big) = \Delta f(x_0+h) - \Delta f(x_0),$$

$$\nabla f(x_0) = f(x_0) - f(x_0 - h), \qquad\qquad\qquad (7.1)$$
$$\nabla^2 f(x_0) = \nabla\big(\nabla f(x_0)\big) = \Delta f(x_0) - \Delta f(x_0-h),$$

$$E f(x_0) = f(x_0+h),$$
$$E^2 f(x_0) = E\big(Ef(x_0)\big) = Ef(x_0+h) = f(x_0+2h),$$
$$E^n f(x_0) = f(x_0+nh).$$

---

* Fourth derivative of tan x is 8 sec^2 x + tan^3 x + 16 sec^4 x tan x.   The maximum value, at $x = 1.0$, is 395.9.   Its minimum value is 10.1 at $x = 0.4$.

There are the obvious relationships:

$$\Delta f(x_0) = Ef(x_0) - f(x_0) = (E - 1)f(x_0),$$
$$\nabla f(x_0) = f(x_0) - E^{-1}f(x_0) = (1 - E^{-1})f(x_0). \tag{7.2}$$

We abstract from Eq. (7.2) the symbolic operator relations

$$\Delta = E - 1,$$
$$\nabla = 1 - E^{-1}. \tag{7.3}$$

Equations (7.3) are really meaningless, for neither side is defined. What they signify is that the effect of Δ when operating on a function is the same as the effect of operating with $(E-1)$, and that ∇ and $(1-E^{-1})$ have the same effect on a function, though all these quantities are without significance standing alone. We must apply them to a function to interpret them.

Since all the operators represented by Δ, ∇, and E are linear operators (the effect on a linear combination of functions is the same as the linear combination of the operator acting on the functions), the laws of algebra are obeyed in relationships between them. This means we can manipulate the relations of Eq. (7.3) by algebraic transformations, and then interpret the results by letting them operate on a function. For example, by raising $\Delta = E-1$ to the nth power, we have

$$\Delta^n = (E-1)^n = E^n - nE^{n-1} + \binom{n}{2}E^{n-2} - \binom{n}{3}E^{n-3} + \ldots,$$

so

$$\Delta^n f(x_0) = \left(E^n - nE^{n-1} + \binom{n}{2}E^{n-2} - \ldots\right)f(x_0)$$
$$= f(x_0+nh) - nf(x_0+(n-1)h) + \binom{n}{2}f(x_0+(n-2)h) - \ldots,$$

or

$$\Delta^n f_0 = f_n - nf_{n-1} + \binom{n}{2}f_{n-2} - \ldots. \tag{7.4}$$

Eq. (7.4) is a proof of the alternating sign, binomial coefficient formula given in Section 1.

We can develop interesting relations between Δ and ∇ such as

$$E\nabla = E(1 - E^{-1}) = E - 1 = \Delta,$$
$$E^n \nabla^n = \nabla^n E^n = \Delta^n,$$
$$\Delta^n f_0 = \nabla^n E^n f_0 = \nabla^n f_n.$$

This illustrates the fact that a given difference entry in a table can be interpreted as either a forward or backward difference of the appropriate f-values. This was already obvious by inspecting a difference table.

We can very simply derive the Newton-Gregory forward formula:

$$E = 1 + \Delta, \qquad E^s = (1 + \Delta)^s,$$
$$f_s = E^s f_0 = (1 + \Delta)^s f_0 = \left(1 + s\Delta + \binom{s}{2}\Delta^2 + \binom{s}{3}\Delta^3 + \cdots\right)f_0$$
$$= f_0 + s\Delta f_0 + \binom{s}{2}\Delta^2 f_0 + \binom{s}{3}\Delta^3 f_0 + \cdots.$$

The Newton-Gregory backward formula is similarly easy:

$$E^{-1} = 1 - \nabla, \qquad E^s = (1 - \nabla)^{-s},$$
$$f_s = E^s f_0 = (1 - \nabla)^{-s} f_0 = \left(1 + s\nabla + \binom{s+1}{2}\nabla^2 + \binom{s+2}{3}\nabla^3 + \cdots\right)f_0$$
$$= f_0 + s\nabla f_0 + \binom{s+1}{2}\nabla^2 f_0 + \binom{s+2}{3}\nabla^3 f_0 + \cdots$$
$$= f_0 + s\Delta f_{-1} + \binom{s+1}{2}\Delta^2 f_{-2} + \binom{s+2}{3}\Delta^3 f_{-3} + \cdots. \tag{7.5}$$

In the last relation of Eq. (7.5) we have expressed the differences as forward differences to agree with our lozenge diagram notation, using $\nabla^n = E^{-n}\Delta^n$.

8. INTERPOLATION WITH NONUNIFORMLY SPACED x-VALUES

When the x-values are not evenly spaced, our methods of writing an interpolating polynomial fail. A parallel development using the concept of divided differences is possible, but perhaps the simplest approach is the Lagrangian polynomial. Data where the x-values are not equispaced often occur as the result of experimental observations, or when historical data is examined.

Suppose we have a table of data, of x- and $f(x)$-values:

x	$f(x)$
x_1	f_1
x_2	f_2
x_3	f_3
x_4	f_4

Here we do not assume uniform Δx, nor even do we need the x-values arranged in any order. Through these four data pairs we can pass a cubic. The Lagrangian form for this is

$$P_3(x) = \frac{(x - x_2)(x - x_3)(x - x_4)}{(x_1 - x_2)(x_1 - x_3)(x_1 - x_4)} f_1 + \frac{(x - x_1)(x - x_3)(x - x_4)}{(x_2 - x_1)(x_2 - x_3)(x_2 - x_4)} f_2$$
$$+ \frac{(x - x_1)(x - x_2)(x - x_4)}{(x_3 - x_1)(x_3 - x_2)(x_3 - x_4)} f_3 + \frac{(x - x_1)(x - x_2)(x - x_3)}{(x_4 - x_1)(x_4 - x_2)(x_4 - x_3)} f_4.$$

Note that it is made up of four terms, each of which is a cubic in x, hence the sum is a cubic. The pattern of each term is to form the numerator as a product of linear factors of the form $(x - x_i)$, omitting one x_i in each term, the omitted value being used to form the denominator by replacing x in each of the numerator factors. In each term, we multiply by the f_i corresponding to the x_i omitted in the numerator factors.

Example. Interpolate for $f(2.3)$ from the table:

x	$f(x)$
1.1	10.6
1.7	15.2
3.0	20.3

With only three pairs of data, a quadratic is the highest degree polynomial possible. It is

$$P_2(x) = \frac{(x - 1.7)(x - 3.0)}{(1.1 - 1.7)(1.1 - 3.0)}(10.6) + \frac{(x - 1.1)(x - 3.0)}{(1.7 - 1.1)(1.7 - 3.0)}(15.2)$$
$$+ \frac{(x - 1.1)(x - 1.7)}{(3.0 - 1.1)(3.0 - 1.7)}(20.3).$$

At $x = 2.3$, we get $P_2(2.3) = 18.38$.

The arithmetic in this method is tedious, unless a desk calculator with transfer features is available.   The recently developed electronic desk calculators are especially convenient for this type of computation.

We develop the error term in the same way as in Section 6.   Write $E(x)$ to exhibit the fact that the error of $P_n(x)$ is zero at the $n + 1$ values of x that are fitted exactly:

$$E(x) = f(x) - P_n(x) = (x - x_1)(x - x_2) \cdots (x - x_{n+1})g(x).$$

The auxiliary function $W(t)$ is

$$W(t) = f(t) - P_n(t) - (t - x_1)(t - x_2) \cdots (t - x_{n+1})g(x).$$

$W(t) = 0$ for $t = x_1, x_2, \cdots, x_{n+1}$, also at $t = x$, for a total of $n + 2$ zeros. Hence $W'(t)$ has n zeros, $W''(t)$ has $n - 1$ zeros, $\cdots$, $W^{(n+1)}(t)$ has one zero. Let ξ be the value of t at which $W^{(n+1)}(t) = 0$.   Hence

$$W^{(n+1)}(\xi) = 0 = f^{(n+1)}(\xi) - 0 - (n + 1)!g(x),$$

$$g(x) = \frac{f^{(n+1)}(\xi)}{(n + 1)!}.$$

The error is then

$$E(x) = (x - x_1)(x - x_2) \cdots (x - x_{n+1}) \frac{f^{(n+1)}(\xi)}{(n + 1)!}. \tag{8.1}$$

The interval in which ξ lies is between the largest and smallest of the x_i for interpolation.   The error can be bracketed between a maximum and a minimum value only if we have information on the $(n + 1)$st derivative of the actual function $f(x)$.

Table 9.1

x	y	Δy	$\Delta^2 y$	$\Delta^3 y$	$\Delta^4 y$
1.6	2.3756				
		0.8926			
1.9	3.2682		0.2963		
		1.1889		0.1079	
2.2	4.4571		0.4042		0.0365
		1.5931		0.1444	
2.5	6.0502		0.5486		
		2.1417			
2.8	8.1919				

9. INVERSE INTERPOLATION

Suppose we have a table of data such as in Table 9.1, and we are required to find the x-value corresponding to a certain value of the function, say at $y = 5.0$.   We have two approaches.   We can consider the y's to be the independent variable (unevenly spaced) and interpolate for x with a Lagrangian polynomial.   Doing so gives $x = 2.312$.   This technique is straightforward, but in some instances gives poor results, the reason being that x considered as a function of y may not be well approximated by a polynomial.   This may be true even though y itself behaves quite like a polynomial.   (Try inverse interpolation among three or four points on the function $y = x^2$, especially for y-values outside the given range.)

The second method is to write y as a polynomial in x and then use the methods of Chapter 1.   This will generally require us to multiply out the interpolating polynomial so as to express the function in the usual polynomial form.   If we use the Gauss forward polynomial, with $x_0 = 2.2$, we have

$$y_s = P_3(x_s) = 4.4571 + s(1.5931) + \frac{(s)(s-1)}{2}(0.4042) + \frac{(s+1)(s)(s-1)}{6}(0.1444)$$

$$+ \frac{(s+1)(s)(s-1)(s-2)}{24}(0.0365) \tag{9.1}$$

$$= 0.00152s^4 + 0.02103s^3 + 0.2006s^2 + 1.3700s + 4.4571.$$

[We have left the polynomial in terms of s to save work; after determining the value of s corresponding to $y = 5.0$ we get x from $s = (x - 2.2)/0.3$.]   The polynomial of Eq. (9.1) could, of course, be determined through the Lagrangian polynomial, but the arithmetic is much greater.   To complete the problem we must find the root near zero of a fourth-degree polynomial in s, which is rather tedious.

The equivalent of this second technique for inverse interpolation can be accomplished more readily by the method of successive approximations.   Our Gauss forward polynomial is

$$y_s = y_0 + s\Delta y_0 + \frac{s(s-1)}{2}\Delta^2 y_{-1} + \frac{(s+1)(s)(s-1)}{6}\Delta^3 y_{-1}$$

$$+ \frac{(s+1)(s)(s-1)(s-2)}{24}\Delta^4 y_{-2}.$$

We rearrange to solve for s in the second term:

$$s = \frac{1}{\Delta y_0}\left(y_s - y_0 - s(s-1)\frac{\Delta^2 y_{-1}}{2} - (s+1)(s)(s-1)\frac{\Delta^3 y_{-1}}{6}\right.$$

$$\left. - (s+1)(s)(s-1)(s-2)\frac{\Delta^4 y_{-2}}{24}\right). \tag{9.2}$$

The method of successive approximations finds s by first neglecting all the terms in s on the right, to give

$$s_1 = \frac{1}{1.5931}(5.0 - 4.4571) = 0.34.$$

The second approximation is obtained using s_1 on the right side of Eq. (9.2), including now one more term:

$$s_2 = \frac{1}{1.5931}(5.0 - 4.4571 - (0.34)(-0.66)(0.2021)) = 0.369.$$

The next approximation uses s_2 on the right and picks up another term:

$$s_3 = \frac{1}{1.5931}(5.0 - 4.4571 - (0.369)(-0.631)(0.2021)$$
$$- (1.369)(0.369)(-0.631)(0.02407)) = 0.375.$$

In the same fashion, $s_4 = 0.3748$, giving

$$x = 2.2 + (0.3)(-0.3748) = 2.31244.$$

[The data in Table 9.1 are for $y = \sinh x$. Substitution in the hyperbolic sine function gives sinh (2.31244) = 5.00001.]

10. INTERPOLATION IN A COMPUTER PROGRAM

The first of the following programs (Program 1) interpolates using the Lagrangian form of interpolating polynomial. The coefficients of each term are built up as the ratio of $(x - x_j)/(x_i - x_j)$ with j varying from 1 to N except omitting the factor when $i = j$. It is demonstrated with the data of Table 9.1 by solving the inverse interpolation problem. Up to a ninth-degree polynomial can be employed.

Program 2 employs the Newton-Gregory interpolating polynomial, again of degree up to nine, to interpolate in a function tabulated with uniform x-spacing.

Program 1

```
ZZJOB 5                          CSC   001      GERALD, C. F.
ZZFORX5
*LIST PRINTER
C   PROGRAM FOR LAGRANGIAN INTERPOLATION
        DIMENSION X(10), Y(10)
C   READ IN NUMBER OF PAIRS OF DATA, THEN READ DATA PAIRS
        READ 100, N, (X(I), Y(I), I = 1,N)
C   READ IN X VALUE FOR WHICH Y IS DESIRED
        READ 101, XIN
C   COMPUTE
        YOUT = 0.0
        DO 20 I = 1,N
        TERM = Y(I)
        DO 10 J = 1,N
        IF (I-J) 9,10,9
      9 TERM = TERM*(XIN - X(J))/(X(I) - X(J))
     10 CONTINUE
     20 YOUT = YOUT + TERM
C   PRINT OUTPUT
        PRINT 200, N, XIN, YOUT
    100 FORMAT (I2/(2F10.0))
    101 FORMAT (F10.0)
    200 FORMAT (49H USING LAGRANGIAN POLYNOMIAL INTERPOLATION AMONG ,I3,
       1 8H VALUES, / 8H AT X = ,F10.3, 27H THE FUNCTION HAS VALUE OF ,
       2 F10.4)
        CALL EXIT
        END
05
2.3756      1.6
3.2682      1.9
4.4571      2.2
6.0502      2.5
8.1919      2.8
5.0
ZZZZ
USING LAGRANGIAN POLYNOMIAL INTERPOLATION AMONG    5 VALUES,
AT X =      5.000 THE FUNCTION HAS VALUE OF     2.3119
```

The differences are computed as they are needed, and a single array is employed both to compute and store them, for maximum economy of core storage use. The value of x_0 is automatically chosen to put the point of interpolation as near the center of the domain as possible.

Program 2

```
ZZJOB 5                         CSC  001     GERALD, C. F.
ZZFORX5
*LIST PRINTER
C   PROGRAM FOR INTERPOLATION IN UNIFORMLY SPACED TABLE
        DIMENSION Y(100), D(10)
C   READ IN BEGINNING AND ENDING X VALUES, DELTA X, AND Y VALUES.
C   THE N VALUES OF Y ARE PUNCHED 8 PER CARD
        READ 100, XI, XN, H
        N = (XN - XI)/H + 1.
        READ 100, (Y(I), I = 1,N)
C   READ IN X VALUE AT WHICH Y IS TO BE INTERPOLATED AND DEGREE OF
C       POLYNOMIAL TO BE EMPLOYED, M.   MAXIMUM DEGREE IS 10.
        READ 101, X, M
C   FIND PROPER SUBSCRIPT FOR YO SO X IS CENTERED IN THE
C       DOMAIN AS WELL AS POSSIBLE.
        FM = M + 1
        IF (X - XI - FM/2.*H) 1,1,2
      2 IF (XN - X - FM/2.*H) 4,4,3
      1 J = 1
        GO TO 5
      4 J = N - M
        GO TO 5
      3 J = (X - XI) / H - FM/2. + 2.
      5 FJ = J
        XO = XI + (FJ - 1.)*H
        YO = Y(J)
C   COMPUTE DIFFERENCES REQUIRED.
        DO 10 I = 1,M
        D(I) = Y(J + 1) - Y(J)
     10 J = J + 1
        DO 20 J = 2,M
        DO 20 I = J,M
        K = M - I + J
     20 D(K) = D(K) - D(K-1)
C   COMPUTE S VALUE
        S = (X - XO)/H
C   COMPUTE INTERPOLATED Y
        YOUT = YO
        FNUM = S
        DEN = 1.
        DO 30 I = 1,M
        FI = I
        YOUT = YOUT + FNUM / DEN * D(I)
        FNUM = FNUM * (S - FI)
     30 DEN = DEN*(FI + 1.)
C   PRINT OUTPUT
        PRINT 200, M, X, YOUT
        CALL EXIT
    100 FORMAT (8F10.0)
    101 FORMAT (F10.0, I2)
    200 FORMAT (35H USING INTERPOLATING POLYNOMIAL OF , I3,
       1 8H DEGREE, / 8H AT X = , F8.3, 17H, Y HAS VALUE OF , F10.5)
        END
0.50       1.55       0.05
.54630     .61311     .68414     .76020     .84229     .93160     1.0296     1.1383
1.2602     1.3984     1.5574     1.7433     1.9648     2.2345     2.5722     3.0096
3.6021     4.4552     5.7979     8.2381     14.101     48.078
1.32       05
ZZZZ


USING INTERPOLATING POLYNOMIAL OF    5 DEGREE,
AT X =     1.320, Y HAS VALUE OF    3.90475
```

PROBLEMS

Section 1

1. Complete a difference table for the following data:

x	1.20	1.25	1.30	1.35	1.40	1.45	1.50
$f(x)$	0.1823	0.2231	0.2624	0.3001	0.3365	0.3716	0.4055

2. In Problem 1, what degree of polynomial is required to fit exactly to all seven data pairs? What lesser degree polynomial will nearly fit the data? Justify your answer.

3. In Problem 1, assume that the third differences are constant and are equal to 0.0001 (which is about their average value). Working backward from this, extrapolate the table to $x = 1.15$ and to $x = 1.55$. The data in Problem 1 are for ln x. Compare your extrapolated values to a table of logarithms.

4. Look up values for sinh x in a handbook and tabluate for $x = 1.0(0.1)2.0$. (This notation means for $x = 1.0$ to $x = 2.0$ at intervals of 0.1.) Round the values to four decimals and make a difference table, computing differences up to and including the fourth. Assuming that the fourth differences are constant at the average value of those in your table, extrapolate to get values of sinh x at $x = 0.9$ and $x = 2.1$. How do these compare to the handbook values? The value at $x = 1.1$ is 1.33565, to five decimals. A common rule for rounding when the last figure is a five is to round to the nearest even digit (which here would give 1.3356) while another rule always rounds a five upwards. What difference does this make in this problem?

5. Form a difference table for $f(x) = x^3 - 3x^2 + 2x + 1$ for $x = -1(0.2)1$. (Recall that nested multiplication is more efficient, especially on a desk calculator.) Verify from your table the validity of Eq. (1.2).

6. The statement is made in Section 1 that one can recompute an entire difference table given only one entry from each column. Verify by completing this table:

x	y	Δy	$\Delta^2 y$	$\Delta^3 y$	$\Delta^4 y$	$\Delta^5 y$
0	___					
		___				
5	___		0.0013			
		0.0888				
10	___		___	___	0.0002	
		___				−0.0002
15	___		___	___	___	
		___		0.0017		
20	___		___			
		___				
25	0.4663					

7. Find the next two terms of the sequence given that the general term is represented by a quartic polynomial:

$$u_1 = 0, \qquad u_2 = 4, \qquad u_3 = 6,$$
$$u_4 = 3, \qquad u_5 = 2.$$

Section 2

8. Form a difference table through fourth differences:

x	$f(x)$		x	$f(x)$
1	0.7		5	7.1
2	0.8		6	13.2
3	1.5		7	22.3
4	3.4		8	35.0

Suppose, in transcribing the above, the entry for $x = 4$ was incorrectly written as 4.3. What would the difference table then be?

9. Find, by the difference table technique, the errors in the following data, which correspond to equispaced arguments: 4.70, 4.91, 5.20, 5.60, 6.14, 6.85, 7.76, 8.83, 10.30, 11.99, 14.00, 16.36, 19.10.

Section 3

10. Use Newton-Gregory forward interpolating polynomials of degree three to estimate $f(0.158)$ and $f(0.636)$, given the following table. In the first polynomial, choose $x_0 = 0.125$. In the second, let $x_0 = 0.375$.

x	$f(x)$	Δf	$\Delta^2 f$	$\Delta^3 f$	$\Delta^4 f$
0.125	0.79168				
		-0.01834			
0.250	0.77334		-0.01129		
		-0.02963		0.00134	
0.375	0.74371		-0.00995		0.00038
		-0.03958		0.00172	
0.500	0.70413		-0.00823		0.00028
		-0.04781		0.00200	
0.625	0.65632		-0.00623		
		-0.05404			
0.750	0.60228				

11. Add one term to the work of Problem 10 to estimate $f(0.158)$ by a fourth-degree polynomial.

12. Many textbooks say, "For interpolation at a point near the end of a table, we cannot use Newton's forward polynomial." In view of Problem 10, do you agree with this statement? Discuss briefly.

13. Use the data below to find the value of y at $x = 0.58$, using a cubic polynomial that fits the table at x values of 0.3, 0.5, 0.7, and 0.9.

x	y	Δy	$\Delta^2 y$	$\Delta^3 y$
0.1	0.003			
		0.064		
0.3	0.067		0.017	
		0.081		0.002
0.5	0.148		0.019	
		0.100		0.003
0.7	0.248		0.022	
		0.122		0.004
0.9	0.370		0.026	
		0.148		0.005
1.1	0.518		0.031	
		0.179		
1.3	0.697			

14. Convert the cubic polynomial of Problem 13 to standard form:

$$a_1 x^3 + a_2 x^2 + a_3 x + a_4.$$

15. What is the minimum degree polynomial that will exactly fit all seven data pairs of Problem 13? (Answer is *not* sixth degree.)

Section 4

16. Using the data of Problem 10, write a Newton-Gregory forward polynomial that fits the table at x-values of 0.500, 0.625, and 0.750. Then write the Newton-Gregory backward polynomial that fits the same three points. Demonstrate that these are two different forms of the same polynomial by rewriting both in the form $a_1 x^2 + a_2 x + a_3$.

17. Repeat Problem 10 using Newton-Gregory backward polynomials, except choose $x_0 = 0.500$ in the first and $x_0 = 0.750$ in the second polynomial. Are the same results obtained?

18. Many textbooks say, "For interpolation at a point near the beginning of a table, Newton's backward polynomial cannot be used." Discuss. Suppose one wished to increase the degree of the polynomial, say from third to fourth degree? (Cf. Problem 17.)

19. Using the data of Problem 10, estimate $f(0.385)$ by the following. Use quadratic polynomials in each case. (a) Newton-Gregory forward, $x_0 = 0.250$. (b) Newton-Gregory backward, $x_0 = 0.500$. (c) Gauss backward, $x_0 = 0.375$. (d) Stirling, $x_0 = 0.375$. Why are all these results identical?

20. The familiar process of linear interpolation can be represented in terms of either the Newton-Gregory forward, or the Newton-Gregory backward, as either

$$y_0 + s\,\Delta y_0 \quad \text{or} \quad y_0 + s\,\Delta y_{-1}.$$

These two must be the same, of course. Compare the y_0, s, and Δy values that should be used in the two equivalent formulas for the two pairs of values, from which $y(1.8)$ is to be determined:

x	y
1	16
3	12

Section 5

21. For the data of Problem 13, write cubic interpolation polynomials that terminate on the third difference whose value is 0.004. Use the lozenge diagram to write polynomials of the following forms: (a) Newton-Gregory forward, (b) Newton-Gregory backward, (c) Gauss forward, (d) Gauss backward, (e) Bessel.

22. Can a Stirling polynomial be written which is the same as the other polynomials of Problem 21? What about the set of quadratics that terminate on the second difference whose value is 0.022? Which polynomial cannot be included in this set?

23. Use each of the polynomials of Problem 21 to interpolate at $x = 0.92$. For each polynomial, compare in a table the values of y_0, s, $y(0.92)$.

24. Interpolate the data of Problem 13 for $y(0.92)$ using a third-degree Stirling polynomial for which y_0 is 0.370. Should this value be identical to the results of Problem 23?

Section 6

25. Write the error terms of each polynomial of Problem 23. Evaluate the coefficients (the s polynomials).

26. Write the error term of the polynomial of Problem 24. Evaluate the coefficient and compare to the result of Problem 23.

27. Problem 26 demonstrates that the coefficient of the error term is smallest when the point at which the interpolation polynomial is used is centered in the range of fit of the polynomial. On what third difference should the cubics of Problem 23 terminate to give the best polynomial for interpolation at $x = 0.92$? How much is the coefficient of the error term reduced over that in Problem 25?

28. What is the error term for linear interpolation? Show that the coefficient is of maximum magnitude at the midpoint between x_0 and x_1.

29. Problem 28 suggests that the maximum error of linear interpolation is no greater than

$$\frac{(x_1 - x_0)^2}{8} M,$$

where M is the maximum of the absolute value of $f''(x)$ in the interval $[x_0, x_1]$. Use this criterion as an estimate of the maximum error to answer these questions.

(a) A commonly used set of tables gives sinh x to four decimals between $x = 3.0$, and $x = 5.0$ with $\Delta x = 0.1$. What might the error be for linear interpolation near $x = 3.0$? Near $x = 5.0$? How many decimal places of accuracy would you expect in the interpolated results?

(b) How close would the entries need to be in the sinh table as above so the maximum error of linear interpolation is sure to be <0.0001 for the above portion of the table?

30. A table of $\tan x$ with the argument given in radians has $\Delta x = 0.01$. Near $x = \pi/2$, the tangent function increases very rapidly and, hence, linear interpolation is not very accurate. What degree of interpolating polynomial is needed to interpolate at $x = 1.506$ to three-decimal accuracy?

Section 7

31. An operator α is called a *linear* operator if $\alpha(f + g) = \alpha f + \alpha g$ and $\alpha(cf) = c\alpha f$, where $c = $ constant. Show that E, Δ, and ∇ are linear operators based on their definitions by Eq. (7.1).

32. Two operators, α and β, are said to commute if the result of operating on a function is not changed by changing the order of the operations, that is, $\alpha(\beta f) = \beta(\alpha f)$. Show that E, Δ, and ∇ commute.

33. Define D as the differentiation operator. Show that D commutes with E, Δ, and ∇.

34. Show that:

$$\text{(a)} \quad \Delta(f(x)g(x)) = f(x)\Delta g(x) + g(x + h)\Delta f(x)$$
$$\text{(b)} \quad \Delta^n x^n = \nabla^n x^n = \nabla^n E^m x^n = n!.$$

35. Express $\Delta^3 E^{-2} \nabla^2 y_4$ in terms of y_i entries of the table.

36. If $\Delta^n y_s = \nabla^n y_r$, express r in terms of s.

37. $\delta = E^{1/2} - E^{-1/2}$ defines the central difference operator. Show that
$$\delta^n y_i = \Delta^n E^{-n/2} y_i = \nabla^n E^{n/2} y_i.$$

38. Based on the relations of Problem 37, where are the even order central differences located in a difference table? Why is it appropriate to call Stirling's formula a "central difference formula?"

39. Again, using the relations of Problem 37, if $y = f(x)$ is known only as the entries of an evenly spaced table, why can the odd order central differences $\delta^n y_0$ not be computed?

40. Define $\mu = \frac{1}{2}(E^{1/2} + E^{-1/2}) = $ averaging operator. Show that $\mu\delta^n y_0$ can be computed for odd orders. Express $\mu\delta y_0$ and $\mu\delta^3 y_0$ in terms of y_i.

41. Prove the following identities:

$$\text{(a)} \quad E^{1/2} = \mu + \frac{\delta}{2} \qquad \text{(b)} \quad E^{-1/2} = \mu - \frac{\delta}{2} \qquad \text{(c)} \quad \sqrt{1 + \delta^2\mu^2} = 1 + \frac{\delta^2}{2}$$

42. Show that $\mu\delta^3 y_0 = \frac{1}{2}(\Delta^3 y_{-2} + \nabla^3 y_2)$.

43. Show that $\mu\delta = \frac{1}{2}e^{hD} - \frac{1}{2}e^{-hD} = \sinh(hD)$ by first expanding e^{hD} as a Maclaurin series in terms of hD from which it is apparent that, symbolically, $Ey_0 = e^{hD}y_0$.

Section 8

44. Write an interpolating polynomial that passes through each point:

x	0.5	1.2	3.1
y	-3.2	1.6	-1.8

Plot the points and sketch the parabola that passes through them.

45. Given the four points $(1,0)$, $(-2,15)$, $(-1,0)$, $(2,9)$, write the Lagrangian form of the cubic that passes through them. Multiply out each term to express in standard form, $ax^3 + bx^2 + cx + d$.

46. Given that $\ln 2 = 0.69315$ and $\ln 5 = 1.60944$, compute the natural logarithms of each integer from 1 to 10. Compare to a five-place table.

47. Compute the error term for $\ln 9$ as you computed that value in Problem 46.

48. If $e^{0.2}$ is estimated by interpolation among the values of $e^0 = 1$, $e^{0.1} = 1.1052$, and $e^{0.3} = 1.3499$, find the maximum and minimum estimates of the error. Compare with the actual error.

49. Repeat Problem 48, except for extrapolation to get $e^{0.4}$.

50. If the x-values are evenly spaced, the Lagrangian form of interpolating polynomial must be identical to the various interpolating polynomials written with the aid of the lozenge diagram (Fig. 5.1). Show that the error terms are also identical by comparing Eqs. (6.4) and (8.1) for this case.

Section 9

51. For the function $y = x^2$, obviously the points $(1,1)$, $(2,4)$, $(3,9)$ are on its graph. Considering x as a function of y, we have $x = \sqrt{y}$, and for $y = 25$, $x = 5$. But one may also compute x corresponding to $y = 25$ by inverse interpolation from the first three points. Do this by Lagrangian interpolation (extrapolating of course.) What error estimates are available?

52. Repeat Problem 51, except determine x at $y = 0$ by inverse interpolation. Are you surprised that the interpolated x-value is again 0.6? Sketch the curves for $x = \sqrt{y}$ and for the interpolating parabola.

53. Use the method of successive approximations to find the value of x corresponding to $y = 3.0$ from the data of Table 9.1. Compare to $\sinh^{-1} 3.0$.

54. The method of successive approximations can also be used for extrapolation (though with less accuracy.) Use this method of inverse interpolation to estimate x corresponding to $y = 10$ for the data in Table 9.1.

55. The inverse interpolation procedure is often recommended as a process for finding the zeros of functions by making a table for values of $f(x)$ near the zero and interpolating. Compute values for $f(x) = 5x^3 - 3x^2 + 2x - 2$ for $x = -1, 0, 1, 2$ and determine x corresponding to $f(x) = 0$. The poor agreement is because the higher order differences are not small. Repeat, except use $x = 0.6, 0.7, 0.8, 0.9$. The good agreement in the second case is because the third difference is now small.

56. Use inverse interpolation, employing the method of successive approximations, to find the zero of $xe^x - \cos x$ near $x = 0.5$. Space the x-values so that the higher order differences are small.

Section 10

57. Use the program for Lagrangian interpolation to find Γ (1.7) given that Γ (0.5) = $\sqrt{\pi}$, Γ (1.0) = 1.0, Γ (1.5) = $0.5\sqrt{\pi}$, Γ (2.0) = 1.0, Γ (3.0) = 2.0.

58. Extending the gamma function data as follows, what degree of interpolating polynomial is needed to get eight-place accuracy? Use either the Lagrangian or Newton interpolation programs:

x	0.5	1.0	1.5	2.0	2.5	3.0	3.5	4.0	4.5	5.0
$\Gamma(x)$	$\sqrt{\pi}$	1.0	$\frac{1}{2}\sqrt{\pi}$	1.0	$\frac{3}{4}\sqrt{\pi}$	2.0	$\frac{15}{8}\sqrt{\pi}$	6.0	$\frac{105}{16}\sqrt{\pi}$	24.0

59. Write a program similar to the one for Newton-Gregory interpolation but employing Stirling's formula.

3
numerical
integration

It is frequently necessary to integrate a function when it is known only as a tabulation of data. Numerical methods permit us to do this. Furthermore, even when the function is explicitly known, since computers cannot readily be programmed to analytically integrate an arbitrary function, nearly all integrations performed on a computer utilize numerical techniques. In addition to these two reasons for studying this topic, the integral of many functions cannot be expressed in terms of ordinary functions. In such cases, numerical methods permit us to evaluate the integral. A further important application is in the solution of ordinary differential equations.

1. TRAPEZOIDAL RULE

The familiar and simple trapezoidal rule can be considered to be an adaptation of the definition of the definite integral as a sum. To evaluate $\int_a^b f(x)dx$, we subdivide the interval from a to b into n subintervals, as in Fig. 1.1. The area under the curve in each subinterval is approximated by the trapezoid formed by replacing the curve by its secant line drawn between the end points of the curve. The integral is then approximated by the sum of all the trapezoidal areas. There is no necessity to make the subintervals equal in width, but our formula is simpler if this is done. Let h be the constant Δx. Since the area of a trapezoid is its average height times the base, for each subinterval,

$$\int_{x_i}^{x_{i+1}} f(x)dx \doteq \frac{f(x_i) + f(x_{i+1})}{2}(\Delta x) = \frac{h}{2}(f_i + f_{i+1}), \tag{1.1}$$

$$\int_a^b f(x)dx \doteq \sum_{i=1}^{n} \frac{h}{2}(f_i + f_{i+1}) = \frac{h}{2}(f_1 + f_2 + f_2 + f_3 + \cdots + f_n + f_{n+1})$$

$$= \frac{h}{2}(f_1 + 2f_2 + 2f_3 + \cdots + 2f_n + f_{n+1}). \tag{1.2}$$

The formula is beautifully simple, and its ability to be applied to unequally spaced values is of value in finding the integral of an experimentally determined function. It is obvious from Fig. 1.1 that the method is subject to large errors unless the subintervals are small, for replacing a curve by a straight line is hardly accurate.

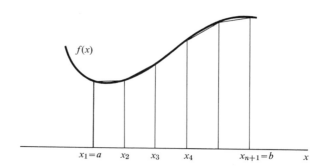

$f(x)$

$x_1 = a$ x_2 x_3 x_4 $x_{n+1} = b$ x **Fig. 1.1**

Table 1.1

x	$f(x)$	x	$f(x)$
1.6	4.953	2.8	16.445
1.8	6.050	3.0	20.086
2.0	7.389	3.2	24.533
2.2	9.025	3.4	29.964
2.4	11.023	3.6	36.598
2.6	13.464	3.8	44.701

Suppose we wished the integral of the function tabulated in Table 1.1 over the interval from $x = 1.8$ to $x = 3.4$. The trapezoidal rule gives

$$\int_{1.8}^{3.4} f(x)dx \doteq \frac{0.2}{2} [6.050 + 2(7.389) + 2(9.025) + 2(11.023) + 2(13.464)$$
$$+ 2(16.445) + 2(20.086) + 2(24.533) + 29.964] = 23.9944.$$

The data in Table 1.1 are for $f(x) = e^x$, so the true value of the integral is $e^{3.4} - e^{1.8} = 23.9144$. We are off in the second decimal.

To find the error term for trapezoidal integration, we express $f(x)$ as a first-degree interpolating polynomial, including an error term, utilizing our work in Chapter 2. We then integrate the polynomial, the integral of the error term giving us the error of our integration formula:

$$f(x_s) = P_1(x_s) + \text{error} = f_0 + s\Delta f_0 + \frac{(s)(s-1)}{2} h^2 f''(\xi),$$

$$\int_{x_0}^{x_0+h} f(x)dx = \int_{x_0}^{x_0+h} (f_0 + s(f_1 - f_0))dx + \tfrac{1}{2}h^2 \int_{x_0}^{x_0+h} (s)(s-1)f''(\xi)dx.$$

We now change variables from x to s. Because

$$s = \frac{1}{h}(x - x_0),$$

$dx = h\,ds$, and the limits become $s = 0$ to $s = 1$:

$$\int_{x_0}^{x_1} f(x)dx = \int_0^1 (f_0 + s(f_1 - f_0))h\,ds + \tfrac{1}{2}h^2\int_0^1 (s^2 - s)f''(\xi)h\,ds$$

$$= h(f_0 + \tfrac{1}{2}(f_1 - f_0)) + \tfrac{1}{2}h^3 f''(\xi_1)(\tfrac{1}{3} - \tfrac{1}{2})$$

$$= \frac{h}{2}(f_0 + f_1) - \frac{1}{12}h^3 f''(\xi_1), \qquad x_0 < \xi < x_0 + h$$

$$= \frac{h}{2}(f_0 + f_1) + \text{error } O(h^3). \tag{1.3}$$

In the integration of the last term we have utilized the *theorem of the mean for integrals*, which is that

$$\int f(x)g(x)\,dx = f(\xi)\int g(x)\,dx$$

provided that $g(x)$ does not change sign on the interval of integration, ξ being found in that interval.   In Eq. (1.3), $s(s - 1)$ does not change sign on [0,1], so the second derivative may be taken outside the integral sign, with ξ being replaced by a (possibly) new value ξ_1, but both ξ and ξ_1 are between x_0 and x_1.   Note that we have reproduced the formula of Eq. (1.1) as well as derived the error term.

Note that the error term can be considered to be the product of a coefficient, $-\frac{1}{12}f''(\xi_1)$, and a power of h, h^3.   The coefficient is vague in that ξ_1 is unknown, and even the derivative function may be unknown, but in any event, it approaches some limiting value as $h \to 0$.   It is conventional to express this as "error $O(h^3)$," read "error of order h cubed", because the most important idea we gain from the error term is that as $h \to 0$, the errors get small proportionally to the cube of h.

This error, it should be emphasized, is the error of only a single step, and is hence called the *local error*.   We normally apply the trapezoidal formula to a series of subintervals to get the integral over a large interval from $x = a$ to $x = b$.   We are interested in the total error, which is called the *global error*.

To develop the formula for global error of the trapezoidal rule, we note that it is the sum of the local errors:

$$\text{global error} = -\tfrac{1}{12}h^3 \left(f''(\xi_1) + f''(\xi_2) + \cdots + f''(\xi_n)\right). \tag{1.4}$$

In Eq. (1.4) each of the values of ξ_i is found in the successive subintervals.   If we assume that $f''(x)$ is continuous on (a,b), there is some value of x in (a,b), say $x = \xi$, at which the value of the sum in Eq. (1.4) is equal to $n \cdot f''(\xi)$.   Since $nh = b - a$, the global error becomes

$$\begin{matrix} \text{Global error of} \\ \text{trapezoidal rule} \end{matrix} = -\frac{1}{12}h^3 n f''(\xi) = \frac{-(b-a)}{12}h^2 f''(\xi) = O(h^2). \tag{1.5}$$

The fact that the global error is $O(h^2)$ while the local error is $O(h^3)$ is reasonable, since, for example, if h is halved, the number of subintervals is doubled, so we add together twice as many errors.

When the function $f(x)$ is known, Eq. (1.5) permits one to estimate the error of numerical integration by the trapezoidal rule.   In applying this equation we bracket the error by calculating with the maximum and the minimum values of $f''(x)$ on the interval $[a, b]$.

2. ROMBERG INTEGRATION

There is a way to improve on the accuracy of the simple trapezoidal rule when the function is known at equispaced intervals, or when it is explicit known so that it can be computed as desired.   This method is known as extrapolation to the limit. We illustrate the method by a reconsideration of the data in Table 1.1.   We saw that the trapezoidal rule gave a value of 23.9944 for $\int_{1.8}^{3.4} f(x)\,dx$, which differs from the true value of 23.9144.   We do not desire to take advantage of our knowledge of what the function is, so we do not make h smaller to improve our estimate of the integral.

Even if we should not make the interval smaller, we can certainly make it larger, by using only every other value, taking $h = 0.4$. If we do this we get

$$I(h = 0.4) \doteq \frac{0.4}{2} (6.050 + 2 (9.025) + 2 (13.464) + 2 (20.086) + 29.964)$$
$$= 24.2328.$$

This value is even more in error than before, which we of course expect with h larger. But we can take advantage of this to nearly eliminate the error of our former computation if we assume that errors of $O(h^2)$ can be interpreted as proportional to h^2, say error $= Ch^2$. Taking C to be a constant is strictly true only in the limit, of course, and at $h = 0.2$ and $h = 0.4$ we are not very near zero. It can be shown,* however, that we make an error of only $O(h^4)$ by assuming the errors are proportional to h^2. We then have, for the two computations above,

$$\text{True value} = \text{computed value} + \text{error},$$
$$\text{True value} = 23.9944 + C(0.2)^2$$
$$\text{True value} = 24.2328 + C(0.4)^2 \tag{2.1}$$

In Eq. (2.1) there are two unknowns, the "true value" and the proportionality constant C, but we can certainly solve for these with two equations. We get, subtracting the second equation from four times the first,

$$\text{True value} = 23.9944 + \tfrac{1}{3}(23.9944 - 24.2328)$$
$$= 23.9149 \text{ (versus 23.9144, exact value)}.$$

We have extrapolated from two inexact values to an improved one. What we have called "true value" is not exact because it used the erroneous assumption that $O(h^2)$ was the same as Ch^2. We can make a further improvement by extrapolation again, if we have two such estimates of $O(h^4)$ accuracy, by assuming their errors can be stated as Ch^4. Such a second estimate can be obtained from the data in Table 1.1 by backing off once more, computing the area with $h = 0.8$, If we do so we get $I(h = 0.8) = 25.1768$. We use this with $I(h = 0.4) = 24.2328$, as shown below.

h	Trapezoidal rule estimate $O(h^2)$ errors	Extrapolated value $O(h^4)$ errors
0.2	23.9944	
0.4	24.2328	23.9149
0.8	25.1768	23.9181

---

* See, for example, Davis and Robinowitz (1967), who show that the error can be expressed as $C_1h^2 + C_2h^4 + C_3h^6 + \cdots$.

As before, we set up relationships

$$\text{True value} = 23.9149 + C(0.2)^4,$$
$$\text{True value} = 23.9189 + C(0.4)^4$$

Solving, we get

$$\text{True value} = 23.9149 + \tfrac{1}{15}(23.9149 - 23.9189)$$
$$= 23.9147.$$

This is now as close as can be determined to the exact value, 23.9144. One further order of extrapolation could be made, for we could start with $h = 1.6$ and by combining with $h = 0.8$ and $h = 0.4$ get another second-order extrapolation. These second-order extrapolations can be shown to be of $O(h^6)$ accuracy and can be combined to an estimate of $O(h^8)$. No further improvement results from this, however, due to the influence of round-off errors in the original data. With only three-decimal accuracy there, it is not surprising to find errors in the fourth decimal of the integrals.

The number of successive improvements which can be made, and hence the order of the error that can be attained, obviously depends on the number of functional values available, or which can be computed when $f(x)$ is a known function. The number of successive improvements which *should* be made may be less than the number which can, in principle, be made because of the effects of round-off error.

We summarize our extrapolation rule, to be used for improving two values for which h varies 2:1:

$$\text{Improved value} = \text{more accurate} + \left(\frac{1}{2^n - 1}\right)(\text{more accurate} - \text{less accurate}). \quad (2.2)$$

The exponent n in the coefficient is the exponent on h in the error term $O(h^n)$ that applies to the two values used for extrapolation. One always knows that the more accurate value is the one computed with smaller h.

When this technique of extrapolation to the limit is applied to integrals of known functions so that values can be computed for smaller values of h, it is known as *Romberg integration*. It is quite popular in computer programs, often being available as a stock subroutine in the computer library. The technique is to compute the integral for an arbitrarily chosen h, and then to compute it again with h halved. If the two values for the integral differ by more than some tolerance value, the value is improved by extrapolation to the limit, and a second integration is computed with h halved again. This is combined with the previous value of the integral to give a second extrapolated value which is compared to the first. If necessary, these two are combined to give a higher order extrapolation. The method is continued until a pair of extrapolated values of the integral agree satisfactorily. A program illustrating this is presented at the end of the chapter. The Romberg method is applicable to a wide class of functions — to any Reimann integrable function, in fact. Smoothness and continuity are not required.

3. LOZENGE DIAGRAMS FOR INTEGRATION

As an alternative to the extrapolation method of Section 2, we can use a higher degree polynomial that approximates to $f(x)$ and integrate it to get other integration formulas of increased accuracy. Integrating over one subdivision (often called one *panel*), we get, using the Newton-Gregory forward polynomial,

$$\int_{x_0}^{x_1} f(x)dx \doteq \int_{s=0}^{s=1} (f_0 + s\,\Delta f_0 + \binom{s}{2}\Delta^2 f_0 + \binom{s}{3}\Delta^3 f_0 + \cdots)h\,ds$$

$$= h\left[sf_0 + \frac{s^2}{2}\Delta f_0 + \left(\frac{s^3}{6} - \frac{s^2}{4}\right)\Delta^2 f_0 + \left(\frac{s^4}{24} - \frac{s^3}{6} + \frac{s^2}{6}\right)\Delta^3 f_0 + \cdots \right]_0^1$$

$$= h(f_0 + \tfrac{1}{2}\Delta f_0 - \tfrac{1}{12}\Delta^2 f_0 + \tfrac{1}{24}\Delta^3 f_0 + \cdots). \tag{3.1}$$

Note that if the formula is truncated after the second term, we get the trapezoidal rule (though expressed in terms of differences). Note also that the next term $-\tfrac{1}{12}\Delta^2 f_0$ will give the local error term of the trapezoidal rule if the second difference is replaced by $h^2 f''(\xi)$, just as we found for interpolation formulas. Indeed,

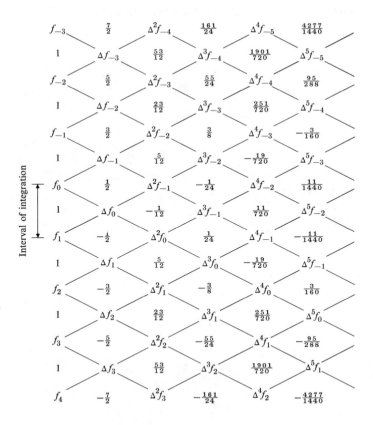

Note: All formulas must be multiplied by h.

Fig. 3.1

this must be so as is seen by comparing Eqs. (3.1) and (1.3), provided that the law of the mean for integrals can be applied.

We would have obtained exactly the same formula, though its form would be different, if we had integrated another interpolating polynomial that also matched the function at the same points, for the definite integral of two identical functions must be the same. This means that we can construct a lozenge diagram for integrals exactly as we have done for interpolating polynomials.

Figure 3.1 shows such a lozenge diagram for integration over one panel, from x_0 to x_1. Figure 3.2 shows a lozenge which pertains to integration over two panels, from x_0 to x_2. Similar diagrams can be constructed for other intervals of integration.

The construction of such lozenges is facilitated by observing that the coefficients which are interspersed among the differences also form a difference table. This, however, is a difference table that reads from right to left, and is also inverted in respect to the order of subtraction. Study of Fig. 5.1, Section 2.5, shows this is true there as well as in Figs. 3.1 and 3.2. We need only one line of coefficients, then, and from these we can complete the entire diagram.

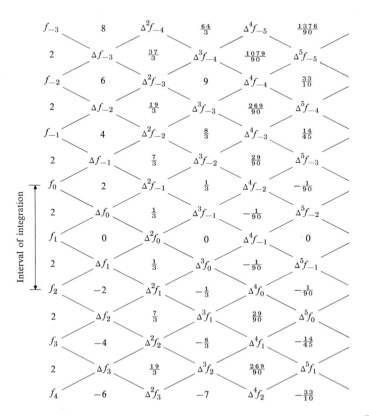

Note: All formulas must be multiplied by h .

Fig. 3.2

4. NEWTON-COTES FORMULAS

The numerical integration formulas that result when interpolating polynomials are integrated over the entire interval at which they match the table are called Newton-Cotes formulas. The first three of these are particularly important. The first one is the trapezoidal rule that we have already discussed in some detail.

The second Newton-Cotes formula is for a quadratic integrated over two panels. Figure 3.2 gives it directly:

$$\int_{x_0}^{x_2} f(x)dx = h(2f_1 + \tfrac{1}{3}\Delta^2 f_0) - \tfrac{1}{90}h^5 f^{iv}(\xi). \tag{4.1}$$

Note that since the coefficient of the $\Delta^3 f$ term was zero, we had to go to the $\Delta^4 f$ column to get the error term. Let us replace $\Delta^2 f_0$ in Eq. (4.1) by its equivalent in terms of f's:

$$\int_{x_0}^{x_2} f(x)dx = h\left(2f_1 + \frac{1}{3}(f_2 - 2f_1 + f_0)\right) - \frac{h^5}{90}f^{iv}(\xi)$$

$$= \frac{h}{3}(f_0 + 4f_1 + f_2) - \frac{h^5}{90}f^{iv}(\xi).$$

This is the very popular Simpson's $\tfrac{1}{3}$ rule. If we apply this to a succession of pairs of panels to evaluate $\int_a^b f(x)dx$, we get

$$\int_a^b f(x)dx = \frac{h}{3}[f_1 + 4f_2 + 2f_3 + 4f_4 + 2f_5 + \cdots + 2f_{n-1} + 4f_n + f_{n+1}]$$

$$- \frac{(b-a)}{90}h^4 f^{iv}(\xi), \qquad x_1 < \xi < x_{n+1}. \tag{4.2}$$

In Eq. (4.2) we have subscripted the x-values so that $x_1 = a$ and $x_{n+1} = b$. In getting the global error we have used the fact that $\sum f^{iv}(\xi_i) = nf^{iv}(\xi)$, which imposes a continuity condition on the fourth derivative. Note that Simpson's $\tfrac{1}{3}$ rule has a global error of $O(h^4)$. Its use requires that we subdivide into an even number of panels.

Example. Apply Simpson's $\tfrac{1}{3}$ rule to the data of Table 1.1 to evaluate $\int_{1.8}^{3.4} f(x)dx$.

Using $h = 0.2$, we get

$$\int_{1.8}^{3.4} f(x)dx \doteq \frac{0.2}{3}[6.050 + 4(7.389) + 2(9.025) + 4(11.023)$$

$$+ 2(13.464) + 4(16.445) + 2(20.086)$$

$$+ 4(24.533) + 29.964]$$

$$= 23.9149.$$

It is no coincidence that this value is the same as the extrapolated value from the trapezoidal rule estimates at $h = 0.2$ and $h = 0.4$. The student should demonstrate that these two procedures always give identical results by working Problem 24.

We can integrate $\int_{1.8}^{3.4} f(x)dx$ using $h = 0.4$ and still apply the $\frac{1}{3}$ rule, getting a second estimate of $O(h^4)$ with h doubled (the value is 23.9181). Extrapolation from these two estimates gives an estimate of $O(h^6)$:$(23.9149 + \frac{1}{15} (23.9149 - 23.9181) = 23.9147)$. Note that the coefficient of the "correction term" which involves the difference between the more accurate and less accurate estimate is $1/(2^{n-1})$ the same as in Eq. (2.2). Since the error term for Simpson's $\frac{1}{3}$ rule can be written as $C_1h^4 + C_2h^6 + C_3h^8 + \cdots$, we can do a Romberg-like set of successive extrapolations.

To get the next Newton-Cotes formula, a cubic integrated over three panels, we must start with the polynomial since we have no lozenge diagram to help us:

$$\int_{x_0}^{x_3} f(x)\,dx \doteq \int_{x_0}^{x_3} P_3(x)\,dx = h\int_{s=0}^{s=3} \left(f_0 + s\,\Delta f_0 + \frac{(s)(s-1)}{2}\Delta^2 f_0 \right.$$
$$\left. + \frac{(s)(s-1)(s-2)}{6}\Delta^3 f_0 \right)ds$$

$$= h\left[sf_0 + \frac{s^2}{2}\Delta f_0 + \left(\frac{s^3}{6} - \frac{s^2}{4}\right)\Delta^2 f_0 + \left(\frac{s^4}{24} - \frac{s^3}{6} + \frac{s^2}{6}\right)\Delta^3 f_0 \right]_0^3$$

$$= h\left[3f_0 + \frac{9}{2}\Delta f_0 + \frac{9}{4}\Delta^2 f_0 + \frac{3}{8}\Delta^3 f_0 \right].$$

If we transform the differences to f's, we get the familiar Simpson's $\frac{3}{8}$ rule,

$$\int_{x_0}^{x_3} f(x)dx \doteq \frac{3h}{8}[f_0 + 3f_1 + 3f_2 + f_3].$$

The local error term of this formula we get by integrating the error term of the cubic polynomial,

$$\text{Local error} = h\int_0^3 \frac{(s)(s-1)(s-2)(s-3)}{24} h^4 f^{iv}(\xi)ds = -\frac{3}{80}h^5 f^{iv}(\xi), \quad x_0 < \xi < x_3.$$

Here the polynomial in s does change sign in the interval but it can still be shown that the error term is proportional to some value of the fourth derivative in the interval. The global error, through the same argument as before, is $O(h^4)$. Note that even though a cubic was used here, in contrast to a quadratic in deriving Simpson's $\frac{1}{3}$ rule, that the error terms are both of $O(h^5)$ locally.* In fact, since the coefficient here is about three times as large, Simpson's $\frac{1}{3}$ rule is often the more accurate. One then wonders if the $\frac{3}{8}$ rule should ever be applied.

As we have seen, Simpson's $\frac{1}{3}$ rule requires the interval to be subdivided into an even number of panels. When applied to tabulated data, this is often impossible. In such cases, one can combine the two rules, applying the $\frac{1}{3}$ rule as far as possible and completing the integration by picking up three panels with the

---

* This is true for all Newton-Cotes formulas from even-degree polynomials. Their global error terms are of order two more than the degree.

$\frac{3}{8}$ rule.   An alternative method for an odd number of panels is to integrate over one panel by the trapezoidal rule.   This involves a penalty of an $O(h^2)$ error over that subinterval.

5. DERIVATION OF FORMULAS BY SYMBOLIC METHODS

In terms of the stepping operator E,

$$f(x_s) = f_s = E^s f_0.$$

Multiplying by $dx = h\,ds$ and integrating from x_0 to x_1 ($s = 0$ to $s = 1$), we obtain

$$\int_{x_0}^{x_1} f(x)\,dx = h\int_0^1 E^s f_0\,ds = \left[\frac{hE^s}{\ln E}f_0\right]_0^1 = \frac{h(E-1)}{\ln E}f_0.$$

Let $E = 1 + \Delta$ and expand $\ln(1 + \Delta)$ as a power series:

$$\ln(1 + \Delta) = \Delta - \tfrac{1}{2}\Delta^2 + \tfrac{1}{3}\Delta^3 - \tfrac{1}{4}\Delta^4 + \cdots.$$

On dividing Δ by this series, we get

$$\int_{x_0}^{x_1} f(x)\,dx = \frac{h\Delta}{\Delta - \tfrac{1}{2}\Delta^2 + \tfrac{1}{3}\Delta^3 - \tfrac{1}{4}\Delta^4 + \cdots}f_0$$

$$= h(f_0 + \tfrac{1}{2}\Delta f_0 - \tfrac{1}{12}\Delta^2 f_0 + \tfrac{1}{24}\Delta^3 f_0 - \cdots). \tag{5.1}$$

The coefficients are considerably easier to get by this technique in comparison to the term by term integration of Section 3.

We can similarly get a line of coefficients for the lozenge diagram (Fig. 3.2) by integrating over two panels

$$\int_{x_0}^{x_2} f(x)\,dx = \left[\frac{hE^s}{\ln E}f_0\right]_0^2 = \frac{h(E^2-1)}{\ln E}f_0 = \frac{h(E+1)(E-1)}{\ln E}f_0.$$

Again letting $E = 1 + \Delta$ so $E - 1 = \Delta$ and $E + 1 = 2 + \Delta$, and dividing Δ by the series for $\ln(1 + \Delta)$, we get

$$\int_{x_0}^{x_2} f(x)\,dx = h(2 + \Delta)(f_0 + \tfrac{1}{2}\Delta f_0 - \tfrac{1}{12}\Delta^2 f_0 + \tfrac{1}{24}\Delta^3 f_0 - \cdots)$$

$$= h(2f_0 + \Delta f_0 - \tfrac{1}{6}\Delta^2 f_0 + \tfrac{1}{12}\Delta^3 f_0 - \cdots$$

$$+ \Delta f_0 + \tfrac{1}{2}\Delta^2 f_0 - \tfrac{1}{12}\Delta^3 f_0 + \cdots)$$

$$= h(2f_0 + 2\Delta f_0 + \tfrac{1}{3}\Delta^2 f_0 + 0 - \cdots).$$

The coefficients of Fig. 3.2 are hence computed from those in Fig. 3.1 quite readily.   The method can obviously be extended.

We now present still another interesting method of deriving formulas that can be applied to a variety of situations including the development of integration formulas.   It may be called the method of *undetermined coefficients*.   We express the formula as a sum of $n + 1$ terms with unknown coefficients, and then evaluate the coefficients by requiring that the formula be exact for all polynomials of degree n or less.   This method is exploited more fully in Chapter 5.   We illustrate it here by finding Simpson's $\frac{1}{3}$ rule by this other technique.   Express the integral as a weighted sum of three equispaced function values:

$$\int_{-1}^{1} f(x)\ dx = af(-1) + bf(0) + cf(+1). \tag{5.2}$$

The symmetrical interval of integration simplifies the arithmetic. We stipulate that the function is to be evaluated at three equally spaced intervals, the two end values and the midpoint. Since the formula contains three terms, we require it to be correct for all polynomials of degree 2 or less. If that is true, it certainly must be true for the three special cases of $f(x) = x^2$, $f(x) = x$, and $f(x) = 1$. We rewrite Eq. (5.2) three times, applying each definition of $f(x)$ in turn:

$$f(x) = 1, \qquad \int_{-1}^{1} dx \quad = 2 = a(1) + b(1) + c(1) = a + b + c,$$

$$f(x) = x, \qquad \int_{-1}^{1} x\ dx = 0 = a(-1) + b(0) + c(1) = -a + c,$$

$$f(x) = x^2, \qquad \int_{-1}^{1} x^2\ dx = \tfrac{2}{3} = a(1) + b(0) + c(1) = a + c.$$

Solving the three equations simultaneously gives $a = \tfrac{1}{3}$, $b = \tfrac{4}{3}$, $c = \tfrac{1}{3}$. Here the spacing between points was one; obviously the integral is proportional to $\Delta x = h$. We then get Simpson's $\tfrac{1}{3}$ rule:

$$\int_{-h}^{h} f(x)\ dx = h[\tfrac{1}{3}f(-h) + \tfrac{4}{3}f(0) + \tfrac{1}{3}f(h)].$$

6. GAUSSIAN QUADRATURE

Our previous formulas for numerical integration were all predicated on evenly spaced x-values, which were specified. With a formula of three terms, then, there were three parameters, the coefficients (weighting factors) applied to each of the functional values. Gauss observed that if we remove the requirement that the function be evaluated at predetermined x-values, a three-term formula will contain six parameters (the three x-values are now unknowns, plus the three weights) and should correspond to an interpolating polynomial of degree 5. Formulas based on this principle are called *Gaussian quadrature formulas.* They can be applied only when $f(x)$ is explicitly known, so that it can be evaluated at any desired value of x.

We shall determine the parameters in the simple case of a two-term formula, containing four unknown parameters:

$$\int_{-1}^{1} f(t) \doteq af(t_1) + bf(t_2).$$

The method is the same as that illustrated in the previous section, by determining unknown parameters. We use a symmetrical interval of integration to simplify the arithmetic, and call our variable t. (This agrees with the notation of most authors. Since the variable of integration is only a dummy variable, its name is unimportant.) Our formula is to be valid for any polynomial of degree 3, hence it will hold if $f(t) = t^3$, $f(t) = t^2$, $f(t) = t$, and $f(t) = 1$:

$$f(t) = t^3, \qquad \int_{-1}^{1} t^3 \, dt = 0 = at_1^3 + bt_2^3,$$

$$f(t) = t^2, \qquad \int_{-1}^{1} t^2 \, dt = \tfrac{2}{3} = at_1^2 + bt_2^2,$$

$$f(t) = t, \qquad \int_{-1}^{1} t \, dt = 0 = at_1 + bt_2,$$

$$f(t) = 1, \qquad \int_{-1}^{1} dt = 2 = a + b.$$

Multiplying the third equation by t_1^2, and subtracting from the first, we have

$$0 = 0 + b(t_2^3 - t_2 t_1^2) = b(t_2)(t_2 - t_1)(t_2 + t_1). \tag{6.1}$$

We can satisfy Eq. (6.1) by either $b = 0$, $t_2 = 0$, $t_1 = t_2$ or $t_2 = -t_1$. Only the last of these possibilities is satisfactory, the others being invalid, or else reduce our formula to only a single term, so we choose $t_1 = -t_2$. We then find that

$$a = b = 1,$$

$$t_2 = -t_1 = \sqrt{\tfrac{1}{3}} = 0.5773,$$

$$\int_{-1}^{1} f(t) \, dx \doteq f(-0.5773) + f(0.5773).$$

It is remarkable that adding these two values of the function gives the exact value for any cubic polynomial over the interval from -1 to 1.

Suppose our limits of integration are from a to b, and not -1 to 1 for which we derived this formula. To use the tabulated Gaussian quadrature parameters we must change the interval of integration to $(-1, 1)$ by a change of variable. If

$$x = \frac{(b-a)t + b + a}{2}, \qquad \text{so} \qquad dx = \left(\frac{b-a}{2}\right) dt,$$

then

$$\int_{a}^{b} f(x) \, dx = \frac{b-a}{2} \int_{-1}^{1} f\!\left(\frac{(b-a)t + b + a}{2}\right) dt.$$

The parameters for Gaussian formulas of higher order than the simple one we have derived lead to a set of simultaneous equations in exactly the same way. The set of equations that results from writing the definition of $f(t)$ as a succession of polynomials is not so easily solved, however. The theory of orthogonal polynomials is employed; the values of t that satisfy the equations are the roots of the Legendre polynomials. We are content to tabulate a few of the results in Table 6.1. More extensive tables are available.*

---

* Love (1966) gives values for up to 200-term formulas.

Table 6.1

Number of terms	Values of t	Weighting factor	Valid up to degree
2	−0.57735027	1.0	3
	0.57735027	1.0	
3	−0.77459667	0.55555555	5
	0.0	0.88888889	
	0.77459667	0.55555555	
4	−0.86113631	0.34785485	7
	−0.33998104	0.65214515	
	0.33998104	0.65214515	
	0.86113631	0.34785485	

Example.   Evaluate $I = \int_{0.2}^{1.5} e^{-x^2} \, dx$ using the three-term Gaussian formula

Change variable: let $x = \dfrac{(1.5 - 0.2)t + 1.5 + 0.2}{2} = 0.65t + 0.85$,

$$I = \frac{1.5 - 0.2}{2} \int_{-1}^{1} e^{-(0.65t + 0.85)^2} \, dt$$

$$= 0.65[0.555 \cdots e^{-(0.65(-0.774\ldots)+0.85)^2} + 0.888 \cdots e^{-(0.65(0.0)+0.85)^2}$$

$$+ 0.555 \cdots e^{-(0.65(0.774\ldots)+0.85)^2}]$$

$$= 0.6586 \quad \text{(cf. to correct value 0.65882).}$$

7. COMPUTER PROGRAMS

Two integration programs are given as examples as to how the computer can employ the algorithms of this chapter.   In Program 1, Simpson's $\frac{1}{3}$ rule is used to find the integral of a function defined by an arithmetic statement function, and a different function can be integrated by changing the card that defines the function.   [Of course, the Hollerith statement in the output line should also be changed when $f(x)$ is not the integrand of the sine integral.]   For the sine integral, which is defined as

$$\int_{0}^{B} \frac{\sin x}{x} \, dx,$$

the upper limit B and the number of panels N are input parameters.   Since N must be even for Simpson's $\frac{1}{3}$ rule, this fact is first checked by comparing the values of N and twice NHALF, where the latter is computed by a fixed point divide.

Program 1

```
ZZJOB 5                        CSC   001      GERALD, C. F.
ZZFORX5
*LIST PRINTER
C   PROGRAM FOR SIMPSONS RULE EVALUATION OF SINE INTEGRAL BETWEEN X = 0 AND X = B
        F(X) = SINF(X)/X
C   READ UPPER LIMIT OF INTEGRATION AND NUMBER OF PANELS, N GREATER THAN 2
        READ 100, B, N
C   CHECK THAT NUMBER OF PANELS IS EVEN
        NHALF = N/2
        IF (N - 2*NHALF) 99,1,99
    99 PRINT 200
        CALL EXIT
C   COMPUTE DELTA X.   ADD FIRST, SECOND, AND LAST VALUES TO SUM
     1 XN = N
        DELX = B/XN
        SUM = 1.0 + 4.*F(DELX) + F(B)
C   ADD REST OF VALUES TO SUM
        X = 2.*DELX
        DO 10 I = 2,NHALF
        SUM = SUM + 2.*F(X) + 4.*F(X+DELX)
    10 X = X + 2.*DELX
C   COMPUTE INTEGRAL AND OUTPUT THE VALUE
        SININT = DELX/3.*SUM
        PRINT 201, B, SININT, N
        CALL EXIT
   100 FORMAT (F10.0, I10)
   200 FORMAT (60H WITH ODD NUMBER OF PANELS, SIMPSONS 1/3 RULE DOES NOT
      1APPLY )
   201 FORMAT (41H VALUE OF SINE INTEGRAL FROM X = 0 TO X = , F6.2,
      1 3H IS , F10.7, 5H WITH , I4, 7H PANELS )
        END
5.0                    100
ZZZZ


VALUE OF SINE INTEGRAL FROM X = 0 TO X =   5.00 IS 1.5499307 WITH 100 PANELS
```

Note that if N should be odd, NHALF will be less than exactly one-half of N because the remainder is truncated in the fixed point division.

After the first, second, and last terms in Simpson's rule sum are accumulated, the pattern of the remaining terms is repetitive so that a DO loop can readily accumulate the rest of the summation. After multiplying by $\Delta x/3$, the integral according to Eq. (4.2) results.

The second integration program (Program 2) uses Romberg integration with 10 subintervals arbitrarily chosen as the starting point. Again the sine integral is computed in the example. After a first approximation has been found with 10 subintervals, the value of Δx is halved, an improved approximation is found, and an extrapolated value is determined utilizing Eq. (2.2) with $n = 2$ the first time. If the difference between the integral with 20 panels and the extrapolated value exceeds an error tolerance, TOL, the interval halving, extrapolation, and comparing continues until either the error tolerance is met, or else eight halvings have been made, whereupon the process is abandoned. The output of the program is a table of successive approximations and corresponding successive extrapolations, together with a statement of the final result.

Program 2

```
ZZJOB 5                          CSC   001      GERALD, C. F.
ZZFORX5
*LIST PRINTER
C   PROGRAM FOR ROMBERG INTEGRATION
C   PROGRAM BEGINS WITH TRAPEZOIDAL INTEGRATION WITH 10 SUBINTERVALS
C   INTERVALS ARE HALVED AND RESULTS ARE EXTRAPOLATED UP TO EIGHTH ORDER
C      OF EXTRAPOLATION.
C   MAXIMUM NUMBER OF SUBINTERVALS USED IN PROGRAM IS 2560.
C   DEFINE INTEGRAND WITH AN ARITHMETIC STATEMENT FUNCTION.
        FCN(X) = SINF(X)/X
        DIMENSION TRAP(9,9)
C   READ UPPER LIMIT AND TOLERANCE AND COMPUTE FIRST INTEGRAL.
        READ 100, B, TOL
        H = B/10.
        SUM = 1. + FCN(B)
        X = 0.0
        DO 10 I = 2,10
        X = X + H
     10 SUM = SUM + FCN(X)*2.
        TRAP(1,1) = H/2.*SUM
C   RECOMPUTE INTEGRAL WITH H HALVED, EXTRAPOLATE AND TEST.  REPEAT UP TO 8 TIMES
        DO 20 I = 1,8
        H = H/2.
        X = H
        K = 10.*2.**I
        DO 30 J = 2,K,2
        SUM = SUM + FCN(X)*2.
     30 X = X + H + H
        TRAP(1,I+1) = H/2.*SUM
        DO 40 II = 1,I
     40 TRAP(II+1,I+1) = TRAP(II,I+1) + 1./(4.**II - 1.)*(TRAP(II,I+1) -
       1 TRAP(II,I))
        IF (ABSF(TRAP(I+1,I+1) - TRAP(I,I+1))- TOL) 50,50,20
     20 CONTINUE
C   IF TOLERANCE NOT MET AFTER 8 EXTRAPOLATIONS, PRINT CALCULATED VALUES AND NOTE
        PRINT 200, B
        DO 60 K = 1,9
     60 PRINT 203, (TRAP(J,K), J = 1,K)
        PRINT 201, TRAP(9,9), TOL
        CALL EXIT
C   PRINT RESULTS WHEN TOLERANCE IS MET.
     50 PRINT 200, B
        I = I + 1
        DO 70 L = 1,I
     70 PRINT 203, (TRAP(J,L), J = 1,L)
        PRINT 202, TRAP(I,I), K, TOL
        CALL EXIT
    100 FORMAT (F10.0, E10.0)
    200 FORMAT (// 38H VALUE OF SINE INTEGRAL BETWEEN 0 AND , F6.2,
       1 24H BY ROMBERG INTEGRATION / )
    201 FORMAT (/ 10H VALUE IS , F10.7, 31H BUT DID NOT MEET TOLERANCE OF
       1 E14.7)
    202 FORMAT (/ 10H VALUE IS , F10.7, 2H, / I5,
       1 49H SUBINTERVALS WERE REQUIRED TO MEET TOLERANCE OF , E14.7)
    203 FORMAT (1H ,8F10.7)
        END
15.0            2.E-7
ZZZZ


VALUE OF SINE INTEGRAL BETWEEN 0 AND  15.00 BY ROMBERG INTEGRATION

1.6077255
1.6156591 1.6183036
1.6175654 1.6182008 1.6181940
1.6180369 1.6181940 1.6181936 1.6181936

VALUE IS  1.6181936,
   80 SUBINTERVALS WERE REQUIRED TO MEET TOLERANCE OF  2.0000000E-07
```

PROBLEMS

Section 1

1. The following values of a function are given.

x	$f(x)$	x	$f(x)$
1.0	1.543	1.5	2.352
1.1	1.668	1.6	2.577
1.2	1.811	1.7	2.828
1.3	1.971	1.8	3.107
1.4	2.151		

Find $\int_{1.0}^{1.8} f(x)\, dx$, using the trapezoidal rule with

     (a) $h = 0.1$,   (b) $h = 0.2$,   (c) $h = 0.4$.

2. The function tabulated in Problem 1 is cosh x. What are the errors of the estimates (a), (b), (c)? How closely are the errors proportional to h^2? What other errors besides that of Eq. (1.5) are present?

3. Compute

$$\int_0^1 \frac{\sin x}{x}\, dx$$

by the trapezoidal rule, subdividing the interval [0,1] into (a) one, (b) two, (c) four subintervals. Carry five decimals. This integral is not expressible in terms of ordinary functions, but is important enough to be tabulated under the name *sine integral*. Compare your answers to the tabulated value.

4. In Section 1, the value of $\int_{1.8}^{3.4} e^x\, dx$ is computed as 23.9944 using the trapezoidal rule with $h = 0.2$, and is in error by -0.08. Use Eq. (1.5) to find maximum and minimum values for the error.

5. If one wished to compute $\int_{1.8}^{3.4} e^x\, dx$ using the trapezoidal rule, and wished to be certain of five-decimal accuracy (error ≤ 0.000005), how small must h be?

6. Use the trapezoidal rule to compute $\int_1^2 dx/x$, choosing h small enough to give three-decimal accuracy. Compare to the analytical value.

7. An advantage of the trapezoidal rule is that the spacing does not have to be uniform. Derive a formula analogous to Eq. (1.2) for unevenly spaced data which is known at $x_0, x_1, \cdots, x_n$. Use $h_i = x_i - x_{i-1}$ as the symbol for the ith interval.

8. Find $\int_0^2 f(x)\,dx$:

x	$f(x)$	x	$f(x)$
0	1.0000	1.08	0.3396
0.12	0.8869	1.43	0.2393
0.53	0.5886	2.00	0.1353
0.87	0.4190		

9. An integral with infinite limits can be approximated numerically if it is convergent. For example, the exponential integral $Ei(x)$ can be evaluated by taking the upper limit, U, sufficiently large in

$$Ei(x) = \int_x^\infty \frac{e^{-v}}{v}\,dv \doteq \int_x^U \frac{e^{-v}}{v}\,dv.$$

We know that U is "sufficiently large" when the additional contributions of making U larger are negligible. Estimate $Ei\,(0.5)$ using the trapezoidal rule. Note that one can use larger subintervals as v increases. Compare to the tabular value of 0.5598.

10. Apply the method suggested in Problem 9 to show that

$$\int_{-\infty}^\infty \frac{1}{\sqrt{2\pi}} e^{-x^2/2}\,dx = 1.$$

(The values of the integrand are available as ordinates of the standard normal curve.)

Section 2

11. Extrapolate the individual answers of Problem 1 to get estimates of improved accuracy. What are the orders of the errors of these extrapolations?

12. Use extrapolation to the limit to evaluate $\int_0^1 f(x)\,dx$, getting a result with error $O(h^6)$:

x	$f(x)$	x	$f(x)$
0	0.3989	0.50	0.3521
0.25	0.3867	0.75	0.3011
		1.00	0.2420

13. Use Romberg integration (successive extrapolations with h halved each time) to evaluate $\int_1^2 dx/x$. Carry six decimals and continue until no change in the fifth place occurs. Compare to the analytical value $\ln 2 = 0.69315$.

14. Use Romberg integration to evaluate

$$\int_0^2 \frac{dx}{\sqrt{1 + x^3}}$$

to four decimals.

Section 3

15. Use the lozenge diagram (Fig. 3.1), to write an integration formula in terms of back-ward differences of f_0 (these backward differences are found on a diagonal sloping upward from f_0), when the integral is taken over the next interval from x_0, i.e.,

$$\int_{x_0}^{x_1} f(x) \, dx.$$

This formula will be useful in Chapter 6.

16. Another formula to be used in Chapter 6 is for $\int_{x_0}^{x_1} f(x) \, dx$ in terms of backward differences of f_1. Write this formula.

17. If the two formulas of Problems 15 and 16 are each terminated on fourth differences, find the respective error terms. Which has the smaller coefficient? How much smaller than the other one?

18. Using the data of Problem 1, make a difference table, carrying it out to fourth difference. The lozenge of Fig. 3.2 then permits one to find the integral over two panels with an (Oh^6) error. Summing four such double panel integrals gives $\int_{1.0}^{1.8} f(x) \, dx$. Compare your answer to results of Problem 11.

19. If the method of Problem 18 were repeated, except by summing one-panel integration formulas which include differences up to the fourth, which may be derived from Fig. 3.1, would the exact same result always be obtained? Justify your answer.

20. In Section 3, it is stated that the same integration formula will result by integration of any of the interpolation formulas (with proper limits of integration, of course). This implies that the interspersed coefficients in the lozenge diagrams should be reproduced as well. Confirm this by computing some of the coefficients in Figs. 3.1 and 3.2 starting with (a) Newton-Gregory backward, (b) Gauss forward, (c) Stirling interpolation formulas.

Section 4

21. Repeat Problem 1 using Simpsons $\frac{1}{3}$ rule with (a) $h = 0.1$, (b) $h = 0.2$.

22. Use the error expression of Eq. (4.2) to estimate the maximum and minimum errors to be expected in Problem 21. [$f(x)$ in Problem 1 is cosh x.] Why do the actual errors not fall within this range?

23. Evaluate $\int_0^{1.0} e^x \, dx$ by Simpson's $\frac{1}{3}$ rule choosing h small enough to guarantee five decimal accuracy. How large can h be?

24. In the text it is stated that the results of using Simpson's $\frac{1}{3}$ rule, and the extrapolated result from two applications of the trapezoidal rule (with h changed by 2:1) are identical. Show this will always be true.

25. Compute
$$\int_0^1 \frac{\sin x}{x}\, dx$$
using Simpson's $\frac{1}{3}$ rule with $h = 0.5$ and with $h = 0.25$.   Extrapolate the results. What is the order of the error for the extrapolated result?

26. Repeat Problem 23, except use Simpson's $\frac{3}{8}$ rule.

27. Apply a combination of the two Simpson rules to evaluate the integral from $x = 3$ to $x = 6.5$.   Note that this is a technique to handle an odd number of panels:

x	$f(x)$	x	$f(x)$
3.0	0.33906	5.0	−0.32758
3.5	0.13738	5.5	−0.34144
4.0	−0.06604	6.0	−0.27668
4.5	−0.23106	6.5	−0.15384

28. In combining the two Simpson rules, as in Problem 27, one has a choice as to where in the interval the group of three panels is selected to be integrated by Simpson's $\frac{3}{8}$ rule.   What will ordinarily be the best choice to make the total error smallest?   Why is such a choice not certain to give the smallest total error?

29. The fact that Simpson's $\frac{1}{3}$ rule has an error term dependent on the third derivative and not the second (as is also true for the $\frac{3}{8}$ rule) means that these integrations are exact for any cubic polynomial.   However, the $\frac{1}{3}$ rule is based on fitting a quadratic through three equispaced points.   The implication of this is that the area under any cubic, from $x = a$ to $x = b$, is exactly the same as the area under the parabola that intersects the cubic at the two end points, and also at $x = \frac{1}{2}(a + b)$.   Prove this.

Section 5

30. Use the symbolic method to determine a line of coefficients for three-panel integration, and from this develop a lozenge diagram for integration from x_0 to x_3.

31. Integration from x_0 to x_1, as in Eq. (5.1), is an arbitrary choice taken for convenience only.   If the limits were taken from x_{-1} to x_0, a similar formula for one-panel integration would result, but different coefficients in the lozenge diagram will be obtained.   Carry out this computation and check your line of coefficients against the lozenge diagram (Fig. 3.1).

32. Perform a similar computation to Problem 31, except get a two-panel integration formula from x_{-1} to x_1.   Check the results against Fig. 3.2.

33. Use the method of undetermined coefficients to derive (a) the trapezoidal rule, (b) Simpson's $\frac{3}{8}$ rule.

Section 6

34. Evaluate
$$\int_0^1 \frac{\sin x}{x}\, dx$$
by a three-term Gaussian quadrature formula.

35. By computing with Gaussian quadrature formulas of increasing complexity, determine how many terms are needed to evaluate $\int_{1.8}^{3.4} e^x \, dx$ to five-decimal accuracy. (The exact value of the integral is 23.9144526.)

36. Repeat Problem 14 using Gaussian quadrature.

37. An n-term Gaussian quadrature formula assumes that a polynomial of degree $2n$-1 is used to fit the function between $x = a$ and $x = b$. Does this mean that the error term is the same as for the integration formulas from Fig. 3.1 which include differences up to the $2n$-1 order? Justify your answer.

38. Referring to Problem 37, what kind of lozenge diagrams would give error terms that are equivalent to Gaussian quadrature? Discuss how the symbolic operator method of Section 5 might be used to get these lozenge coefficients.

Section 7

39. Use the Romberg integration program to evaluate the following integrals:

(a) $\dfrac{2}{\sqrt{\pi}} \displaystyle\int_0^1 e^{-x^2} \, dx$, tolerance $= 10^{-6}$

(b) $\displaystyle\int_0^5 e^{-x} \, dx$, tolerance $= 10^{-7}$

(c) $\displaystyle\int_0^{\pi/2} \dfrac{d\theta}{\sqrt{1 - \sin^2 \theta}}$, tolerance $= 2 \times 10^{-8}$

(d) $\displaystyle\int_0^{\pi/2} \sqrt{1 - 0.5 \sin^2 \theta} \, d\theta$, tolerance $= 2 \times 10^{-8}$

40. Repeat Problem 39 except use the program for Simpson's rule with 100 panels in each case.

41. Write a program that will apply a combination of Simpson's $\frac{1}{3}$ and $\frac{3}{8}$ rules to a set of equispaced tabular data. If the number of panels is even, use the $\frac{1}{3}$ rule throughout. If odd, use the $\frac{3}{8}$ rule on either the first three or last three panels as chosen by the user.

42. Rewrite the program for Simpson's rule so that the integral is recomputed with the number of panels doubled each time, the estimates of the integral compared, and unless a certain tolerance is met, the number of panels doubled again. Is there a way to avoid recomputing values of the function which are to be reused in the next summation?

43. Write a program that will integrate a function using Gaussian quadrature, with two-, three-, or four-term formulas available as desired.

4
numerical
differentiation

We are interested in methods of finding the derivatives of functions which are known only as a tabulation of data so that the rate of change of such functions can be approximated. We are also interested in finding derivatives by non-analytical means, even if the function is known, for application to computer programs. The methods of this chapter are fundamental to the numerical solution of partial differential equations.

1. DERIVATIVES FROM INTERPOLATING POLYNOMIALS

If a function is reasonably well approximated by an interpolating polynomial, the slope of the function should also be approximated by the slope of the poly-nomial, although we shall discover that the error of estimating the slope is greater than the error of estimating the function.

We begin with a Newton-Gregory forward polynomial:

$$f(x_s) \doteq P_n(x_s) = f_0 + s\Delta f_0 + \binom{s}{2}\Delta^2 f_0 + \cdots + \binom{s}{n}\Delta^n f_0. \tag{1.1}$$

The error term of Eq. (1.1) is, by our rule for changing the next term,

$$\text{Error of } P_n(x_s) = \binom{s}{n+1}h^{n+1}f^{(n+1)}(\xi), \qquad x_0 < \xi < x_n. \tag{1.2}$$

Differentiating Eq. (1.1), remembering that all the Δ-terms are constants (numbers from the difference table), we have

$$f'(x_s) \doteq P'_n(x_s) = \frac{d}{dx}\left(P_n(x_s)\right) = \frac{d}{ds}\left(P_n(x_s)\right)\frac{ds}{dx} = \frac{d}{ds}\left(P_n(x_s)\right)\frac{1}{h}$$
$$= \frac{1}{h}\left[\Delta f_0 + \tfrac{1}{2}(s-1+s)\Delta^2 f_0 + \tfrac{1}{6}\left((s-1)(s-2)+s(s-2)+s(s-1)\right)\Delta^3 f_0 + \cdots\right].$$
$$\tag{1.3}$$

The derivatives of the factorials of s rapidly become algebraically complicated. We get some simplification if we let $s = 0$, giving us the derivative corresponding to x_0 (where $s = 0$), however. If we let $s = 0$, Eq. (1.3) becomes

$$f'(x_0) \doteq \frac{1}{h}[\Delta f_0 - \tfrac{1}{2}\Delta^2 f_0 + \tfrac{1}{3}\Delta^3 f_0 - \tfrac{1}{4}\Delta^4 f_0 + \cdots \pm \frac{1}{n}\Delta^n f_0]. \tag{1.4}$$

The error in Eq. (1.3) is obtained by differentiating the error term given in Eq. (1.2):

$$\text{Error of } P'_n(x_s) = h^{n+1}f^{(n+1)}(\xi)\left[\frac{d}{ds}\binom{s}{n+1}\right]\frac{1}{h} + \binom{s}{n+1}h^{n+1}\frac{d}{dx}\left[f^{(n+1)}(\xi)\right]. \tag{1.5}$$

The second term here cannot be evaluated, for the way that ξ varies with x is unknown, but when we let $s = 0$ this second term vanishes because

$$\binom{s}{n+1} = \frac{1}{(n+1)!}(s)(s-1)\cdots(s-n) = 0$$

Table 1.1

x	y	Δy	$\Delta^2 y$	$\Delta^3 y$	$\Delta^4 y$
1.3	3.669				
		0.813			
1.5	4.482		0.179		
		0.992		0.041	
1.7	5.474		0.220		0.007
		1.212		0.048	
1.9	6.686		0.268		0.012
		1.480		0.060	
2.1	8.166		0.328		0.012
		1.808		0.072	
2.3	9.974		0.400		
		2.208			
2.5	12.182				

at $s = 0$. We then need only evaluate the first term of Eq. (1.5). Since

$$\frac{d}{ds}\binom{s}{n+1} =$$

$$\frac{(s-1)(s-2)\cdots(s-n)+s(s-2)(s-3)\cdots(s-n)+\cdots+s(s-1)\cdots(s-n+1)}{(n+1)!},$$

at $s = 0$, only the first term of the numerator remains, and Eq. (1.5) becomes

$$\text{Error of } P'_n(x_0) = h^{n+1}f^{(n+1)}(\xi)\left[(-1)^n\frac{n!}{(n+1)!}\right]\left(\frac{1}{h}\right) = \frac{(-1)^n}{n+1}h^n f^{(n+1)}(\xi).$$

$$(1.6)$$

Note that even though the interpolating polynomial gives the function exactly at $s = 0$, our derivative formula is subject to error at that point, unless $f^{(n+1)}(x_0) = 0$. In comparing Eq. (1.6) and (1.4) we see that the error term of the derivative can be obtained by changing the $\Delta^n f_0$ in the next term beyond the last one included into $h^n f^{(n)}(\xi)$ as we have grown accustomed to with our interpolating and integration formulas.

Example. Use the data in Table 1.1 to estimate the derivative of y at $x = 1.7$. Use $h = 0.2$, and compute using two, three, or four terms of the formula.

With two terms $y'(1.7) = \dfrac{1}{0.2}\left(1.212 - \tfrac{1}{2}(0.268)\right) = 5.390,$

With three terms $y'(1.7) = \dfrac{1}{0.2}\left(1.212 - \tfrac{1}{2}(0.268) + \tfrac{1}{3}(0.060)\right) = 5.490,$

With four terms $y'(1.7) = \dfrac{1}{0.2}\left(1.212 - \tfrac{1}{2}(0.268) + \tfrac{1}{3}(0.060) - \tfrac{1}{4}(0.012)\right)$

$$= 5.475.$$

The data tabulated in Table 1.1 are for $y = e^x$, rounded to three decimals.*
Since the derivative of e^x is also e^x, we see that the error in the derivative is quite
small with four terms, comparing 5.475 to 5.474. We can anticipate this; since
the fourth differences in Table 1.1 do not vary greatly, the function is represented
reasonably well by a fourth-degree polynomial.

2. SYMBOLIC DERIVATION OF FORMULAS

We can get formulas for higher derivatives by further differentiating Eq. (1.3),
and differentiating Eq. (1.5) will give the error terms. For example,

$$f''(x_s) \doteq P''_n(x_s) = \frac{1}{h^2} \frac{d^2}{ds^2} P_n(x_s)$$

$$\doteq \frac{1}{h^2} \left[\tfrac{1}{2}(2)\Delta^2 f_0 + \tfrac{1}{6}((s-2)+(s-1)+(s-2)+s+(s-1)+s)\Delta^3 f_0 + \cdots \right].$$

At $s = 0$,

$$f''(x_0) \doteq \frac{1}{h^2} [\Delta^2 f_0 - \Delta^3 f_0 + \cdots].$$

There is no simple pattern for the coefficients, and the error terms have complicated coefficients as well. We have seen before that symbolic methods have
given formulas with relative ease. It is true for derivative formulas also:

$$E = 1 + \Delta,$$

$$y_s = E^s y_0,$$

$$y'_s = \frac{d}{dx}(E^s y_0) = \frac{1}{h}\frac{d}{ds}(E^s y_0) = \frac{1}{h}(\ln E) E^s y_0. \tag{2.1}$$

At $s = 0$,

$$y'_0 = \frac{1}{h}(\ln E)y_0 = \frac{1}{h}\ln(1 + \Delta)y_0$$

$$= \frac{1}{h}[\Delta y_0 - \tfrac{1}{2}\Delta^2 y_0 + \tfrac{1}{3}\Delta^3 y_0 - \tfrac{1}{4}\Delta^4 y_0 + \cdots]. \tag{2.2}$$

In writing Eq. (2.2), we have used the Maclaurin expansion for $\ln(1 + \Delta)$.
If we use the symbol D for the derivative operator,

$$Dy_0 = \frac{1}{h}\ln(1 + \Delta)y_0.$$

---

* The table incidentally illustrates the effect of rounding errors in the fourth differences.
A more accurate table would show fourth differences of 0.0088, 0.0108, 0.0132.

We abstract the equivalence between operators:

$$D = \frac{1}{h} \ln(1 + \Delta),$$

$$D^2 = \frac{1}{h^2} \ln^2(1 + \Delta), \tag{2.3}$$

$$D^n = \frac{1}{h^n} \ln^n(1 + \Delta).$$

Equations (2.3) show that we can get formulas for higher derivatives by multiplying the series in Eq. (2.1) by itself. The second derivative is given by

$$D^2 y_0 = \frac{1}{h^2} [\Delta - \tfrac{1}{2}\Delta^2 + \tfrac{1}{3}\Delta^3 - \tfrac{1}{4}\Delta^4 + \cdots]^2 y_0$$

$$= \frac{1}{h^2} [\Delta^2 - \Delta^3 + \tfrac{11}{12}\Delta^4 - \tfrac{5}{6}\Delta^5 + \cdots] y_0$$

$$= \frac{1}{h^2} (\Delta^2 y_0 - \Delta^3 y_0 + \tfrac{11}{12}\Delta^4 y_0 - \tfrac{5}{6}\Delta^5 y_0 + \cdots). \tag{2.4}$$

Formulas for derivatives of higher order can be similarly obtained by multiplication of the series in Eq. (2.2).

Equations (2.3) lead to an important generalization for the estimation of derivatives. The first term of all the formulas in terms of differences has only Δ^n. Hence, as a first approximation,

$$D^n y = \frac{1}{h^n} \Delta^n y + \text{error } O(h).$$

Example. Use formula (2.4) to estimate y'' (1.7) from Table 1.1 using terms through Δ^3. Also estimate the error.

$$y''(1.7) = \frac{1}{(0.2)^2} (0.268 - 0.060)$$

$$= 5.200 \quad \text{(compare to 5.474, exact answer).}$$

$$\text{Error} = \frac{1}{h^2} \left(-\tfrac{11}{12} h^4 y^{iv}(\xi) \right), \qquad 1.7 < \xi < 2.1,$$

from the next term in the formula. Since $y = e^x$, $y^{iv} = e^x$.

$$\text{Max value of error} = \tfrac{11}{12}(0.2)^2(e^{2.1}) = 0.298,$$
$$\text{Min value of error} = \tfrac{11}{12}(0.2)^2(e^{1.7}) = 0.201.$$

Compare to actual error of 0.274.

In the absence of knowledge of the function, we could have used the fourth differences to estimate $h^4 f^{iv}$. If the differences vary greatly, this is hazardous, however. In this instance we would estimate the error as about 0.23 if we used 0.010 (the average value of $\Delta^4 y$) as an estimate of $h^4 f^{iv}(\xi)$.

3. LOZENGE DIAGRAM FOR DERIVATIVES

Since there is a variety of equivalent forms of the interpolating polynomial, we may obtain a variety of forms of derivative formulas by differentiating the various polynomials. They are interrelated, of course, as we have seen for integration formulas. It is convenient to exhibit this interrelation by a lozenge diagram as in Fig. 3.1. We can build up the lozenge diagram from Eq. (1.4) if we remember that the interspersed coefficients make a second difference table, rotated 180 degrees, and interlaced with the difference table of the function.

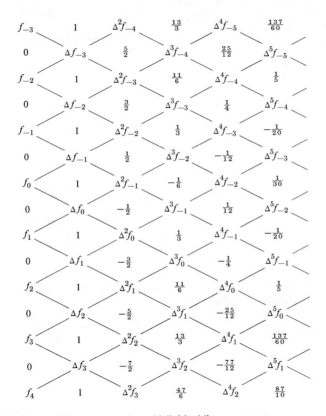

Note: All formulas must be multiplied by $1/h$.

Fig. 3.1

Equation (2.4) serves to start a lozenge diagram for the second derivative at $x = x_0$ in a similar fashion. The result is Fig. 3.2.

We use the lozenge diagrams to write formulas for differentiation of a tabulated function in the same way we used similar diagrams to write interpolation and integration formulas. Any path through the table may be chosen; the coefficients

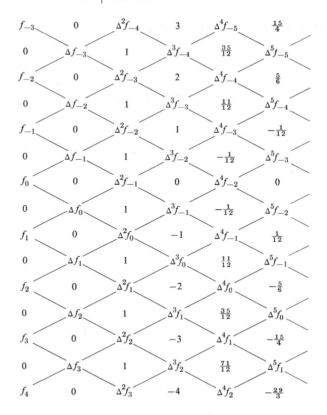

Note: All formulas must be multiplied by $1/h^2$.

Fig. 3.2

are then given by the rules discussed in Section 5, Chapter 2. Suppose we use a horizontal path beginning at f_0:

$$\frac{df}{dx}\bigg|_{x=x_0} = \frac{1}{h}\left[\frac{\Delta f_0 + \Delta f_{-1}}{2} + (0)\,\Delta^2 f_{-1} + \left(-\tfrac{1}{6}\right)\frac{\Delta^3 f_{-1} + \Delta^3 f_{-2}}{2} + \cdots\right]$$

$$= \frac{1}{h}\left[\frac{(f_1 - f_0) + (f_0 - f_{-1})}{2} + O(h^3)\right] = \frac{f_1 - f_{-1}}{2h} + O(h^2). \quad (3.1)$$

Using Fig. 3.2 to find the second derivative, again following a horizontal path, we obtain

$$\frac{d^2 f}{dx^2}\bigg|_{x=x_0} = \frac{1}{h^2}\left[\Delta^2 f_{-1} + 0 + \left(-\tfrac{1}{12}\right)\Delta^4 f_{-2} + \cdots\right]$$

$$= \frac{1}{h^2}\left[(f_1 - 2f_0 + f_{-1}) + O(h^4)\right]$$

$$= \frac{f_1 - 2f_0 + f_{-1}}{h^2} + O(h^2). \quad (3.2)$$

Equations (3.1) and (3.2) are formulas of special importance. Note that even though they are only one-term formulas, they have errors of $O(h^2)$, in contrast to the $O(h)$ error for Eq. (2.4), when only one term is used. These $O(h^2)$ formulas are known as *central difference approximations*. Their good error terms occur because the point at $x = x_0$ is at the center of the range of x-values at which the interpolating polynomial matches the table.

The central difference approximations extend directly to higher derivatives as well. For the even-order derivatives we have

$$f^{(n)}(x)\Big|_{x=x_0} = \frac{1}{h^n}\Delta^n f_{-n/2} + O(h^2), \qquad n \text{ even.}$$

The difference required is always found in the table on the horizontal line through x_0 and f_0.

For odd-order derivatives, the expression involves the average of differences found just above and just below the horizontal line through x_0:

$$f^{(n)}(x)\Big|_{x=x_0} = \frac{\Delta^n f_{(-n+1)/2} + \Delta^n f_{(-n-1)/2}}{2h^n} + O(h^2), \qquad n \text{ odd.} \qquad (3.3)$$

4. EXTRAPOLATION TECHNIQUES

There is an alternative way of successively improving the accuracy of our estimates of the derivative which is equivalent to adding terms as given by the lozenge diagram. This is the technique of successive extrapolations, and resembles the method for Romberg integration. It has special utility in computer programs for differentiation of arbitrary functions.

Table 4.1

x	$f(x)$	x	$f(x)$
2.0	0.42298	2.6	0.25337
2.1	0.40051	2.7	0.22008
2.2	0.37507	2.8	0.18649
2.3	0.34718	2.9	0.15290
2.4	0.31729	3.0	0.11963
2.5	0.28587		

We illustrate the method first for a function known only from a table, such as Table 4.1 We desire the derivative $f'(x)$ at $x = 2.5$. The central difference formula, Eq. (3.1), gives

$$f'(x)\Big|_{x=2.5} \doteq \frac{f_1 - f_{-1}}{2h} = \frac{0.25337 - 0.31729}{2(0.1)} = -0.3196.$$

This estimate has an error $O(h^2)$.

We can also compute the derivative by using values at $x = 2.7$ and $x = 2.3$, for which the spacing is 0.2, and get

$$f'(x)\Big|_{x=2.5} \doteq \frac{0.22008 - 0.34718}{2(0.2)} = -0.3178.$$

The error here is also $O(h^2)$, but now h is twice as large as in the previous calculation.

In order to combine these two estimates and extrapolate to a more accurate estimate, we examine in greater detail the nature of the error term of Eq. (3.1). Using the lozenge diagram, Fig. 3.1, we have

$$f'_0 = \frac{1}{h}\left(\frac{\Delta f_{-1} + \Delta f_0}{2} - \frac{1}{6}\frac{\Delta^3 f_{-2} + \Delta^3 f_{-1}}{2} + \frac{1}{30}\frac{\Delta^5 f_{-3} + \Delta^5 f_{-2}}{2} - \cdots\right). \quad (4.1)$$

If we truncate after the first term, we get an error of $O(h^2)$, but Eq. (3.3) shows that

$$f'''_0 = \frac{\Delta^3 f_{-2} + \Delta^3 f_{-1}}{2h^3} + O(h^2).$$

We may write equivalently,

$$-\frac{1}{6}\frac{\Delta^3 f_{-2} + \Delta^3 f_{-1}}{2} = -\frac{1}{6}h^3 f'''_0 + O(h^5) = Ch^3 + O(h^5).$$

In this expression we merely recognize that the third derivative of f evaluated at x_0 has a fixed value. Equation (4.1) can hence be written as

$$f'_0 = \frac{1}{h}\frac{\Delta f_{-1} + \Delta f_0}{2} + Ch^2 + O(h^4) + \tfrac{1}{30}h^4 f^v(\xi)$$

$$= \frac{1}{h}\frac{\Delta f_{-1} + \Delta f_0}{2} + Ch^2 + O(h^4).$$

We see, just as for the integration formulas, that we make only an order h^4 error in assuming the error of Eq. (3.1) is proportional to h^2. We then combine the two estimates of the derivative, -0.3196 with $h = 0.1$ and -0.3178 with $h = 0.2$, as follows:

$$f'(2.5) = \text{more accurate} + \frac{1}{2^2 - 1}(\text{more} - \text{less accurate})$$

$$= -0.3196 + \frac{1}{2^2 - 1}(-0.3196 + 0.3178)$$

$$= -0.3203.$$

This extrapolation method can be extended since the first-order extrapolations with error $O(h^4)$ can be shown to have errors of the form $Ch^4 + O(h^6)$. (Observe that this follows from the alternately zero coefficients on the horizontal path through Fig. 3.1, so this principle extends to all orders of extrapolation. Errors due to rounding of the original data eventually dominate, however, so the extrapolation technique must be abandoned at some point.)

For the data in Table 4.1 we can extend to one more order of extrapolation.

h	Initial estimate	First-order extrapolation	Second-order extrapolation
0.1	−0.31960		
		−0.32022	
0.2	−0.31775		−0.32020
		−0.32050	
0.4	−0.30951		

The last entry was calculated as

$$-0.32022 + \frac{1}{2^4 - 1}\,(-0.32022 + 0.32050) = -0.32020.$$

An extra guard figure was carried to minimize round-off in the computations.

Since the lozenge for second derivatives, Fig. 3.2, also shows alternate coefficients to be zero, approximations for the second derivative can be extrapolated in exactly the same way, the order of the error increasing to h^4, h^6, h^8, etc., at each step.

It is left as an exercise for the student to show that this extrapolation method is the equivalent to passing a higher degree interpolating polynomial through the data (with the point $x = x_0$ at the center of the region of fit for each of these), and then finding the derivative by central difference formulas.

The application of this extrapolation method in a computer program is important because successive estimates, with h decreasing, become more and more accurate, and hence provide a built-in measure of accuracy. The method is generally employed to differentiate a function of known form, rather than a tabulated function. The basic scheme is to compute the derivative at two values of h, one being one-half the value of the other, and to extrapolate as in the above example. If a significant change is made by the extrapolation, compared to the more accurate of the original estimates, h is halved again and a second-order extrapolation is made. Again, if a significant improvement results, another computation is made with h halved still once more and a third-order extrapolation is made. The process is continued until the improvement is less than some criterion.

Since round-off affects calculation by computers just as it does hand computations, eventually these errors will control the accuracy of successive values, so the student is cautioned against indiscriminate use of extrapolation. It is generally good practice to put a limit to the number of repetitions of the process.

5. A PROGRAM FOR DIFFERENTIATION

Program 1 evaluates the first and second derivatives of a function (defined by an arithmetic statement function) by finite difference methods. The central difference formulas, Eqs. (3.1) and (3.2), are employed and a tabulation is printed showing how the values of the approximation improve as Δx is made smaller.   The values of the second derivative, whose analytical value is 2, show an oscillatory behavior due to round-off.   The first derivative values converge monotonically to zero.

Program 1

```
ZZJOB 5                          CSC  001     GERALD, C. F.
ZZFORX5
*LIST PRINTER
C   PROGRAM TO COMPUTE DERIVATIVES BY FINITE DIFFERENCE FORMULAS
C   DEFINE THE FUNCTION BY AN ARITHMETIC STATEMENT FUNCTION
      F(X) = X**2*EXPF(X)
C   READ IN X VALUE AND INITIAL VALUE OF DELTA X
      READ 100, X, H
C   PRINT HEADING
      PRINT 200, X
C   COMPUTE DERIVATIVES AND PRINT.  DECREASE H AND REPEAT
      FO = F(X)
      DO 10 I = 1,10
      FR = F(X + H)
      FL = F(X - H)
      DX = (FR - FL)/(2.*H)
      DDX = (FR - 2.*FO + FL)/H**2
      PRINT 201, H, DX, DDX
   10 H = 0.1*H
      CALL EXIT
  100 FORMAT (2F10.0)
  200 FORMAT (52H DERIVATIVES COMPUTED BY CENTRAL DIFFERENCE FORMULAS /
     1 8H AT X = , F5.2 // 4X, 1HH, 13X, 2HDX, 17X, 3HDDX /)
  201 FORMAT (1H ,E8.1, 5X, E14.7, 5X, E14.7)
      END
0.0        0.1
ZZZZ
DERIVATIVES COMPUTED BY CENTRAL DIFFERENCE FORMULAS
AT X =   0.00

      H              DX                    DDX

1.0E-01        1.0016675E-02         2.0100083E+00
1.0E-02        1.0000150E-04         2.0000999E+00
1.0E-03        9.9995000E-07         2.0000009E+00
1.0E-04        1.0000000E-08         2.0000000E+00
1.0E-05        9.9500000E-11         1.9999999E+00
1.0E-06        9.5000000E-13         1.9999999E+00
1.0E-07        5.0000000E-15         1.9999999E+00
1.0E-08        0.0000000E-99         2.0000000E+00
1.0E-09        0.0000000E-99         2.0000000E+00
1.0E-10        0.0000000E-99         2.0000000E+00
```

PROBLEMS

Section 1

1. The following table is for sin θ, with θ given in degrees.   Find $d(\sin\theta)/d\theta$, at $\theta = 22°$, correct to four decimals.   How many terms of Eq. (1.4) are required?   Discuss your findings.

θ	$\sin\theta$	θ	$\sin\theta$
20	0.34202	25	0.42262
21	0.35837	26	0.43837
22	0.37461	27	0.45399
23	0.39073	28	0.46947
24	0.40674		

2. The following table is for $(1 + \log x)$. Determine estimates of $d(1 + \log x)/dx$ at $x = 0.15, 0.19, 0.23$ using (a) one term, (b) two terms, and (c) three terms of Eq. (1.4). By comparing to the analytical values, determine the errors of each estimate.

x	$1 + \log x$	Δ	Δ^2	Δ^3
0.15	0.1761			
		0.0543		
0.17	0.2304		−0.0059	
		0.0484		0.0009
0.19	0.2788		−0.0050	
		0.0434		0.0011
0.21	0.3222		−0.0039	
		0.0395		0.0006
0.23	0.3617		−0.0033	
		0.0362		0.0005
0.25	0.3979		−0.0027	
		0.0335		0.0002
0.27	0.4314		−0.0025	
		0.0310		0.0005
0.29	0.4624		−0.0020	
		0.0290		
0.31	0.4914			

3. Write expressions for the errors in each computation of Problem 2, by properly interpreting Eq. (1.6). From these expressions find upper and lower bounds for each of the computations.

4. If one wished the derivative of $(1 + \log x)$ at $x = 0.29$, the table in Problem 2 would be inadequate for this, using Eq. (1.4), because the necessary forward differences cannot be computed. A formula in terms of backward differences can be derived, however, by differentiating the Newton-Gregory backward interpolation polynomial. Do this, and use three terms to obtain $d(1 + \log x)/dx$ at $x = 0.29$ from the data given in table of Problem 2.

5. Derive the error term for the formula of Problem 4, and show that the "next-term rule" applies.

Section 2

6. Use the symbolic method to derive a formula for (a) the third derivative and (b) the fourth derivative of $f(x)$ at $x = x_0$ from a difference table.

7. The Newton-Gregory backward interpolating formula can be developed by expanding the symbolic relation

$$f_s = E^s f_0 = (1 - \nabla)^{-s} f_0.$$

Using this representation of the interpolation formula, derive the derivative formula required in Problem 4 by the symbolic technique.

8. Suppose the nature of the function tabulated in Problem 2 were unknown. How could estimates of the errors of each computation in that problem be determined? Compare with the error bounds determined in Problem 3.

9. If one uses the nth difference as an estimate of the quantity $h^n f^n(\xi)$, what rule of thumb results regarding the error of the derivative formulas? What is this rule of thumb for integration formulas? Interpolation formulas?

Section 3

10. If one begins with Stirling's interpolation formula (Eq. 5.1, Chapter 2), differentiates, and then evaluates at $x = x_0$, one obtains a derivative formula in terms of central differences. Show that when this is done, the coefficients are the same as given by a horizontal path through the lozenge diagram, Fig. 3.1, starting at f_0.

11. If one differentiates Stirling's formula twice and then sets $s = 0$, a central difference formula for the second derivative results. Show that the coefficients so obtained match those on a horizontal path from f_0 in Fig. 3.2.

12. Make a lozenge diagram, similar to Figs. 3.1 and 3.2, for the third derivative.

13. Equations (3.1) and (3.2) show that the one-term central difference formulas for derivatives are more accurate than forward difference formulas. Using the data tabulated in Problem 2, compute $d(1 + \log x)/dx$ at $x = 0.23$ using formulas up to (a) first, (b) second, and (c) third differences. What are the errors of each computation?

14. If the difference table of Problem 2 is extended to the right, differences up to the eighth order can be calculated. The lozenge diagram, Fig. 3.1, can also be extended to a similar order of differences. Discuss whether the accuracy of estimates of the derivative of the logarithm function will be improved by such extensions.

15. Using the data of Problem 1 for $\sin \theta$, evaluate the second derivative at $\theta = 22°$ as accurately as possible. Over what range of values for θ will the basic interpolating polynomial fit that is the precursor of this second derivative computation?

16. The following table is for the first-order Bessel function $J_1(x)$. Prepare tables for $J_1'(x)$ and $J_1''(x)$ for $x = 0(0.1)1$.

x	$J_1(x)$	x	$J_1(x)$
0	0.0000	0.6	0.2867
0.1	0.0499	0.7	0.3290
0.2	0.0995	0.8	0.3688
0.3	0.1483	0.9	0.4059
0.4	0.1960	1.0	0.4401
0.5	0.2423		

17. An alternative way of deriving error terms for derivative formulas is through Taylor series expansions. For example, the error term of Eq. (3.1) can be obtained by ex-

panding $f_1 = f(x_0 + h)$ and $f_{-1} = f(x_0 - h)$ each about the point $x = x_0$, and then combining the two series. Show that the error term of Eq. (3.1) is $-\frac{1}{6}h^2 f''(\xi)$.

18. Evaluate the error term for Eq. (3.2) by the Taylor series method as described in Problem 17.

19. For unevenly spaced data, a derivative expression analogous to Eq. (3.1) is

$$f'(x)\Big|_{x=x_0} \doteq \frac{f(x_1) - f(x_{-1})}{x_1 - x_{-1}}.$$

In this case $x_1 - x_0 \neq x_0 - x_{-1}$. Take $x_1 - x_0 = h$ and $x_0 - x_{-1} = \alpha h$, $\alpha \neq 1$. Show by the Taylor series method that the error is now $O(h)$ and not $O(h^2)$ unless $\alpha = 1$.

Section 4

20. Use the extrapolation method to determine $f'(0.23)$ for the data of Problem 2, to an $O(h^6)$ extrapolation.

21. Determine $d^2(\sin \theta)/d\theta^2$ at $\theta = 24°$ by the extrapolation method.

22. Show that the first-order extrapolation for $f'(x_0)$ with Δx-values differing by 2:1 is the same as the formula

$$f'_0 = \frac{1}{h}\left(\frac{\Delta f_{-1} + \Delta f_0}{2} - \frac{1}{6}\frac{\Delta^3 f_{-2} + \Delta^3 f_{-1}}{2}\right),$$

where h is the smaller of the two Δx-values.

23. Show that a second-order extrapolation for $f'(x_0)$ is the equivalent of using central differences through $\Delta^5 f$.

24. Repeat Problems 22 and 23 for $f''(x_0)$.

25. Develop a first-order extrapolation formula if the Δx-values (a) are in the ratio 3:1 rather than 2:1, (b) if the ratio is r.

Section 5

26. Use the computer program of the text to compute $f'(x_0)$ and $f''(x_0)$ when:
   (a) $f(x) = \sinh x$, $x_0 = 0$
   (b) $f(x) = \sqrt{x^3/(2 - x)}$, $x_0 = 1$ (cissoid of Diocles)
   (c) $f(x) = e^x \sin x$, $x_0 = \pi/2$

27. Modify the program of the text to decrease H to half its previous value each time, instead of to one-tenth, and cause the program to terminate when neither derivative changes in value by more than some input value TOL. Test with the functions in Problem 26.

28. Further modify the program of Problem 27 to incorporate the extrapolation technique of Section 4.

5
formulas
by method of
undetermined
coefficients

The previous chapters have developed formulas for integration and differentiation of functions by replacing the actual function with a polynomial that agrees at a number of points (a so-called *interpolating polynomial*), and then integrating or differentiating the polynomial. These formulas are valuable if one wishes to write computer programs for integration and differentiation, but the most important use is for solving differential equations numerically. Because computers calculate their functional values rather than interpolating in a table, there is less interest today in interpolating polynomials than in earlier times. Hence, there is reason to present an alternative method of deriving the formulas for derivatives and integrals which are needed to solve differential equations. This chapter does this, and consequently provides a shorter path to that interesting application of numerical analysis.

We shall call the method that we employ in this chapter the *method of undetermined coefficients*. Basically, we impose certain conditions on a formula of desired form and use these conditions to determine values of the unknown coefficients in the formula. Hamming (1962) presents the method in considerable detail.

1. DERIVATIVE FORMULAS BY METHOD OF UNDETERMINED COEFFICIENTS

Since the derivative of a function is the rate of change of the function relative to changes in the independent variable, we should expect that formulas for the derivative would involve differences between function values in the neighborhood of the point where we wish to evaluate the derivative. It is in fact possible to approximate the derivative as a linear combination of such function values. While one can argue that the best formula should use function values very near to the point in question, the practically important formulas impose the restriction that only function values at equally spaced x-values are to be used.

For example, we can write a formula for the first derivative in terms of $n + 1$ equispaced points:

$$f'(x_0) = c_0 f(x_0) + c_1 f(x_1) + c_2 f(x_2) + \cdots + c_n f(x_n),$$
$$x_{i+1} - x_i = h = \text{constant}. \tag{1.1}$$

The more terms that we employ, the greater accuracy we shall expect, since more information about the function is being fed in. We shall evaluate the coefficients in the equation by requiring that the formula be exact whenever the function is a polynomial of degree n or less. (We shall find throughout this chapter that the method of undetermined coefficients uses this criterion to develop formulas. It has validity because any function that is continuous on an interval can be approximated to any specified precision by a polynomial of sufficiently high degree. Using polynomials to replace the function also greatly simplifies the work, in contrast to replacing with other functions.)

Let us illustrate the method of undetermined coefficients with a simple case. We shall simplify the notation by defining

$$f_i = f(x_i).$$

If we write the derivative in terms of only two function values, we would have

$$f_0' = c_0 f_0 + c_1 f_1. \tag{1.2}$$

We require the formula to be exact if $f(x)$ is a polynomial of degree 1 or less. (The maximum degree is always one less than the number of undetermined constants.) Hence, since the formula is to be exact if $f(x)$ is any first-degree polynomial, it must be exact if either $f(x) = x$ or $f(x) = 1$. We write Eq. (1.2) for both these cases:

$$\begin{array}{llll} f(x) = x, & f'(x) = 1, & 1 = c_0(x_0) + c_1(x_0 + h), \\ f(x) = 1, & f'(x) = 0, & 0 = c_0(1) + c_1(1). \end{array} \tag{1.3}$$

We solve Eqs. (1.3) simultaneously to get $c_0 = -1/h$, $c_1 = 1/h$. Consequently,

$$f'(x_0) \doteq \frac{f_1 - f_0}{h}. \tag{1.4}$$

The dot over the equal sign in Eq. (1.4) is to remind us that the formula is only an approximation. It is exact if $f(x)$ is a polynomial of degree 1, but not exact if a polynomial of higher degree, or some transcendental function.

Similarly, a three-term formula can be derived by replacing the function with x^2 and x and 1 in

$$f_0' = c_0 f_0 + c_1 f_1 + c_2 f_2.$$

The set of equations to solve is

$$\begin{array}{l} 2x_0 = c_0(x_0)^2 + c_1(x_0 + h)^2 + c_2(x_0 + 2h)^2, \\ 1 = c_0(x_0) + c_1(x_0 + h) + c_2(x_0 + 2h), \\ 0 = c_0(1) + c_1(1) + c_2(1). \end{array} \tag{1.5}$$

The arithmetic is simplified by letting $x_0 = 0$. That this is valid is readily seen. Imagine the graph of $f(x)$ versus x. The derivative we desire is the slope of the curve at the point where $x = x_0$, which obviously is unchanged by a translation of the axes. Taking $x_0 = 0$ is the equivalent of a translation of axes so that the origin corresponds to x_0. Equations (1.5) become, with this change,

$$\begin{array}{l} 0 = c_0(0) + c_1(h)^2 + c_2(2h)^2 \\ 1 = c_0(0) + c_1(h) + c_2(2h) \\ 0 = c_0(1) + c_1(1) + c_2(1) \end{array}$$

Solving, we get $c_0 = -3/2h$, $c_1 = 2/h$, $c_2 = -1/2h$; so a three-term formula for the derivative is

$$f_0' \doteq \frac{-3f_0 + 4f_1 - f_2}{2h}. \tag{1.6}$$

We could extend these formulas to include more and more terms, but after a little reflection we can conclude that the original form, Eq. (1.1), is not the best to use. It utilizes only functional values to one side of the point in question, while it would be better to utilize information from both sides of x. After all, the limit in the definition of the derivative is two-sided, and further, we utilize closer and hence more pertinent information by going to both the left and the right.

We expect an improvement over Eq. (1.6) if we begin with

$$f_0' = c_{-1}f_{-1} + c_0f_0 + c_1f_1, \tag{1.7}$$

where $f_{-1} = f(x_0 - h)$. Adopting the simplification of letting $x_0 = 0$ as before, and taking x^2, x, and 1 for $f(x)$, we have to solve

$$
\begin{aligned}
0 &= c_{-1}(-h)^2 + c_0(0) + c_1(h)^2, \\
1 &= c_{-1}(-h) + c_0(0) + c_1(h), \\
0 &= c_{-1}(1) + c_0(1) + c_1(1).
\end{aligned} \tag{1.8}
$$

Completing the algebra, we get $c_{-1} = -1/2h$, $c_0 = 0$, $c_1 = 1/2h$.

The following equation is particularly important:

$$f_0' \doteq \frac{f_1 - f_{-1}}{2h}. \tag{1.9}$$

In the next section we will compare its accuracy with Eqs. (1.4) and (1.6). Because the point where the derivative is evaluated is centered among the function values whose differences appear in the formula, it is called a *central difference approximation*. Higher order central difference approximations to the first derivative, utilizing five or seven or more points, can be derived by this same procedure.

We now apply the method of undetermined coefficients to higher derivatives. We shall discuss only the central difference approximations since they are the more widely used. In terms of values both to the right and left of x_0,

$$f_0'' = c_{-1}f_{-1} + c_0f_0 + c_1f_1.$$

Taking $f(x) = x^2$, x, and 1, we get the relations ($x_0 = 0$),

$$
\begin{aligned}
2 &= c_{-1}(-h)^2 + c_0(0) + c_1(h)^2, \\
0 &= c_{-1}(-h) + c_0(0) + c_1(h), \\
0 &= c_{-1}(1) + c_0(1) + c_1(1)
\end{aligned}
$$

The resulting formula is

$$f_0'' \doteq \frac{f_{-1} - 2f_0 + f_1}{h^2}. \tag{1.10}$$

Equation (1.10), like Eq. (1.9), is particularly useful.

In the same way, we can derive formulas for the third and fourth derivatives. Derivatives beyond these do not generally appear in applied problems. We must use a minimum of four and five terms, however, since only polynomials of degree 3 and 4 have nonzero third or fourth derivatives. For the third-degree formula, complete symmetry with four points is impossible; five-term formulas for both derivatives are therefore given. We present the results only, leaving the derivations as an exercise:

$$f_0''' = \frac{-f_{-2} + 2f_{-1} - 2f_1 + f_2}{2h^3}, \tag{1.11}$$

$$f_0^{iv} = \frac{f_{-2} - 4f_{-1} + 6f_0 - 4f_1 + f_2}{h^4}. \tag{1.12}$$

2. ERROR TERMS FOR DERIVATIVE FORMULAS

In the previous section, we used the method of undetermined coefficients to derive several formulas for the first derivative of a function, utilizing function values at equispaced x-values. These were, using the notation that $f_i = f(x_i)$,

$$f_0' \doteq \frac{f_1 - f_0}{h}, \tag{2.1}$$

$$f_0' \doteq \frac{-3f_0 + 4f_1 - f_2}{2h}, \tag{2.2}$$

$$f_0' \doteq \frac{f_1 - f_{-1}}{2h}. \tag{2.3}$$

While each of these formulas is not exact, as suggested by the $\doteq$ symbol, we argued heuristically that the error should decrease in each succeeding one. We now wish to develop expressions for the errors.

We begin with the Taylor series expansion of $f(x_1) = f(x_0 + h)$ in terms of $x_1 - x_0 = h$:

$$f(x_1) = f(x_0) + hf'(x_0) + \tfrac{1}{2}h^2 f''(x_0) + \tfrac{1}{6}h^3 f'''(x_0) + \cdots.$$

Changing to our subscript notation, and truncating after the term in h, with the usual error term, we have

$$f_1 = f_0 + hf_0' + \tfrac{1}{2}h^2 f''(\xi), \qquad x_0 < \xi < x_0 + h. \tag{2.4}$$

Solving for f_0', we have

$$f_0' = \frac{f_1 - f_0}{h} - \frac{1}{2}hf''(\xi), \qquad x_0 < \xi < x_0 + h, \tag{2.5}$$

so that the error term is $-\tfrac{1}{2}hf''(\xi)$ for the derivative formula, Eq. (2.1).

In Eq. (2.5), the error term involves the second derivative of the function evaluated at a place which is known only within a certain interval. In the majority of applications for numerical differentiation, not only is the point of evaluation uncertain, but the function $f(x)$ is also unknown. If we do not know $f(x)$, we can hardly expect to know its derivatives. We do know, however, that the error involves the first power of $h = x_{i+1} - x_i$, and, in fact, the only way we can change the error is to change h. Making h smaller will decrease h, and in the limit as $h \to 0$ the error will go to zero. Further, as $h \to 0$, $x_1 \to x_0$, and the value of ξ is squeezed into a smaller and smaller interval. In other words, $f''(\xi)$ approaches a fixed value, specifically $f''(x_0)$, as h goes to zero.

The special importance of h in the error term is denoted by a special notation in numerical analysis, the *order relation*. We say the error of Eq. (2.5) is "of order h" and write

$$\text{error} = O(h), \qquad \text{when} \qquad \lim_{h \to 0}(\text{error}) = ch,$$

where c is a fixed value.

To develop the error term for Eq. (2.2), we proceed similarly, except we write expansions for both f_1 and f_2. It is also necessary to carry terms through h^3

because the second derivatives cancel, resulting in an error term involving the third derivative:

$$f_1 = f_0 + hf_0' + \tfrac{1}{2}h^2f_0'' + \tfrac{1}{6}h^3f'''(\xi_1), \qquad x_0 < \xi_1 < x_0 + h,$$
$$f_2 = f_0 + 2hf_0' + \tfrac{1}{2}(2h)^2f_0'' + \tfrac{1}{6}(2h)^3f'''(\xi_2), \qquad x_0 < \xi_2 < x_0 + 2h.$$

Note that the values of ξ_1 and ξ_2 may not be identical. If we multiply the first equation by 4, the second by -1, and add $-3f_0$ to their sum, we get

$$-3f_0 + 4f_1 - f_2 = (-3 + 4 - 1)f_0 + 2hf_0' + (2-2)f_0'' + \frac{4h^3}{6}\Big(f'''(\xi_1) - 2f'''(\xi_2)\Big).$$

Solving for f_0', we get

$$f_0' = \frac{-3f_0 + 4f_1 - f_2}{2h} + \frac{1}{3}h^2\Big(2f'''(\xi_2) - f'''(\xi_1)\Big), \qquad (2.6)$$

$$x_0 < \xi_1 < x_0 + h, \quad x_0 < \xi_2 < x_0 + 2h.$$

The last term of Eq. (2.6) is the error term. As $h \to 0$, the two values of ξ approach the same value. Consequently the error term approaches $\tfrac{1}{3}h^2f'''(x_0)$. We then conclude the error of Eq. (2.2) is $O(h^2)$.

For the error of Eq. (2.3), we proceeded similarly. We leave the development as an exercise to show that

$$f_0' = \frac{f_1 - f_{-1}}{2h} - \frac{1}{6}h^2f'''(\xi), \qquad x_0 - h < \xi < x_0 + h, \qquad (2.7)$$

$$= \frac{f_1 - f_{-1}}{2h} + O(h^2).$$

The error terms of both Eqs. (2.6) and (2.7) are $O(h^2)$, but the coefficient in Eq. (2.7) is only half the magnitude of the coefficient in Eq. (2.6). The progressive increase in accuracy we anticipated is confirmed.

By similar arguments, we can show that the formulas for third and fourth derivatives, Eqs. (1.11) and (1.12), both have errors $O(h^2)$.

3. INTEGRATION FORMULAS BY METHOD OF UNDETERMINED COEFFICIENTS

A numerical integration formula will estimate the value of the integral of the function by a formula involving the values of the function at a number of points in or near the interval of integration. Figure 3.1 illustrates the general situation.

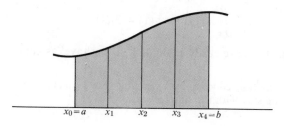

Fig. 3.1

If we desire to evaluate $\int_a^b f(x)\,dx$, it is obvious that if we could find some average value of the function on the interval $[a,b]$, the integral would be

$$\int_a^b f(x)\,dx = (b - a)f_{av}.$$

It seems reasonable to assume that f_{av} could be approximated by a linear combination of function values within or near the interval.* We therefore write, similarly to our procedure for derivatives,

$$\int_a^b f(x)\,dx = c_0 f(x_0) + c_1 f(x_1) + \cdots + c_n f(x_n),$$

where the coefficients c_i are to be determined. The points x_i at which the function is to be evaluated can also be left undetermined, but, for the formulas which we need to derive, we shall again impose the restriction that they are equally spaced with the value of $x_{i+1} - x_i = \Delta x = h = $ constant. We shall impose the further restriction that the boundaries of the interval of integration, a and b, coincide with two of the x_i-values.

We start with a simple case. Suppose we wished to express the integral in terms of just two functional values, specifically $f(a)$ and $f(b)$. Our formula takes the form

$$\int_a^b f(x)\,dx = c_0 f(a) + c_1 f(b). \tag{3.1}$$

Obviously, the values of c_0 and c_1 depend on $f(x)$, and probably on the values of a and b as well. The method of undetermined coefficients assumes that $f(x)$ can be approximated by a polynomial over the interval $[a,b]$ and determines c_0 and c_1 so that Eq. (3.1) is exact for all polynomials of a certain maximum degree or less.

Equation (3.1) has only two arbitrary constants, so we would not expect that it could be made exact for polynomials of degree higher than the first, since more than two parameters appear in second or higher degree polynomials. We therefore force Eq. (3.1) to be exact when $f(x)$ is replaced by either a first-degree polynomial or one of zero degree (a constant function). If it is to be exact for all first-degree polynomials, it must be exact for the very simple function $f(x) = x$. If it is to be exact when the function is any constant, it must hold if $f(x) = 1$. We express these two relations:

$$\int_a^b x\,dx = c_0(a) + c_1(b),$$

$$\int_a^b (1)dx = c_1(1) + c_1(1). \tag{3.2}$$

The integrals of Eqs. (3.2) are easily evaluated, so we have, as conditions that the

---

* One could generalize the concept by extending this to include derivatives of the function as well, but we stay with the simpler case.

constants must satisfy,

$$\int_a^b x \, ds = \frac{x^2}{2}\Big|_a^b = \frac{b^2}{2} - \frac{a^2}{2} = c_0 a + c_1 b,$$

$$\int_a^b dx = x\Big|_a^b = b - a = c_0 + c_1.$$

Solving these equations gives $c_0 = (b - a)/2$, $c_1 = (b - a)/2$.
Consequently,

$$\int_a^b f(x) \doteq \frac{b-a}{2}\Big(f(a) + f(b)\Big). \tag{3.3}$$

This formula is the *trapezoidal rule*, and was derived in Chapter 3 as Eq. (1.1) through an entirely different approach. We have put a dot over the equality sign in Eq. (3.3) to remind ourselves that the relation is only approximately true, for the function $f(x)$ cannot ordinarily be replaced without error by a first-degree polynomial. It is intuitively obvious that unless the interval $[a,b]$ is very small, the error will be considerable.

We can reduce the error in the above formula by using a higher degree polynomial to replace $f(x)$, but we would then need to take a linear combination of more than two function values. In fact, we shall have to preserve a balance between the number of undetermined coefficients in the formula and the number of parameters in the polynomial. The number of coefficients hence must be one greater than the degree of the polynomial.

We now look at the next case, a three-term formula corresponding to replacing $f(x)$ by a quadratic. For three terms, using $x_0 = a$ and $x_2 = b$, with x_1 at the midpoint,

$$\int_a^b f(x) \, dx = c_0 f(a) + c_1 f\left(\frac{a+b}{2}\right) + c_2 f(b).$$

The formula must be exact if $f(x) = x^2$, or $f(x) = x$, or $f(x) = 1$, so

$$\int_a^b x^2 \, dx = \frac{b^3}{3} - \frac{a^3}{3} = c_0 a^2 + c_1\left(\frac{a+b}{2}\right)^2 + c_2 b^2,$$

$$\int_a^b x \, dx = \frac{b^2}{2} - \frac{a^2}{2} = c_0 a + c_1\left(\frac{a+b}{2}\right) + c_2 b,$$

$$\int_a^b dx = b - a = c_0 + c_1 + c_2.$$

The solution is $c_0 = c_2 = \frac{1}{3}(b - a)$, $c_1 = \frac{4}{3}(b - a)$. Hence

$$\int_a^b f(x) \, dx \doteq \frac{b-a}{6}\Big(f(a) + 4f\left(\frac{a+b}{2}\right) + f(b)\Big). \tag{3.4}$$

A more common form of Eq. (3.4) is found by writing $b - a = 2h$ (the interval from a to b is subdivided into two panels) and substituting x_0 for a, x_2 for b, and x_1 for the midpoint. We then have

$$\int_{x_0}^{x_0+2h} f(x)\ dx \doteq \frac{h}{3}\ (f_0 + 4f_1 + f_2). \tag{3.5}$$

Formula (3.5) is Simpson's $\frac{1}{3}$ rule, a particularly popular formula. Again, it is not exact as indicated by the dot over the equality sign.

The application of Eqs. (3.3) and (3.5) over an extended interval is straight-forward. It is intuitively apparent that the error in these formulas will be large if the interval is not small. (We discuss these errors quantitatively in a later section.) To apply these to a large interval of integration and still keep control over the error, we break the interval into a large number of small subintervals and add together the formulas applied to the subintervals. When this is done we get the *extended trapezoidal rule*

$$\int_a^b f(x)\ dx \doteq \frac{b-a}{2n}[f(x_0) + 2f(x_1) + 2f(x_2) + \cdots + 2f(x_{n-1}) + f(x_n)], \tag{3.6}$$

and the extended Simpson's $\frac{1}{3}$ rule

$$\int_a^b f(x)\ dx \doteq \frac{b-a}{3(2n)}\ [f(x_0) + 4f(x_1) + 2f(x_2) + 4f(x_3) + \cdots$$

$$+ 2f(x_{2n-2}) + 4f(x_{2n-1}) + f(x_{2n})]. \tag{3.7}$$

These forms of the Trapezoid rule (Eq. 3.6) and of Simpson's $\frac{1}{3}$ rule (Eq. 3.7) are widely used in computer programs for integration. By applying them with n increasing, the error can be made arbitrarily small.* For Simpson's rule, ob-serve that the number of subintervals must be even.

4. INTEGRATION FORMULAS USING POINTS OUTSIDE THE INTERVAL

In studying numerical methods to solve differential equations we shall have need for some specific integral formulas which involve function values computed at points outside the interval of integration. Figures 4.1, 4.2, and 4.3 sketch three special cases of importance. In Fig. 4.1 the curve that passes through the four points whose abscissas are $x_{n-3}, x_{n-2}, x_{n-1},$ and x_n is extrapolated to x_{n+1}, and we desire the integral only over the extrapolated interval. Figure 4.2 presents the case where the curve passes through four points, and the integral is taken only over the last panel. In Figure 4.3 we consider a case where a curve that fits at three points is extrapolated in both directions and integration is taken over four panels.

In the derivations of this section it will be convenient to adopt the notation that $f_i = f(x_i)$, so that subscripts on the function indicate the x-value at which it is evaluated.

---

* Except for round-off error effects which eventually will dominate since they are not decreased by small subdivision of the interval and, in fact, may increase as the number of computations increases.

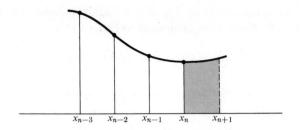

Fig. 4.1

For the first case (Fig. 4.1) we desire a formula of the form

$$\int_{x_n}^{x_{n+1}} f(x)\,dx = c_0 f_{n-3} + c_1 f_{n-2} + c_2 f_{n-1} + c_3 f_n.$$

With four constants, we can make the formula exact when $f(x)$ is any polynomial of degree 3 or less. Accordingly, we replace $f(x)$ successively by x^3, x^2, x, and 1 to evaluate the coefficients.

It is apparent that the formula must be independent of the actual x-values. To simplify the equations, let us shift the origin to the point $x = x_n$; our integral is then taken over the interval from 0 to h, where $h = x_{n+1} - x_n$:

$$\int_0^h f(x)\,dx = c_0 f(-3h) + c_1 f(-2h) + c_2 f(-h) + c_3 f(0).$$

Carrying out the computations by replacing $f(x)$ with the particular polynomials, we have

$$\frac{h^4}{4} = c_0(-3h)^3 + c_1(-2h)^3 + c_2(-h)^3 + c_3(0),$$

$$\frac{h^3}{3} = c_0(-3h)^2 + c_1(-2h)^2 + c_2(-h)^2 + c_3(0),$$

$$\frac{h^2}{2} = c_0(-3h) + c_1(-2h) + c_2(-h) + c_3(0),$$

$$h = c_0(1) + c_1(1) + c_2(1) + c_3(1).$$

(4.1)

After completion of the algebra we get

$$\int_{x_n}^{x_{n+1}} f(x)\,dx \doteq \frac{h}{24}(-9f_{n-3} + 37f_{n-2} - 59f_{n-1} + 55f_n).$$

(4.2)

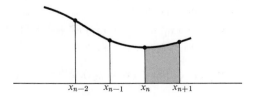

Fig. 4.2

For the second case, illustrated by Fig. 4.2, we again translate the origin to x_n to simplify.   The set of equations analogous to Eqs. (4.1) is

$$\frac{h^4}{4} = c_0(-2h)^3 + c_1(-h)^3 + c_2(0) + c_3(h)^3,$$

$$\frac{h^3}{3} = c_0(-2h)^2 + c_1(-h)^2 + c_2(0) + c_3(h)^2,$$

$$\frac{h^2}{2} = c_0(-2h) + c_1(-h) + c_2(0) + c_3(h),$$

$$h = c_0(1) + c_1(1) + c_2(1) + c_3(1).$$

These give values of the constants in the integration formula:

$$\int_{x_n}^{x_{n+1}} f(x)\, dx \doteq \frac{h}{24} (f_{n-2} - 5f_{n-1} + 19f_n + 9f_{n+1}). \qquad (4.3)$$

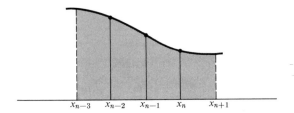

$x_{n-3} \qquad x_{n-2} \qquad x_{n-1} \qquad x_n \qquad x_{n+1}$

Fig. 4.3

The third case, Fig. 4.3, where extrapolation of a quadratic in both directions is involved, leads to the equation

$$\int_{x_{n-3}}^{x_{n+1}} f(x)\, dx \doteq \frac{4h}{3} (2f_{n-2} - f_{n-1} + 2f_n). \qquad (4.4)$$

Setting up the equations for this case is left as an exercise for the student.

The particular methods for solving differential equations that use the formulas derived in this section are known as *Adam's method* and *Milne's method*.

5. ERROR OF INTEGRATION FORMULAS

In each of the formulas that we derived above, we used the symbol $\doteq$ to remind us that the formulas are not exact unless $f(x)$ is in fact a polynomial of a certain degree or less.   It is important to derive expressions for the error.

We first determine the error term for the Trapezoidal rule over one panel, Eq. (3.3).   Begin with a Taylor expansion:

$$F(x_1) = F(x_0) + F'(x_0)\, h + \tfrac{1}{2} F''(x_0)\, h^2 + \tfrac{1}{6} F'''(\xi_1)\, h^3,$$
$$x_0 < \xi_1 < x_1 = x_0 + h. \qquad (5.1)$$

Let us define $F(x) = \int_a^x f(t)\, dt$ so that $F'(x) = f(x)$, $F''(x) = f'(x)$, $F'''(x) = f''(x)$.

Equation (5.1) becomes, with this definition of $F(x)$,

$$\int_a^{x_1} f(x)\, dx = \int_a^{x_0} f(x)\, dx + f(x_0)\, h + \tfrac{1}{2} f'(x_0)\, h^2 + \tfrac{1}{6} f''(\xi_1)\, h^3.$$

On rearranging, and using subscript notation, we get

$$\int_a^{x_1} f(x)\, dx - \int_a^{x_0} f(x)\, dx = \int_{x_0}^{x_1} f(x)\, dx = f_0 h + \tfrac{1}{2} f_0' h^2 + \tfrac{1}{6} f''(\xi_1)\, h^3. \quad (5.2)$$

We now replace the term f_0' by the forward difference approximation with error term from Section 5.2:

$$f_0' = \frac{f_1 - f_0}{h} - \tfrac{1}{2} h f''(\xi_2), \qquad x_0 < \xi_2 < x_1.$$

Equation (5.2) becomes

$$\int_{x_0}^{x_1} f(x)\, dx = f_0 h + \tfrac{1}{2} h(f_1 - f_0) - \tfrac{1}{4} h^3 f''(\xi_2) + \tfrac{1}{6} h^3 f''(\xi_1).$$

We can combine the last two terms into $-\tfrac{1}{12} h^3 f''(\xi)$, where ξ also lies in $[x_0, x_1]$. Hence,

$$\int_{x_0}^{x_1} = \frac{h}{2}(f_0 + f_1) - \frac{1}{12} h^3 f''(\xi).$$

The error of the Trapezoidal rule over one panel is then $O(h^3)$. This is called the *local error*. Since we normally use a succession of applications of this rule over the interval $[a,b]$ to evaluate $\int_a^b f(x)\, dx$, by subdividing into $n = (b - a)/h$ panels, we need to determine the so-called global error for such intervals of integration. The global error will be the sum of the local errors in each of the n panels:

$$\text{Global error} = -\tfrac{1}{12} h^3 f''(\xi_1) - \tfrac{1}{12} h^3 f''(\xi_2) - \cdots - \tfrac{1}{12} h^3 f''(\xi_n)$$
$$= -\tfrac{1}{12} h^3 [f''(\xi_1) + f''(\xi_2) + \cdots + f''(\xi_n)].$$

Here the subscripts on ξ indicate the panel in whose interval the value lies. If $f''(x)$ is continuous throughout the interval of integration,

$$\sum_{i=1}^{n} f''(\xi_i) = n f''(\xi), \qquad a < \xi < b.$$

Using $n = (b - a)/h$, we get

$$\text{Global error} = -\frac{1}{12} h^3 \left(\frac{b - a}{h} \right) f''(\xi) = -\frac{b - a}{12} h^2 f''(\xi) = O(h^2).$$

Therefore the global error is $O(h^2)$, even though the local error is $O(h^3)$.

Similar treatment shows that the local error for Simpson's $\tfrac{1}{3}$ rule is

$$- \tfrac{1}{90} h^5 f^{iv}(\xi) = O(h^5)$$

and the global error is

$$- \tfrac{1}{180} (b - a) h^4 f^{iv}(\xi) = O(h^4).$$

One can also show that the local error of each of the integration formulas in Section 4 is $O(h^5)$.

We summarize the results:

Trapezoidal Rule

local error

$$\int_a^b f(x)\, dx = \frac{h}{2}\, [f_0 + 2f_1 + 2f_2 + \cdots + 2f_{n-1} + f_n] - \frac{h^3}{12}\, f''(\xi)$$

Global error $= -\dfrac{(b-a)}{12}\, h^2 f''(\xi).$

Simpson's $\frac{1}{3}$ Rule

local error

$$\int_a^b f(x)\, dx = \frac{h}{3}\, [f_0 + 4f_1 + 2f_2 + 4f_3 + \cdots + 2f_{n-2} + 4f_{n-1} + f_n] - \frac{h^5}{90}\, f^{iv}(\xi)$$

Global error $= -\dfrac{(b-a)}{180}\, h^4 f^{iv}(\xi).$

Rule for Milne's Method

local error

$$\int_{x_{n-3}}^{x_{n+1}} f(x)\, dx = \frac{4h}{3}\, [2f_{n-2} - f_{n-1} + 2f_n] + \tfrac{28}{90}\, h^5 f^{iv}(\xi).$$

Rules for Adam's Method

local error

$$\int_{x_n}^{x_{n+1}} f(x)\, dx = \frac{h}{24}\, [-9f_{n-3} + 37f_{n-2} - 59f_{n-1} + 55f_n] + \tfrac{251}{720}\, h^5 f^{iv}(\xi),$$

local error

$$\int_{x_n}^{x_{n+1}} f(x)\, dx = \frac{h}{24}\, [f_{n-2} - 5f_{n-1} + 19f_n + 9f_{n+1}] - \tfrac{19}{720}\, h^5 f^{iv}(\xi).$$

PROBLEMS

Section 1

1. Derive formulas similar to Eqs. (1.4) and (1.6) for $f'(x_0)$, the derivative at x_0, but in terms of f-values to the left of x_0, e.g., evaluate the constants in the relations $f'_0 = c_0 f_0 + c_{-1} f_{-1}$ and $f'_0 = c_0 f_0 + c_{-1} f_{-1} + c_{-2} f_{-2}$. [Take the x-values as equally spaced, and $\Delta x = h$.]

2. When more information about the function is used to compute the derivative, a more accurate formula results. Hence the formula $f'_0 = c_0 f_0 + c_1 f_1 + c_2 f_2 + c_3 f_3$ is more accurate than either Eq. (1.4) or (1.6). Find the constants assuming equispaced x-values.

3. When values on both sides are used in a derivative formula, the accuracy is improved over a one-sided formula. Derive a formula more accurate than the four-term for-

mulas of Problems 1 and 2 by evaluating the constants in $f_0' = c_{-1}f_{-1} + c_0f_0 + c_1f_1 + c_2f_2$.

4. Find formulas for the second derivative f_0'' in terms of

(a) f_0, f_1, f_2 (b) f_{-1}, f_0, f_1, f_2

(c) f_{-2}, f_{-1}, f_1, f_2 (d) f_{-3}, f_{-1}, f_0, f_2

Note in (c) and (d) that uniform spacing is not required.

5. Use the unevenly spaced points f_L, f_0, f_R to write a formula for f_0'', where $x_0 - x_L = h$ and $x_R - x_0 = \alpha h$.

6. Show that the solution to Problem 5 reduces to Eq. (1.10) when $\alpha = 1$.

7. (a) Derive Eq. (1.11). (b) Derive Eq. (1.12).

8. The following data are for the function $f(x) = \sin x$. Estimate the value of the derivative at $x = 1.0$ using:

(a) Eq. (1.4)

(b) Eq. (1.6)

(c) Eq. (1.9)

(d) The formulas derived in Problem 1

(e) The formula derived in Problem 2

(f) The formula derived in Problem 3.

In each case determine the error by comparison to the true value, 0.5403:

x	y	x	y
0.7	0.6442	1.1	0.8912
0.8	0.7174	1.2	0.9320
0.9	0.7833	1.3	0.9636
1.0	0.8415	1.4	0.9854

9. For the same tabular values of $\sin x$ given in Problem 8, determine $f''(x)$ at $x = 1.1$ using:

(a) Eq. (1.10)

(b) The formulas derived in Problem 4.

In each case, determine the magnitude of the error.

10. From a table of natural logarithms obtain values of $\ln x$ at $x = 0.46, 0.48, 0.50, 0.52,$ and 0.54. From these values estimate the value of the first, second, third, and fourth derivatives of the function at $x = 0.50$. Compare to the exact values.

11. The following data are from an experimental test. Use the formula of Problem 5 to estimate the second derivative at the middle x-value. What value would one assign to the first derivative at this point?

x	Function
0.87	18.402
0.92	19.307
0.94	19.006

Section 2

12. Show that the formula

$$f_0' \doteq \frac{1}{h}(f_0 - f_{-1})$$

has an error $O(h)$, and that the formula

$$f_0'' \doteq \frac{1}{h^2}(f_2 - 2f_1 + f_0)$$

also has error $O(h)$.

13. Derive Eq. (2.7).

14. (a) Find the error term for Eq. (1.11).   (b) Find the error term for Eq. (1.12).

15. Show that the error for f_0'' in terms of unevenly spaced values on either side of x_0 (Problem 5) has $O(h)$ error unless $\alpha = 1$, when it has $O(h^2)$ error.

16. Observe that in finding an expression for the error term of a formula by manipulation of Taylor series expansions, we also get the formula as a by-product.   Thus this is an alternative technique for deriving derivative formulas.   Use this method to derive Eq. (1.10) by eliminating y_0' from the expansions of y_1 and y_{-1}, and solving for y_0''.

17. Use the Taylor series technique to derive (a) Eq. (1.11), (b) Eq. (1.12).

18. While the error expression of Eq. (2.4) involves the parameter ξ, which is known only within a certain interval, it can be used to bracket the error of the formula by determining the maximum and minimum values that the second derivative can assume. Compare such upper and lower limits to the error with the actual error in Problem 8(a).   Similarly, use Eq. (2.7) to bound the error of Problem 8(c), and compare to the actual error.

19. Observe that the technique of Problem 18 requires us to know the function before we can set bounds to the error.   However, if enough values of the function are available, one can estimate the second derivative by the use of Eq. (1.10).   Employ this procedure to find bounds on the error as in Problem 18, this time not using your knowledge that the function tabulated in Problem 8 is $\sin x$.

Section 3

20. Derive an integration formula similar to Eq. (3.5) except employing four equispaced values of $f(x)$.   This then gives an approximation to the value of

$$\int_{x_0}^{x_0+3h} f(x)\, dx,$$

which is known as Simpson's $\frac{3}{8}$ rule.

21. Use Eq. (1.7) to approximate the integral of $f(x) = e^x$, from $x = 0$ to $x = 1$.   Use x-values spaced 0.25 apart.   Compare to the exact value of $e^1 - e^0 = 1.7183$.

22. Repeat Problem 21 except with x values spaced 0.1 apart.   The error should decrease as Δx is made smaller as discussed in Section 5.   How much smaller is the error?

23. Repeat Problem 21 except using Simpson's $\frac{1}{3}$ rule (Eq. 3.7). This rule is more accurate, so the error should be less. How much less is it?

24. Derive an extended version of Simpson's $\frac{2}{3}$ rule by applying the formula derived in Problem 20 to an interval $[a,b]$ divided into a number of subintervals which is divisible by 3.

25. Observe that a function known only as an equispaced tabulation can be integrated numerically. Hence, the integral of experimental data can be obtained by these formulas. Find the integral of the following function between $x = 10$ and $x = 16$ (a) using the trapezoidal rule, (b) using Simpson's $\frac{1}{3}$ rule, (c) using Simpson's $\frac{2}{3}$ rule.

x	$f(x)$	x	$f(x)$
10	10.1	14	14.7
11	4.5	15	16.6
12	5.4	16	17.5
13	10.7		

26. Simpson's $\frac{2}{3}$ rule is generally not more accurate than Simpson's $\frac{1}{3}$ rule, and is more complicated. One then might wonder why one uses the rule. Note, however, that Simpson's $\frac{1}{3}$ rule requires an even number of panels, which may not be the case for tabulated data. In such a situation, a combination of the two Simpson rules can be applied. Use the two rules together to evaluate the integral of $f(x)$ as given in Problem 25 between $x = 10$ and $x = 15$.

Section 4

27. Evaluate $\int_0^1 e^x \, dx$ using values at $x = 0, -1, -2, -3$ by means of Eq. (4.2), and compare to the exact value.

28. Evaluate $\int_0^1 e^x \, dx$ using values at $x = 1, 0, -1, -2$ by means of Eq. (4.3). Compare to the results of Problem 27, and observe that extrapolation of the function is relatively less accurate.

29. Derive Eq. (4.4).

30. Apply Eq. (4.4) to evaluate $\int_0^1 e^x dx$ using values of the integrand at $x = 0.25, 0.50,$ 0.75. Note that answer is less accurate than in Problem 23 even though three-term formulas with $\Delta x = 0.25$ are used in both cases. This is caused by the extrapolation in Eq. (4.4).

Section 5

31. Derive the expression for the local error of Simpson's $\frac{1}{3}$ rule to obtain

$$\text{Local error} = -\frac{1}{90} h^5 f^{iv}(\xi).$$

32. Using the expression for the local error of Simpson's $\frac{1}{3}$ rule, derive the expression for the global error:

$$\text{Global error} = \frac{-(b-a)}{180} h^4 f^{iv}(\xi).$$

33. The local error for Simpson's $\frac{3}{8}$ rule is $-(3h^5/80)f^{iv}(\xi)$. Use this to derive the global error and observe that it is twice the magnitude of the error in the $\frac{1}{3}$ rule, provided that the fourth derivative is relatively constant on the interval $[a,b]$.

34. While the exact value of ξ is not known in the error formulas of Section 5, it must take on values within the interval $[a,b]$. Hence, if $f(x)$ is a known function, we can find bounds for the error by finding maximum and minimum values for the derivative on $[a,b]$. Use this method to bracket the errors of Problems 21, 22, 23, and 30. Compare to the actual errors. Note that the effects of round-off of the original data as well as the effects of carrying only a limited number of decimals in the calculations are not included in the error terms. For this reason, the actual error may sometimes lie outside the predicted upper and lower bounds.

6
numerical
solution of
ordinary
differential equations

Because the rate of change of a quantity is generally more easily determined from the various factors which influence a given physical system than the values of the quantity itself, most scientific and engineering laws are expressed as differential equations. We mean by the term "differential equation" a relationship involving derivatives of the function as well as the function and the independent variable. For example,

$$\frac{du}{dt} = -0.27(u - 60)^{5/4}$$

is an equation describing (approximately) the rate of change of temperature u of a body losing heat by natural convection with constant temperature surroundings. This is termed a first-order differential equation because the highest order derivative is the first.

If the equation contains derivatives of nth order, it is said to be an nth order differential equation. For example, a second-order equation describing the oscillation of a weight acted upon by a spring, with resistance to motion proportional to the square of the velocity, might be

$$\frac{d^2x}{dt^2} + 4\left(\frac{dx}{dt}\right)^2 + 0.6x = 0,$$

where x is the displacement and t is time.

The solution to a differential equation is the function which satisfies the differential equation and which also satisfies certain initial conditions on the function. In solving a differential equation analytically, one usually finds a general solution containing arbitrary constants and then evaluates the arbitrary constants so that the expression agrees with the initial conditions. For an nth order equation, n independent initial conditions must usually be known. The analytical methods are limited to certain special forms of equations; elementary courses normally treat only linear equations with constant coefficients when the degree of the equation is higher than the first. Neither of the above examples is linear.

Numerical methods have no such limitations to only standard forms. We obtain the solution as a tabulation of the values of the function at various values of the independent variable, however, and not as a functional relationship. We also must pay a price for our ability to solve practically any equation in that we must recompute the entire table if the initial conditions are changed.

Our procedure will be to explore several methods of solving first-order equations, and then show how these same methods can be applied to systems of simultaneous first-order equations and to higher order differential equations. We shall use for our typical first-order equation the form

$$\frac{dy}{dx} = f(x,y),$$

$$y(x_0) = y_0.$$

1. TAYLOR SERIES METHOD

The first method we discuss is not strictly a numerical method, but it is sometimes used in conjunction with the numerical schemes, is of general applicability, and serves as an introduction to the other techniques we shall study. Consider the example problem

$$\frac{dy}{dx} = x + y, \qquad y(0) = 1. \tag{1.1}$$

(This particularly simple example is chosen to illustrate the method so that the student can readily check the computational work. The analytical solution, $y = 2e^x - x - 1$, is obtained immediately by application of the standard methods, and will be compared with our numerical results to show exactly the error at any step.)

We develop the relation between y and x by finding the coefficients of the Taylor series:

$$y(x) = y(x_0) + y'(x_0)(x - x_0)$$
$$+ \frac{y''(x_0)}{2!}(x - x_0)^2$$
$$+ \frac{y'''(x_0)}{3!}(x - x_0)^3 + \cdots.$$

Since $y(x_0)$ is our initial condition, $y(0) = 1$, with $x_0 = 0$, the first term is known. (Since the expansion is about the point $x = 0$, our Taylor series is actually a Maclaurin series in this example.)

We get the coefficient of the second term by substituting $x = 0$, $y = 1$ into the equation for the first derivative, Eq. (1.1):

$$y'(x_0) = y'(0) = 0 + 1 = 1.$$

We get equations for the second and higher order derivatives by successively differentiating the equation for the first derivative. Each of these is evaluated corresponding to $x = 0$ to get the various coefficients:

$$\begin{aligned} y''(x) &= 1 + y', & y''(0) &= 1 + 1 = 2, \\ y'''(x) &= y'', & y'''(0) &= 2, \\ y^{iv}(x) &= y''', & y^{iv}(0) &= 2, \\ \text{etc.} & & y^{(n)}(0) &= 2. \end{aligned}$$

[Getting the derivatives is deceptively easy in this example. The student should compare this to the function $f(x,y) = x/y$, with $y(1) = 1$.]

We then write our series solution for y, letting $x = h$ be the value at which we wish to determine y:

$$y(h) = 1 + h + h^2 + \tfrac{1}{3}h^3 + \tfrac{1}{12}h^4 + \text{error}.$$

The solution to our differential equation, $dy/dx = x + y$, $y(0) = 1$, is then given by the table

x	y	y, analytical
0	1.000	1.0000
0.1	1.1103	1.1103
0.2	1.2428	1.2428
0.3	1.3997	1.3997
0.4	1.5835 (?)	1.5836
0.5	1.7969 (?)	1.7974

As we computed the last two entries, there was some doubt in our minds as to their accuracy without using more terms of the Taylor series, because the successive terms were decreasing less and less rapidly. We need more terms than we have calculated to get four-decimal-place accuracy.

The error term of the Taylor series after the h^4 term,

$$\text{error} = \frac{y^{(v)}(\xi)}{5!} h^5, \qquad 0 < \xi < h,$$

cannot be computed because evaluating the derivative at $x = \xi$ is impossible with ξ unknown, and even bounding it in the interval $[0,h]$ is impossible because the lower derivatives are known only at $x = 0$ and not at $x = h$.

Numerical analysis is sometimes termed an art instead of a science because, in situations like this, the number of Taylor series terms to be included is a matter of judgment and experience. We normally truncate the Taylor series when the contribution of the last term is negligible to the number of decimal places to which we are working. However, this is correct only when the succeeding terms become small rapidly enough — in some cases the sum of the many neglected small terms is significant.

The Taylor series method can be readily applied to higher order equations as the next example shows. Here the difficulty of getting successive derivatives is more realistic. We desire the value of y at $x = 0.2$:

$$\frac{d^2y}{dx^2} = y^2 e^x - \frac{dy}{dx}, \qquad y(0) = 2, \quad y'(0) = -1.$$

$$y''(x) = y^2 e^x - y',$$
$$y'''(x) = 2yy'e^x + y^2 e^x - y'',$$
$$y^{iv}(x) = 2y'y'e^x + 2yy''e^x$$
$$+ 2yy'e^x + y^2 e^x - y'''.$$

$$y''(0) = 4(1) + 1 = 5,$$
$$y'''(0) = 2(2)(-1)(1) + 4(1) - 5 = -5,$$
$$y^{iv}(0) = 2(-1)(-1)(1) + 2(2)(5)(1) + 2(2)(-1) + 2(2)(-1)(1) + 4(1) + 5 = 23.$$

Substituting the derivative values in the Taylor series, we get

$$y(0.2) = 2 - 0.2 + \tfrac{5}{2}(0.2)^2 - \tfrac{5}{6}(0.2)^3 + \tfrac{23}{24}(0.2)^4 + \text{error}$$
$$= 1.902 + \text{error}.$$

Here there is real doubt in our mind as to the accuracy of the answer. The last term contributed only 0.0015, but we have no assurance that the succeeding terms will be small. (Actually the next derivative has the value -5, but the general term is impossible to write.) In fact, in most cases, even the convergence interval is uncertain for a Taylor series developed from a differential equation.

2. EULER AND MODIFIED EULER METHODS

As we have seen, the Taylor series method is awkward to apply if the various derivatives are complicated, and the error is difficult to determine. An even more significant criticism in this computer age is that taking derivatives of arbitrary functions cannot easily be written into a computer program. We look for an approach that is not subject to these disadvantages.

One thing we do know about the Taylor series. The error will be small if the step size h (the interval beyond x_0 where we evaluate the series) is small. In fact, if it is small enough, only a few terms are needed for good accuracy. The Euler method may be thought of as following this idea to the extreme for first-order differential equations. Suppose we choose h small enough that we may truncate after the first derivative term. Then

$$y(x_0+h) = y(x_0) + hy'(x_0) + \frac{y''(\xi)}{2}h^2, \qquad x_0 < \xi < x_0 + h.$$

We have written the usual form of the error term for the truncated Taylor series.

It will of course be necessary to use this method iteratively, advancing the solution to $x = x_0 + 2h$ after $y(x_0 + h)$ has been found, then to $x = x_0 + 3h$, etc. Adopting the subscript notation for the y-values and representing the error by the order relation, we may write the algorithm for the Euler method as

$$y_{n+1} = y_n + hy'_n + O(h^2)\,\text{error.*} \tag{2.1}$$

As an example, we apply this to the simple equation

$$\frac{dy}{dx} = x + y, \qquad y(0) = 1,$$

where the computations can be done mentally. It is convenient to arrange the

---

* This is the order of error for one step only, the "local error." As detailed in a later section, over many steps the error becomes $O(h)$. Such accumulated error is termed the *global error*.

work in a table.   Take $h = 0.02$:

x_n	y_n	y_n'	hy_n'
0	1.0000	1.0000	0.0200
0.02	1.0200	1.0400	0.0208
0.04	1.0408	1.0808	0.0216
0.06	1.0624	1.1224	0.0224
0.08	1.0848	1.1648	0.0233
0.10	1.1081		

(1.1103 analytical, error is 0.0022)

Comparing this to the analytical answer $y(0.10) = 1.1103$, we see that there is only two-decimal-place accuracy, even though we have advanced the solution only five steps.   To gain four-decimal accuracy, we must reduce the error at least 22-fold.   Since the global error is about proportional to h, we will need to reduce the step about 20-fold to 0.004.

The trouble with this most simple method is its lack of accuracy, requiring an extremely small step size.   Figure 2.1 suggests how we might improve this method with little additional effort.

In the simple Euler method, we use the slope at the beginning of the interval y_n' to determine the increment to the function, but this is always wrong.   After all, if the slope of the function were constant, the solution is the obvious linear relation.   We need to use an average slope over the interval if we hope to estimate the change in y.

Suppose we do this, using the arithmetic average of the slopes at the beginning and end of the interval:

$$y_{n+1} = y_n + h\frac{y_n' + y_{n+1}'}{2}. \tag{2.2}$$

We are unable to employ Eq. (2.2) immediately, because since the derivative is a function of both x and y, we cannot evaluate y_{n+1}' with y_{n+1} unknown.   The modified Euler method surmounts the difficulty by estimating, or "predicting," a value of y_{n+1} by the simple Euler relation, Eq. (2.1), and uses this value to compute y_{n+1}', giving an improved estimate ("corrected" value) for y_{n+1}.   Since the y_{n+1}' was computed using the predicted value, of less accuracy, one generally re-

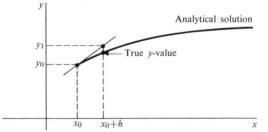

Fig. 2.1

corrects the y_{n+1} value as many times as will make a significant difference. (If more than two or three recorrections are required, it is usually more efficient to reduce the step size.)

We will illustrate the modified Euler method, also termed the Euler predictor-corrector method, on the same problem previously treated. A table is convenient:

$$\frac{dy}{dx} = x + y, \qquad y(0) = 1, \quad h = 0.02.$$

x_n	y_n	y'_n	hy'_n	y_{n+1}	y'_{n+1}	y'_{av}	hy'_{av}
0	1.0000	1.0000	0.0200	1.0200	1.0400	1.0200	0.0204
				1.0204*	1.0404	1.0202	0.0204
0.02	1.0204	1.0404	0.0208	1.0412	1.0812	1.0608	0.0212
				1.0416	1.0816	1.0610	0.0212
0.04	1.0416	1.0816	0.0216	1.0632	1.1232	1.1024	0.0220
				1.0636	1.1236	1.1026	0.0221
				1.0637	1.1237	1.1027	0.0221
0.06	1.0637	1.1237	0.0225	1.0862	1.1662	1.1449	0.0229
				1.0866	1.1666	1.1451	0.0229
0.08	1.0866	1.1666	0.0233	1.1099	1.2099	1.1883	0.0238
				1.1104	1.2104	1.1885	0.0238
0.10	1.1104						

(1.1103 analytical value)

The answer agrees within 1 in the fourth decimal place. We have done more work than in the simple Euler method, but certainly not the 20 times more that would have been needed with that method to attain four-decimal-place accuracy. Hence the modified method is more efficient.

We can find the error of the modified Euler method by comparing with the Taylor series,

$$y_{n+1} = y_n + y'_n h + \tfrac{1}{2}y''_n h^2 + \frac{y'''(\xi)}{6}h^3, \qquad x_n < \xi < x_n + h.$$

Replace the second derivative by the forward difference approximation for y'', $(y'_{n+1} - y'_n)/h$, which has error of $O(h)$, and write the error term as $O(h^3)$:

$$y_{n+1} = y_n + h\left(y'_n + \frac{1}{2}\left[\frac{y'_{n+1} - y'_n}{h} + O(h)\right]h\right) + O(h^3),$$

$$y_{n+1} = y_n + h(y'_n + \tfrac{1}{2}y'_{n+1} - \tfrac{1}{2}y'_n) + O(h^3),$$

$$y_{n+1} = y_n + h\left(\frac{y'_n + y'_{n+1}}{2}\right) + O(h^3).$$

---

* It is convenient to use this column for both the predicted and corrected values of y_{n+1} The first entry in any x-row is the predicted value. Corrected and recorrected values follow.

This shows that the error of one step of the modified Euler method is $O(h^3)$. This is the "local error." There is an accumulation of errors from step to step, so that the error over the whole range of application, the so-called "global error," is $O(h^2)$. This seems intuitively reasonable since the number of steps into which the interval is subdivided is proportional to $1/h$; hence the order of error is reduced to $O(h^2)$ on the continuing application. We treat the accumulation of errors more fully in a later section.

3. RUNGE-KUTTA METHODS

A further advance in efficiency, i.e., obtaining the most accuracy per unit of computational effort, can be secured with a group of methods due to the German mathematicians Runge and Kutta. The fourth-order Runge-Kutta methods are widely used in computer solutions to differential equations. The development of this technique is algebraically complicated.*

To convey some idea of how the methods are developed, we show the derivation of a second-order method. We write the increment to y as a weighted average of two estimates of Δy, k_1 and k_2. For the equation $dy/dx = f(x,y)$,

$$
\begin{aligned}
y_{n+1} &= y_n + ak_1 + bk_2, \\
k_1 &= hf(x_n, y_n), \\
k_2 &= hf(x_n + \alpha h, y_n + \beta k_1).
\end{aligned} \tag{3.1}
$$

The Runge-Kutta methods always use as the first estimate of Δy the simple Euler estimate; the other estimates are taken with x and y stepped up the fractions α and β of h and of the earlier estimate of Δy, k_1. Our problem is to devise a scheme of choosing the four parameters, a, b, α, β. We do this by making Eq. (3.1) agree as well as possible with the Taylor series expansion, in which the y-derivatives are written in terms of f,

$$
y_{n+1} = y_n + hf(x_n, y_n) + (h^2/2) f'(x_n, y_n) + \cdots .
$$

An equivalent form, since $df/dx = f_x + f_y\,dy/dx = f_x + f_y f$, is

$$
y_{n+1} = y_n + hf_n + h^2(\tfrac{1}{2}f_x + \tfrac{1}{2}f_y f)_n. \tag{3.2}
$$

We now rewrite Eq. (3.1) by substituting the definitions of k_1 and k_2:

$$
y_{n+1} = y_n + ahf(x_n, y_n) + bhf\big(x_n + \alpha h, y_n + \beta hf(x_n, y_n)\big). \tag{3.3}
$$

To make the last term of Eq. (3.3) comparable to Eq. (3.2), we expand $f(x, y)$ in a Taylor series in terms of x_n, y_n, remembering that f is a function of two variables,† retaining only first derivative terms,

$$
f\big(x_n + \alpha h, y_n + \beta hf(x_n, y_n)\big) = (f + f_x \alpha h + f_y \beta hf)_n. \tag{3.4}
$$

On the right side of both Eqs. (3.2) and (3.4), f and its partial derivatives are all to be evaluated at (x_n, y_n).

---

* The development is given rather completely in Kopal (1955) pp. 195-213.
† The appendix will remind the student of this expansion.

Substituting from Eq. (3.4) into Eq. (3.3), we have

$$y_{n+1} = y_n + ahf_n + bh(f + f_x\alpha h + f_y\beta hf)_n,$$

or, rearranging,

$$y_{n+1} = y_n + (a + b)hf_n + h^2(\alpha bf_x + \beta bf_y f)_n. \tag{3.5}$$

Equation (3.5) will be identical to Eq. (3.2) if

$$a + b = 1, \qquad \alpha b = \tfrac{1}{2}, \qquad \beta b = \tfrac{1}{2}.$$

Note that there are only three equations that need to be satisfied by the four unknowns. We can choose one value arbitrarily (with minor restrictions); hence we have a set of second-order methods. For example, taking $a = \tfrac{2}{3}$, we have $b = \tfrac{1}{3}$, $\alpha = \tfrac{3}{2}$, $\beta = \tfrac{3}{2}$. Other choices give other sets of parameters that agree with the Taylor series expansion. If one takes $a = \tfrac{1}{2}$, the other variables are $b = \tfrac{1}{2}$, $\alpha = 1$, $\beta = 1$. This last set of parameters gives the modified Euler algorithm that we have previously discussed.

The fourth-order Runge-Kutta method is most widely used and is derived in similar fashion. Greater complexity results from having to compare terms through h^4, and gives a set of eleven equations in thirteen unknowns. The set of eleven equations can be solved with two unknowns being chosen arbitrarily. The most commonly used set of values lead to the algorithm

$$y_{n+1} = y_n + \tfrac{1}{6}(k_1 + 2k_2 + 2k_3 + k_4),$$
$$k_1 = hf(x_n, y_n),$$
$$k_2 = hf(x_n + \tfrac{1}{2}h, y_n + \tfrac{1}{2}k_1),$$
$$k_3 = hf(x_n + \tfrac{1}{2}h, y_n + \tfrac{1}{2}k_2),$$
$$k_4 = hf(x_n + h, y_n + k_3).$$

As an example, we again solve $dy/dx = x + y$, $y(0) = 1$, this time taking $h = 0.1$:

$$k_1 = 0.1(0 + 1) = 0.10000,$$
$$k_2 = 0.1(0.05 + 1.05) = 0.11000,$$
$$k_3 = 0.1(0.05 + 1.055) = 0.11050,$$
$$k_4 = 0.1(0.10 + 1.1105) = 0.12105,$$
$$y(0.1) = 1.0000 + \tfrac{1}{6}(0.10000 + 0.20000 + 0.22100 + 0.12105) = 1.11034.$$

This agrees to five decimals with the analytical result, and illustrates a further gain in accuracy with less effort than our example of Section 2.

The local error term for the fourth-order Runge-Kutta is $O(h^5)$; the global error would be about $O(h^4)$. It is computationally more efficient than the modified Euler method because, while four evaluations of the function are required per step rather than two (or three), the steps can be many-fold larger for the same accuracy.

4. MILNE'S METHOD

To get further efficiency improvement over the Runge-Kutta method, we must take a different attack. Suppose, instead of only the single starting value $y(x_0) = y_0$, given by the initial condition of our problem, we have several. These might have been computed by the Runge-Kutta method, or possibly by the Taylor series method. We can employ quadrature formulas to integrate as follows:

$$\frac{dy}{dx} = f(x,y),$$

$$\int_{x_{n-3}}^{x_{n+1}} \left(\frac{dy}{dx}\right) dx = \int_{x_{n-3}}^{x_{n+1}} f(x,y)\, dx = \int_{x_{n-3}}^{x_{n+1}} P_2(x)\, dx, \qquad (4.1)$$

$$y_{n+1} - y_{n-3} = \frac{4h}{3}(2f_n - f_{n-1} + 2f_{n-2}) + \frac{28}{90} h^5 y^v(\xi_1), \qquad x_{n-3} < \xi_1 < x_{n+1}.$$

We integrate the function $f(x,y)$ by replacing it with a quadratic interpolating polynomial that fits at the three points, where $x = x_n$, x_{n-1}, and x_{n-2}, and integrating according to the methods of Chapter 3.* Note that we extrapolate in the integration by one panel both to the left and to the right of the region of fit. Hence the error is not as small, because of the extrapolation, as it would be with only interpolation.

With this value of y_{n+1} one can calculate f_{n+1} reasonably accurately. In Milne's method, we use Eq. (4.1) as a predictor formula and then correct with

$$\int_{x_{n-1}}^{x_{n+1}} \left(\frac{dy}{dx}\right) dx = \int_{x_{n-1}}^{x_{n+1}} f(x,y)\, dx = \int_{x_{n-1}}^{x_{n+1}} P_2(x)\, dx, \qquad (4.2)$$

$$y_{n+1,c} - y_{n-1} = \frac{h}{3}(f_{n+1} + 4f_n + f_{n-1}) - \frac{h^5}{90} y^v(\xi_2), \qquad x_{n-1} < \xi_2 < x_{n+1}.$$

Note the changed range of integration and the smaller coefficient in the error term because the polynomial is not extrapolated. f_{n+1} is calculated using y_{n+1} from the predictor formula. The integration formula is the familiar Simpson's $\frac{1}{3}$ rule, since we integrate a quadratic over two panels.

We illustrate with our familiar simple problem, $dy/dx = x + y$, $y(0) = 1$. From Section 1, we take the four values calculated by Taylor series:

	x	y	$f(x,y) = x + y$	
From Taylor Series	0	1.0000	1.0000	
	0.1	1.1103	1.2103	
	0.2	1.2428	1.4428	
	0.3	1.3997	1.6997	
	0.4	(1.5836)	(1.9836)	predictor value
		1.5836	1.9836	corrector value
	0.5	(1.7974)	(2.2974)	predictor value
		1.7974	2.2974	corrector value

Analytical value at $x = 0.5$ is 1.7974

* We also derived the formula in Chapter 5 by the method of undetermined coefficients, Eq. (4.4).

For this example, the predictor and corrector values agree, and give the analytical answer to four decimal places. This means that the value of h could have been chosen larger. With the set of values available, h cannot be increased without additional computations, but if we had seven equally spaced values, we could double h by taking only every other one and still have four values to move ahead from.

Normally, the value of y_{n+1} from the predictor and the corrector do not agree. Consideration of the error terms of Eqs. (4.1) and (4.2) suggests that the true value should usually lie between the two values and closer to the corrector value. While ξ_1 and ξ_2 are not necessarily the same value, they lie in similar intervals. If one assumes that the values of $y^v(\xi_1)$ and $y^v(\xi_2)$ are equal, the error in the corrector formula is $\frac{1}{28}$ times the error in the predictor formula. Hence the difference between the predictor and corrector formula is about 29 times the error in the corrected value. This is frequently used as a criterion of accuracy for Milne's method.*

Milne's method is simple and has a good error term, $O(h^5)$ for local error. It is subject to an instability problem in certain cases, however, wherein the errors do not tend to zero as h is made smaller. This unexpected phenomenon is discussed below. Because of possible instability, another method, that of Adams, is more widely used than Milne's method.

5. INSTABILITY IN MILNE'S METHOD

It is sufficient to show the instability of Milne's method for one simple case. We shall be able to draw the necessary conclusions from this. Consider the differential equation, $dy/dx = Ay$, where A is a constant. The general solution is $y = ce^{Ax}$. If $y(x_0) = y_0$ is the initial condition which the solution must satisfy, $c = y_0 e^{-Ax_0}$. Hence, letting y_n be the value of the function when $x = x_n$, the analytical solution is

$$y_n = y_0 e^{A(x_n - x_0)}. \tag{5.1}$$

If we solve the differential equation by the method of Milne, we have, from the corrector formula,

$$y_{n+1} = y_{n-1} + \frac{h}{3}(y'_{n+1} + 4y'_n + y'_{n-1}).$$

Letting $y'_n = Ay_n$, from the original differential equation, and rearranging, we get

$$y_{n+1} = y_{n-1} + \frac{h}{3}(Ay_{n+1} + 4Ay_n + Ay_{n-1}),$$

$$\left(1 - \frac{hA}{3}\right)y_{n+1} - \frac{4hA}{3}y_n - \left(1 + \frac{hA}{3}\right)y_{n-1} = 0. \tag{5.2}$$

---

* A further criterion as to whether or not the single correction that is normally used in Milne's method is adequate is discussed in Section 7.

We would like to solve this equation for y_n in terms of y_0 to compare to Eq. (5.1).   Equation (5.2) is a second-order linear difference equation which can be solved in an analogous manner to that for differential equations.   The solution is

$$y_n = C_1 Z_1^{\ n} + C_2 Z_2^{\ n}, \tag{5.3}$$

where Z_1, Z_2 are the roots of the quadratic

$$\left(1 - \frac{hA}{3}\right) Z^2 - \frac{4hA}{3} Z - \left(1 + \frac{hA}{3}\right) = 0. \tag{5.4}$$

(The student should check that Eq. (5.3) is a solution of Eq. (5.2) by direct substitution.)   For simplification, let $hA/3 = r$; the roots of Eq. (5.4) are then

$$Z_1 = \frac{4r + \sqrt{12r^2 + 4}}{2(1 - r)},$$

$$Z_2 = \frac{4r - \sqrt{12r^2 + 4}}{2(1 - r)}. \tag{5.5}$$

We are interested in comparing the behavior of Eq. (5.3) with Eq. (5.1) as the step size h gets small.   As $h \rightarrow 0$, $r \rightarrow 0$, and $r^2 \rightarrow 0$ even faster.   Neglecting the $12r^2$ terms in comparison to the constant 4 under the radical in Eq. (5.5) gives

$$Z_1 \doteq \frac{4r + 2}{2(1 - r)} = \frac{2r + 1}{1 - r} = 1 + 3r + O(r^2) = 1 + Ah + O(h^2),$$

$$Z_2 \doteq \frac{4r - 2}{2(1 - r)} = \frac{2r - 1}{1 - r} = -1 + r + O(r^2) = -\left(1 - \frac{Ah}{3}\right) + O(h^2). \tag{5.6}$$

The last results are obtained by dividing the fractions.   We now compare Eq. (5.6) with the Maclaurin series of the exponential function,

$$e^{hA} = 1 + hA + O(h^2),$$

$$e^{-hA/3} = 1 - \frac{hA}{3} + O(h^2).$$

We see that, for $h \rightarrow 0$,

$$Z_1 \doteq e^{-hA}, \qquad Z_2 \doteq e^{-hA/3}.$$

Hence the Milne solution is represented by

$$y_n = C_1 (e^{hA})^n + C_2 (e^{-hA/3})^n = C_1 e^{A(x_n - x_0)} + C_2 e^{-A(x_n - x_0)/3}. \tag{5.7}$$

In Eq. (5.7), we have used $x_n - x_0 = nh$.   The solution consists of two parts. The first term obviously matches with the analytical solution, Eq. (5.1).   The second term, called a *parasitic term*, will die out as x_n increases if A is a positive constant, but if A is negative, it will grow exponentially with x_n.   Note that we get this peculiar behavior independent of h; smaller step size is of no benefit in eliminating the error.

6. ADAMS-MOULTON METHOD

A method which does not have the same instability problem as the Milne method, but is about as efficient, is the Adams-Moulton method. It also assumes a set of starting values already calculated by some other technique.* Similar to the derivation of Milne's method, we integrate the derivative function by representing it as a polynomial. Here we take a cubic through four points, from x_{n-3} to x_n, and integrate over one step, from x_n to x_{n+1}. For $dy/dx = f(x,y)$,

$$\int_{x_n}^{x_{n+1}} \left(\frac{dy}{dx}\right) dx = \int_{x_n}^{x_{n+1}} f(x,y)\, dx = \int_{x_n}^{x_{n+1}} P_3(x)\, dx,$$

$$y_{n+1} - y_n = h(f_n + \tfrac{1}{2}\Delta f_{n-1} + \tfrac{5}{12}\Delta^2 f_{n-2} + \tfrac{3}{8}\Delta^3 f_{n-3}) + \tfrac{251}{720} h^5 y^v(\xi), \quad (6.1)$$

$$x_{n-3} < \xi < x_{n+1}.$$

The integration formula comes directly from the lozenge diagram for one panel integration in Chapter 3. Alternatively, it can be developed by writing $P_3(x)$ as a Newton-Gregory backwards interpolating polynomial (fitting at x_n, x_{n-1}, x_{n-2}, and x_{n-3}) and integrating. Use of Eq. (6.1) to compute y_{n+1} is called the Adams-Bashforth method, but more accurate results are obtained by correcting y_{n+1} before calculating the next step. Using Eq. (6.1) as a predictor formula, we can compute a nearly correct value of f_{n+1}. If we now approximate $f(x,y)$ as a cubic that fits over the range from x_{n-2} to x_{n+1}, and integrate from x_n to x_{n+1}, we will not be extrapolating the polynomial, and will have a more favorable error term. The result is

$$y_{n+1} - y_n = h(f_{n+1} - \tfrac{1}{2}\Delta f_n - \tfrac{1}{12}\Delta^2 f_{n-1} - \tfrac{1}{24}\Delta^3 f_{n-2}) - \tfrac{19}{720} h^5 y^v(\xi_1),$$

$$x_{n-2} < \xi_1 < x_{n+1}. \quad (6.2)$$

The Adams-Moulton method consists of using Eq. (6.1) as a predictor and Eq. (6.2) as a corrector. We illustrate it with the same example as before, $dy/dx = x + y$, $y(0) = 1$:

	x	y	f	Δf	$\Delta^2 f$	$\Delta^3 f$
	0	1.00000	1.00000			
				0.21034		
Starting values	0.1	1.11034	1.21034		0.02213	
				0.23247		0.00231
	0.2	1.24281	1.44281		0.02444	
				0.25691		(0.00257)
	0.3	1.39972	1.69972		(0.02701)	
				(0.28392)		
predicted	0.4	(1.58364)	(1.98364)			
corrected		1.58365				

---

* Methods such as modified Euler or Runge-Kutta are known as *single step*, or *starting* methods. The methods of Milne and of Adams are called *multistep* methods.

By the predictor formula,

$$y(0.4) = 1.39972 + 0.1\left(1.69972 + \tfrac{1}{2}(0.25691) + \tfrac{5}{12}(0.02444) + \tfrac{3}{8}(0.00231)\right)$$
$$= 1.58364.$$

f at $x = 0.4$ is then computed and the difference table calculated.   The corrector formula then gives

$$y(0.4) = 1.39972 + 0.1\left(1.98364 - \tfrac{1}{2}(0.28392) - \tfrac{1}{12}(0.02701) - \tfrac{1}{24}(0.00257)\right)$$
$$= 1.58365.$$

The corrected value almost agrees to five decimals with the predicted value. Comparing error terms of Eq. (6.1) and (6.2), and assuming that the two fifth derivative values are equal, we see that the true value should lie between the predicated and corrected values, with the error in the corrected value being about

$$\frac{19}{251 + 19} \quad \text{or} \quad \frac{1}{14.2}$$

times the difference between the predicted and corrected values.   A frequently used criterion for accuracy of the Adams-Moulton method with four starting values is that the corrected value is not in error by more than one in the last place if the difference between predicted and corrected values is less than 14 in the last decimal place.*

Equations (6.1) and (6.2) are not well suited to computer utilization because calculating and storing the difference tables is wasteful of both time and core. Each of these differences is a linear function of the various f-values, however. Expressing the differences in terms of the f's gives an alternative form for the Adams-Moulton method:

Predictor:

$$y_{n+1} = y_n + \frac{h}{24}(55f_n - 59f_{n-1} + 37f_{n-2} - 9f_{n-3}),$$

Corrector: (6.3)

$$y_{n+1} = y_n + \frac{h}{24}(9f_{n+1} + 19f_n - 5f_{n-1} + f_{n-2}).$$

These same formulas were derived directly by the method of undetermined co-efficients as Eqs. (4.2) and (4.3) of Chapter 5.

These equations are better suited to machine calculation and to digital computer programming.   Without such calculation aids, the large coefficients and alternating signs lead to large round-off errors unless extra guard figures are carried. For slide rule and hand computation, the difference table form of the equation is preferred.

Adams formulas employing more or less than four starting values can be derived in analogous fashion.   The formulas we have given are widely used,

---

* The convergence criterion of Section 7 should also be met.

especially in combination with Runge-Kutta, because both kinds of methods then have local errors of $O(h^5)$.

When the predicted and corrected values agree to as many decimals as the desired accuracy, we can save computational effort by increasing the step size. As mentioned in connection with Milne's method, we can conveniently double the step size after we have seven equispaced values by omitting every second one. When the difference between predicted and corrected values reaches or exceeds the accuracy criterion, we should decrease step size. If we interpolate two additional y-values with a fourth-degree polynomial, where the error will be $O(h^5)$, consistent with the rest of our work, we can readily halve the step size. Convenient formulas for this are

$$y_{n-1/2} = \tfrac{1}{128}[35y_n + 140y_{n-1} - 70y_{n-2} + 28y_{n-3} - 5y_{n-4}],$$
$$y_{n-3/2} = \tfrac{1}{64}[-y_n + 24y_{n-1} + 54y_{n-2} - 16y_{n-3} + 3y_{n-4}].$$
(6.4)

Use of these values with y_n, y_{n-1} gives four values of the function at intervals of $\Delta x = h/2$.

The efficiency of both the Adams-Moulton and Milne methods is about twice that of the Runge-Kutta technique, because only two new function evaluations are needed per step rather than four, and the error terms are similar. In addition, these multistep methods have built in accuracy criteria so that step size control can readily be secured. Warten (1963) gives a method for accuracy estimation and step size control for the Runge-Kutta method.

7. CONVERGENCE CRITERIA

In Section 6, we recorrected in the modified Euler method until no further change in y_{n+1} resulted. Usually this means one more calculation than would otherwise be needed if we could predict whether the recorrection would make a significant change. In the methods of Milne and Adams-Moulton, we usually do not recorrect, but use a value of h small enough that this is unnecessary. We now look at a criterion to say how small h should be in the Milne method, for $dy/dx = f(x,y)$. Let

y_p = value of y_{n+1} from predictor formula,
y_c = value of y_{n+1} from corrector formula,
y_{cc}, y_{ccc}, etc. = values of y_{n+1} if successive recorrections are made,
y_∞ = value to which successive recorrections converge,
$D = y_c - y_p$.

The change of y_c by recorrecting once is

$$y_{cc} - y_c = \left(y_n + \frac{h}{3}(y'_c + 4y'_n + y'_{n-1})\right) - \left(y_n + \frac{h}{3}(y'_p + 4y'_n + y'_{n-1})\right)$$
$$= \frac{h}{3}(y'_c - y'_p).$$
(7.1)

In Eq. (7.1) we have used the subscripts p or c to denote which y-value is used in evaluating the derivative at $x = x_{n+1}$. We now manipulate the difference $(y_c' - y_p')$:

$$y_c' - y_p' = f(x_{n+1}, y_c) - f(x_{n+1}, y_p) = \frac{f(x_{n+1}, y_c) - f(x_{n+1}, y_p)}{(y_c - y_p)}(y_c - y_p)$$

$$= f_y(\xi_1)D, \qquad \xi_1 \text{ between } y_c \text{ and } y_p.$$

Hence

$$y_{cc} - y_c = \frac{hD}{3}f_y(\xi_1)$$

is the difference on recorrecting. If recorrected again,

$$y_{ccc} - y_{cc} = \frac{h}{3}(y_{cc}' - y_c')$$

$$= \frac{h}{3}f_y(\xi_2) \cdot (y_{cc} - y_c) = \frac{h}{3}f_y(\xi_2)\left[\frac{hD}{3}f_y(\xi_1)\right] \qquad (7.2)$$

$$= \left(\frac{h}{3}\right)^2 (f_y(\xi))^2 D, \qquad \xi_2 \text{ between } y_c \text{ and } y_{cc}.$$

In Eq. (7.2), we need to impose on f the restrictions of continuity and that $f(\xi_1)$ and $f(\xi_2)$ have the same sign. ξ lies between the extremes of y_p, y_c, and y_{cc}.

On further recorrections we will have a similar relation. We get y_∞ by adding all the corrections of y_p together:

$$y_\infty = y_p + (y_c - y_p) + (y_{cc} - y_c) + (y_{ccc} - y_{cc}) + \cdots$$

$$= y_p + D + \frac{hf_y(\xi)}{3}D + \left(\frac{(hf_y(\xi))}{3}\right)^2 D + \left(\frac{hf_y(\xi)}{3}\right)^3 D + \cdots.$$

The increment to y_p is a geometric series; so, if the ratio is less than unity,

$$y_\infty = y_p + \frac{D}{1-r}, \qquad r = \frac{hf_y(\xi)}{3}, \qquad \xi \text{ between } y_p \text{ and } y_\infty.$$

Hence, unless

$$|r| = \frac{h|f_y(\xi)|}{3} \doteq \frac{hf_y(x_n, y_n)}{3} < 1,$$

the successive recorrections diverge. Our first convergence criterion is

$$h < \frac{3}{|f_y(x_n, y_n)|}. \qquad (7.3)$$

We wish to have y_c and y_∞ the same to within one in the Nth decimal place:

$$y_\infty - y_c = \left(y_p + \frac{D}{1-r}\right) - (y_p + D) = \frac{rD}{1-r} < 10^{-N}.$$

If $r \ll 1$, the fraction

$$\frac{r}{1-r} \doteq r,$$

and a second convergence criterion which ensures that the first corrected value is adequate, i.e., it will not be changed in the Nth decimal place by further corrections, is

$$D \cdot 10^N < \left| \frac{1}{r} \right| \doteq \frac{3}{h \, | \, f_y(x_n, y_n) \, |}. \tag{7.4}$$

For the Milne method we have the three criteria given below. If all are met, the corrected value should be good to N decimals:

$$\text{Convergence criteria} \quad \begin{cases} h < \dfrac{3}{| \, f_y \, |}, \\[2mm] D \cdot 10^N < \dfrac{3}{h \, | \, f_y \, |}. \end{cases}$$

$$\text{Accuracy criterion} \quad \begin{cases} D \cdot 10^N < 29. \end{cases}$$

A similar set of criteria for the Adams-Moulton method are derived in the same way. They are:

$$h < \frac{24/9}{| \, f_y \, |},$$

$$D \cdot 10^N < \frac{24/9}{h \, | \, f_y \, |}, \tag{7.5}$$

$$D \cdot 10^N < 14.$$

These criteria are for a single first-order equation only. A similar analysis for a system is much more complicated.

8. ERRORS AND ERROR PROPAGATION

Our previous error analyses have examined the error of a single step only, the so-called local error of the methods. Since all practical applications of numerical methods to differential equations involve many steps, the accumulation of these errors, termed the global error, is important. We recognize that there are several sources of error in a numerical calculation.

Original data errors. If the initial conditions are not known exactly (or must be expressed inexactly as a terminated decimal number) the solution will be affected to a greater or lesser degree depending on the sensitivity of the equation. Highly sensitive equations are said to be subject to *inherent instability*.

Round-off errors. Since we can carry only a finite number of decimal places, our computations are subject to inaccuracy from this source, no matter whether we round or whether we chop off. Carrying more decimal places than we require

in the final answer is the normal practice to minimize this, but in lengthy calculations this is a source of error that is most difficult to analyze and control. This
type of error is especially troublesome when two nearly equal quantities are subtracted. Both floating point and fixed point calculations in computers are subject to round-off errors.

Truncation errors of the method. These are the type of errors we have been
discussing, due to using truncated series for approximation in our work, when an
infinite series is needed for exactness. The choice of method is our best control
here, with suitable selection of h.

In addition to these three types of errors, when we solve differential equations
numerically we must worry about the propagation of previous errors through the
subsequent steps. Since we use the end values at each step as the starting values
for the next one, it is as if incorrect original data was distorting the later values.
(Round-off will almost always produce error even if our method were exact.)
This effect we now examine, but only for the very simple case of the Euler method.
This will show how such error studies are made, as well as suggest how difficult
is the analysis of more practical methods.

We consider the first-order equation $dy/dx = f(x,y)$, $y(x_0) = y_0$. Let

$$Y_n = \text{calculated value at } x_n,$$
$$y_n = \text{true value at } x_n,$$
$$e_n = y_n - Y_n = \text{error in } Y_n; \ y_n = Y_n + e_n.$$

By the Euler algorithm,

$$Y_{n+1} = Y_n + hf(x_n, Y_n).$$

By Taylor series,

$$y_{n+1} = y_n + hf(x_n, y_n) + \frac{h^2}{2} y''(\xi_n), \qquad x_n < \xi_n < x_n + h,$$

$$e_{n+1} = y_{n+1} - Y_{n+1} = y_n - Y_n + h\big(f(x_n, y_n) - f(x_n, Y_n)\big) + \frac{h^2}{2} y''(\xi_n)$$

$$= e_n + h\frac{f(x_n, y_n) - f(x_n, Y_n)}{y_n - Y_n}(y_n - Y_n) + \frac{h^2}{2} y''(\xi_n) \tag{8.1}$$

$$= e_n + hf_y(x_n, \eta_n)e_n + \frac{h^2}{2} y''(\xi_n), \qquad \eta_n \text{ between } y_n, Y_n.$$

In Eq. (8.1) we have used the mean value theroem, imposing continuity and
existence conditions on $f(x,y)$ and f_y. We suppose, in addition, that the magnitude
of f_y is bounded by the positive constant K in the region of x,y space in which we
are interested.* Hence,

$$e_{n+1} \leq (1 + hK)e_n + \tfrac{1}{2}h^2 y''(\xi_n). \tag{8.2}$$

---

* This is essentially the same as the Lipschitz condition which will guarantee existence
and uniqueness of a solution.

$y(x_0) = y_0$ is our initial condition which we assume free of error. Since $Y_0 = y_0$, $e_0 = 0$:

$$e_1 \leq (1 + hK)e_0 + \tfrac{1}{2}h^2 y''(\xi_0) = \tfrac{1}{2}h^2 y''(\xi_0),$$
$$e_2 \leq (1 + hK)(\tfrac{1}{2}h^2 y''(\xi_0)) + \tfrac{1}{2}h^2 y''(\xi_1) = \tfrac{1}{2}h^2[(1 + hK)y''(\xi_0) + y''(\xi_1)].$$

Similarly,

$$e_3 \leq \tfrac{1}{2}h^2[(1 + hK)^2 y''(\xi_0) + (1 + hK)y''(\xi_1) + y''(\xi_2)],$$
$$e_n \leq \tfrac{1}{2}h^2[(1 + hK)^{n-1}y''(\xi_0) + (1 + hK)^{n-2}y''(\xi_1) + \cdots + y''(\xi_{n-1})].$$

If $f_y \leq K$ is positive, the truncation error at every step is propagated to every later step after being amplified by the factor $(1 + hf_y)$ each time. Note that as $h \to 0$, the error at any point is just the sum of all the previous errors. If the f_y are negative and of magnitude such that $|hf_y| < 2$, the errors are propagated with diminishing effect.

We now show that the accumulated error after n steps is $O(h)$, i.e., global error of simple Euler is $O(h)$. We assume in addition to the above that y'' is bounded, $|y''(x)| < M$, $M > 0$. Equation (8.2) becomes, after taking absolute values,

$$|e_{n+1}| \leq (1 + hK)|e_n| + \tfrac{1}{2}h^2 M.$$

Compare to the second-order difference equation:

$$Z_{n+1} = (1 + hK)Z_n + \tfrac{1}{2}h^2 M, \tag{8.3}$$
$$Z_0 = 0.$$

Obviously the values of Z_n are at least equal to the magnitudes of $|e_n|$. The solution to Eq. (8.3) is (check by direct substitution)

$$Z_n = \frac{hM}{2K}(1 + hK)^n - \frac{hM}{2K}.$$

The Maclaurin expansion of e^{hK} is

$$e^{hK} = 1 + hK + \frac{(hK)^2}{2} + \frac{(hK)^3}{6} + \cdots,$$

so that

$$1 + hK < e^{hK} \qquad (K > 0),$$

$$Z_n < \frac{hM}{2K}(e^{hK})^n - \frac{hM}{2K} = \frac{hM}{2K}(e^{nhK} - 1)$$

$$= \frac{hM}{2K}(e^{(x_n - x_0)K} - 1) = O(h).$$

It follows that the global error e_n is $O(h)$.

9. SYSTEMS OF EQUATIONS AND HIGHER ORDER EQUATIONS

We have so far treated only the case of a first-order differential equation. Most differential equations which are the mathematical model for a physical problem are of higher order, or a set of simultaneous differential equations. For example,

$$\frac{w}{g}\frac{d^2x}{dt^2} + b\frac{dx}{dt} + kx = f(x,t)$$

represents a vibrating system in which a linear spring with spring constant k restores a displaced mass of weight w against a resisting force whose resistance is b times the velocity. The function $f(x,t)$ is an external forcing function acting on the mass.

An analogous second-order equation describes the flow of electricity in a circuit containing inductance, capacitance, and resistance. The external forcing function in this case represents the applied electromotive force. Compound spring-mass systems and electrical networks can be simulated by a system of such equations.

We show how a higher order differential equation can be reduced to a system of simultaneous first-order equations. These can be solved by an application of the methods previously studied. We treat here initial value problems only, for which the n values of the functions or derivatives are all specified at the same (initial) value of the independent variable. When some of the conditions are specified at one value of the independent variable and others at a second value, we call it a *boundary value problem*. Methods of solving these are discussed in the next chapter.

By solving for the second derivative, we can normally express a second-order equation as

$$\frac{d^2x}{dt^2} = f\left(x,t,\frac{dx}{dt}\right), \qquad x(t_0) = x_0, \qquad x'(t_0) = x_0'. \tag{9.1}$$

The initial value of the function x and its derivative are generally specified. We convert this to a pair of first-order equations by the simple expedient of defining the derivative as a second function. Then, since $d^2x/dt^2 = (d/dt)(dx/dt)$,

$$\frac{dx}{dt} = y, \qquad x(t_0) = x_0,$$

$$\frac{dy}{dt} = f(x,t,y), \qquad y(t_0) = x_0'.$$

The pair of first-order equations is equivalent to the original Eq. (9.1). For higher order, each of the lower derivatives is defined as a new function, giving n first-order equations that correspond to an nth order differential equation. For a system of higher order equations, each is similarly converted, so that a larger set of first-order equations results.

We illustrate the application of the various methods to the pair of first-order equations.

$$\frac{dx}{dt} = xy + t, \qquad x(0) = 1,$$

$$\frac{dy}{dt} = ty + x, \qquad y(0) = -1. \tag{9.2}$$

Taylor series method. We need the various derivatives x', x'', x''', $\cdots$, y', y'', y''', $\cdots$ all evaluated at $t = 0$:

$$x' = xy + t,$$
$$y' = ty + x,$$
$$x'' = xy' + x'y + 1,$$
$$y'' = y + ty' + x',$$
$$x''' = x'y' + xy'' + x''y + x'y',$$
$$y''' = y' + y' + ty'' + x'',$$

etc.,

$$x'(0) = (1)(-1) + 0 = -1,$$
$$y'(0) = (0)(-1) + 1 = 1,$$
$$x''(0) = (1)(1) + (-1)(-1) + 1 = 3,$$
$$y''(0) = -1 + (0)(1) - 1 = -2,$$
$$x'''(0) = -7,$$
$$y'''(0) = 5,$$

etc.,

$$x(t) = 1 - t + \tfrac{3}{2}t^2 - \tfrac{7}{6}t^3 + \tfrac{27}{24}t^4 - \tfrac{124}{120}t^5 + \cdots,$$
$$y(t) = -1 + t - t^2 + \tfrac{5}{6}t^3 - \tfrac{13}{24}t^4 + \tfrac{47}{120}t^5 + \cdots. \tag{9.3}$$

At $t = 0.2$, $x = 0.8521$ and $y = -0.8341$.

Equations (9.3) are the solution to the set (9.2). Note that we need to alternate between the functions in getting the derivatives, i.e., we cannot get $x''(0)$ until $y'(0)$ is known; we cannot get $y'''(0)$ until $x''(0)$ is known.

Euler predictor-corrector. We apply the predictor to each equation; then the corrector can be used. Again note that we work alternately with the two functions.

Take $h = 0.2$. Let p and c subscripts indicate predicted and corrected values, respectively:

$$x_p(0.2) = 1 + 0.2\left((1)(-1) + 0\right) = 0.8,$$
$$y_p(0.2) = -1 + 0.2\left((0)(-1) + 1\right) = -0.8,$$
$$x_c(0.2) = 1 + 0.2\left(\frac{-1 + [(0.8)(-0.8) + 0.2]}{2}\right) = 0.856,$$

$$y_c(0.2) = -1 + 0.2\left(\frac{1 + [(0.2)(-0.8) + 0.856]}{2}\right) = -0.830.$$

In computing $x_c(0.2)$, we used the x_p and y_p. In computing $y_c(0.2)$ after $x_c(0.2)$ is known, we have a choice between x_p and x_c. There is an intuitive feel that one should use x_c with the idea that one should always use the best available values. This does not always hasten convergence, probably due to compensating errors. Here we have used the best values to date. Recorrecting in the obvious manner gives

$$x(0.2) = 0.849,$$
$$y(0.2) = -0.832.$$

We could now advance the solution another step if desired.

Runge-Kutta. Again there is an alternation between the x and y calculations. In applying this method, one always uses the previous k-value in incrementing the function values. The k-values for x are

$$k_1 = 0.2(-1) = -0.2,$$

$$k_2 = 0.2\left[\left(1 - \frac{0.2}{2}\right)\left(-1 + \frac{0.2}{2}\right) + \frac{0.2}{2}\right] = -0.142,$$

$$k_3 = 0.2\left[\left(1 - \frac{0.142}{2}\right)\left(-1 + \frac{0.162}{2}\right) + \frac{0.2}{2}\right] = -0.1508,$$

$$k_4 = 0.2[(1 - 0.1508)(-1 + 0.1674) + 0.2] = -0.1014.$$

The k-values for y are

$$k_1 = 0.2(1) = 0.2,$$

$$k_2 = 0.2\left[\left(\frac{0.2}{2}\right)\left(-1 + \frac{0.2}{2}\right) + \left(1 - \frac{0.2}{2}\right)\right] = 0.162,$$

$$k_3 = 0.2\left[\left(\frac{0.2}{2}\right)\left(-1 + \frac{0.162}{2}\right) + \left(1 - \frac{0.142}{2}\right)\right] = 0.1674,$$

$$k_4 = 0.2[(0.2)(-1 + 0.1674) + (1 - 0.1508)] = 0.1365.$$

Then

$$x(0.2) = 1 + \tfrac{1}{6}(-0.2 + 2(-0.142) + 2(-0.1508) - 0.1014) = 0.8522,$$

$$y(0.2) = -1 + \tfrac{1}{6}(0.2 + 2(0.162) + 2(0.1674) + 0.1365) = -0.8341.$$

The Runge-Kutta values appear to be correct to four decimals, while the modified Euler values are correct to only two. $h = 0.2$ is too large for that method.

Adams-Moulton. After getting four starting values, we proceed in the obvious way, again alternately computing x and then y:

	t	x	x'	t	y	y'
	0	1.0	-1.0	0	-1.0	1.0
Starting	0.1	0.9139	-0.7309	0.1	-0.9092	0.8230
values	0.2	0.8522	-0.5108	0.2	-0.8341	0.6854
	0.3	0.8107	-0.3246	0.3	-0.7705	0.5796
predicted	0.4	(0.7867)	(-0.1638)	0.4	(-0.7167)	(0.5000)
corrected		0.7865			-0.7167	

In the computations we first get predicted values of x and y:

$$x(0.4)_p = 0.8107 + \frac{0.1}{24}\left[55(-0.3246) - 59(-0.5108) + 37(-0.7309) - 9(-1.0)\right]$$

$$= 0.7867,$$

$$y(0.4)_p = -0.7705 + \frac{0.1}{24}\left[55(0.5796) - 59(0.6854) + 37(0.8230) - 9(1.0)\right]$$

$$= -0.7167.$$

After getting x' and y' at $t = 0.4$, using $x(0.4)_p$ and $y(0.4)_p$, we then correct:

$$x(0.4)_c = 0.8107 + \frac{0.1}{24}\left[9(-0.1638) + 19(-0.3246) - 5(-0.5108) + (-0.7309)\right]$$

$$= 0.7865,$$

$$y(0.4)_c = -0.7705 + \frac{0.1}{24}\left[9(0.5000) + 19(0.5796) - 5(0.6854) + (0.8230)\right]$$

$$= -0.7167.$$

The close agreement of predicted and corrected values indicates four-decimal-place accuracy.

10. SPECIAL METHODS

The methods we have discussed are of general utility and are adequate for most applications. For certain special forms of differential equations, other methods have been proposed.

Kunz (1957) gives the following predictor-corrector formulas for second-order differential equations which do not involve the first derivative:

$$y_{n+,p} = y_n + y_{n-2} - y_{n-3} + \frac{h^2}{4}(5y_n'' + 2y_{n-1}'' + 5y_{n-2}'') + \frac{17}{240}h^6 y^{vi}(\xi),$$

$$y_{n+1,c} = 2y_n - y_{n-1} + \frac{h^2}{12}(y_{n+1}'' + 10y_n'' + y_{n-1}'') - \frac{1}{240}h^6 y^{iv}(\xi_1). \tag{10.1}$$

Hamming (1962) gives this and other similar formulas for this special type of second-order equation.

Fox (1962) discusses the methods of Lanczos and of Clenshaw which involve representing the solution as a Chebyshev polynomial. These methods are beyond the present text.

The Numerov method is useful for second-order linear equations which lack the first derivative. Kunz (1957) discusses this. The method has an error term of $O(h^6)$ and requires only two starting values.

11. COMPUTER APPLICATIONS

Program 1 solves a second-order differential equation by the modified Euler method, after reducing it to a pair of first-order equations. The computer output is the solution to the equation

$$\frac{W}{g}\frac{d^2x}{dt^2} + b\frac{dx}{dt} + kx = a \sin \omega t.$$

Program 1

```
ZZJOB 5                    CSC  001    GERALD, C. F.
ZZFORX5
*LIST PRINTER
C SOLUTION TO SECOND ORDER DIFFERENTIAL EQUATION, EULER PRED-CORR METHOD
C DEFINE SECOND DERIVATIVE FUNCTION AND READ PARAMETERS
      FCN(X,Y,T)=(A*SINF(O*T)-B*Y-XK*X)*32.1725/W
      READ 100, W, B, XK, A, O
      READ 100, XO, DXO, DT, TOL, TEND
C PRINT HEADINGS AND FIRST LINE OF OUTPUT
      T=0.
      PRINT 50, T, XO, DXO
C COMPUTE BY PREDICTOR FORMULAS
15    XP=XO+DT*DXO
      YP=DXO+DT*FCN(XO,DXO,T)
C COMPUTE BY CORRECTOR FORMULAS, RECORRECTING UP TO 10 TIMES.
      DO 30 I=1,10
      XC=XO+DT*(DXO+YP)/2.0
      YC=DXO+DT*(FCN(XO,DXO,T)+FCN(XC,YP,T+DT))/2.0
      Z=ABSF(XC-XP)
      IF(Z-TOL)1,1,2
2     XP=XC
30    YP=YC
      T=T+DT
      PRINT 200, T, XP, YP
      IF(T-TEND)20,99,99
1     T=T+DT
      XP=XC
      YP=YC
      PRINT 300, T, XP, YP
C RESET VARIABLES AND CALCULATE NEXT STEP.
20    XO=XP
      DXO=YP
      IF(T-TEND)15,99,99
99    CALL EXIT
100   FORMAT(5F10.0)
50    FORMAT(/ 6H TIME , 7X, 8HDISTANCE , 15X, 8HVELOCITY /
     1 F5.2, 6X, F10.7, 12X, F10.7 / )
200   FORMAT (1H ,F5.2, 6X, F10.7, 12X, F10.7, 17HTOLERANCE NOT MET )
300   FORMAT (1H ,F5.2, 6X, F10.7, 12X, F10.7)
      END
16.0    1.6     24.0    0.6     2.0
1.0    -2.0     0.05    0.0001  2.0
ZZZZ
```

TIME	DISTANCE	VELOCITY
0.00	1.0000000	-2.0000000
.05	.8557082	-3.7716199
.10	.6396284	-4.8715109
.15	.3859001	-5.2775660
.20	.1278021	-5.0463069
.25	-.1057319	-4.2953908
.30	-.2926581	-3.1818665
.35	-.4191903	-1.8794967
.40	-.4801044	-.5574827
.45	-.4781096	.6374180
.50	-.4223449	1.5933858
.55	-.3264980	2.2407386
.60	-.2066513	2.5533821
.65	-.0791891	2.5453363
.70	.0410248	2.2634072
.75	.1420409	1.7773592
.80	.2156988	1.1690166
.85	.2579674	.5216237
.90	.2687752	-.0893631
.95	.2514589	-.6033833
1.00	.2118981	-.9791641
1.05	.1575118	-1.1964095
1.10	.0962304	-1.2549625
1.15	.0355613	-1.1719000
1.20	-.0181638	-.9771774
1.25	-.0603140	-.7084645
1.30	-.0881795	-.4060621
1.35	-.1010276	-.1080150
1.40	-.0998778	.1536661
1.45	-.0871458	.3556626
1.50	-.0661432	.4845028
1.55	-.0406149	.5366831
1.60	-.0142594	.5175885
1.65	.0096785	.4395823
1.70	.0286676	.3197608
1.75	.0410967	.1773170
1.80	.0463163	.0311357
1.85	.0445390	-.1020884
1.90	.0367438	-.2095098
1.95	.0244276	-.2828859
2.00	.0093784	-.3188254

Program 2

```
ZZJOB 5                    CSC  001       GERALD, C. F.
ZZFORX5
*LIST PRINTER
C  SOLUTION TO FIRST ORDER DIFFERENTIAL EQUATION BY RUNGE-KUTTA PLUS ADAMS
C  DEFINE DERIVATIVE FUNCTION
      DERFN(X,Y)=X+Y*EXPF(-Y)
      DIMENSION X(5),Y(5)
C  READ IN PARAMETERS
      READ 100, XO,XF,H,YO
C  COMPUTE FOUR VALUES BY RUNGE-KUTTA AND PRINT OUT
      X(1)=XO
      Y(1)=YO
      DO 10 I=2,4
      FK1=H*DERFN(XO,YO)
      FK2=H*DERFN(XO+.5*H,YO+.5*FK1)
      FK3=H*DERFN(XO+.5*H,YO+.5*FK2)
      FK4=H*DERFN(XO+H,YO+FK3)
      X(I)=XO+H
      Y(I)=YO+(FK1+2.*FK2+2.*FK3+FK4)/6.
      XO=X(I)
   10 YO=Y(I)
      PRINT 60
      PRINT 200, (X(I),Y(I),I=1,4)
C  ADVANCE THE SOLUTION BY ADAMS-MOULTON
      PRINT 70
   30 Y(5)=Y(4)+H*( 55.*DERFN(X(4),Y(4))-59.*DERFN(X(3),Y(3))
     1+37.*DERFN(X(2),Y(2))-9.*DERFN(X(1),Y(1)))/24.
      Y(5)=Y(4)+H*(9.*DERFN(X(4),Y(5))+19.*DERFN(X(4),Y(4))
     1-5.*DERFN(X(3),Y(3))+DERFN(X(2),Y(2)))/24.
      X(5)=X(4)+H
      PRINT 200,X(5),Y(5)
C  RESUBSCRIPT X AND Y VALUES AND REPEAT ADAMS CALCULATION
      DO 20 J=1,4
      X(J)=X(J+1)
   20 Y(J)=Y(J+1)
      IF(X(4)-XF)30,99,99
   99 CALL EXIT
   60 FORMAT(33H STARTING VALUES FROM RUNGE-KUTTA//5X,1HX,12X,1HY/)
   70 FORMAT(/37H CONTINUATION OF SOLN BY ADAMS METHOD/)
  100 FORMAT(4F10.0)
  200 FORMAT(1H ,F10.7,8X,F10.7)
      END
0.0      2.0       0.1        0.0
ZZZZ
```

$$\frac{dy}{dx}=x+y\,e^{-y} \quad \text{for } 0<x<1$$
$$y(0)=0$$

STARTING VALUES FROM RUNGE-KUTTA

X	Y
0.0000000	0.0000000
.1000000	.0051701
.2000000	.0213838
.3000000	.0497096

CONTINUATION OF SOLN BY ADAMS METHOD

X	Y
.4000000	.0911612
.5000000	.1466001
.6000000	.2166289
.7000000	.3015114
.8000000	.4011332
.9000000	.5150156
1.0000000	.6423821
1.1000000	.7822589
1.2000000	.9335892
1.3000000	1.0953367
1.4000000	1.2665649
1.5000000	1.4464872
1.6000000	1.6344891
1.7000000	1.8301294
1.8000000	2.0331268
1.9000000	2.2433389
2.0000000	2.4607380

The parameters and initial conditions are read in.   The value of the function is recorrected until the difference in successive values is less than a certain tolerance. If the tolerance is not met in 10 recorrections, the value is punched with a message to indicate that the tolerance was not met.

Program 2 solves a first-order differential equation by finding four starting values by Runge-Kutta, and then advances the solution by Adams-Moulton.   In the example, the equation $dy/dx = x + ye^{-y}$, $y(0) = 0$ is solved for $0 < x < 1$. By changing the definition of the derivation function, a different equation can be solved.

PROBLEMS

Section 1

1. Solve the differential equation

$$\frac{dy}{dx} = x + y + xy, \qquad y(0) = 1$$

by Taylor series expansion to get the value of y at $x = 0.1$.   Use enough terms to get four-decimal-place accuracy.

2. Use the Taylor series method to find $y(2)$, given

$$\frac{dy}{dx} = x^2 + y^2, \qquad y(1) = 0.$$

How many terms are required for three-decimal accuracy?

3. The general solution to a differential equation normally defines a family of curves. For the differential equation

$$\frac{dy}{dx} = x^2 y^2$$

find, using the Taylor series method, the particular curve that passes through $(1,0)$.

4. Use the Taylor series method to get y at $x = 0.2(0.2)0.6$, given that

$$y'' = xy, \qquad y(0) = 1, \quad y'(0) = 1.$$

5. Find y at $t = 0.5$ if it is known that y and t are related by the second-order differential equation

$$y'' + y' = t \sin y, \qquad y(0) = 0, \quad y'(0) = 1.$$

6. A spring system has resistance to motion proportional to the square of the velocity, and its motion is described by

$$\frac{d^2 x}{dt^2} + 0.1 \left(\frac{dx}{dt}\right)^2 + 0.6x = 0.$$

If the spring is released from a point which is a unit distance above its equilibrium point, $x(0) = 1$, $x'(0) = 0$, use the Taylor series method to write a series expression for the displacement as a function of time, including terms up to t^6.

Section 2

7. Use the simple Euler method to solve Problem 1, with $h = 0.01$. Comparing your result to the value determined by Taylor series, estimate how small h would need to be to obtain four-decimal accuracy.

8. Solve the differential equation

$$\frac{dy}{dx} = \frac{x}{y}, \quad y(0) = 1,$$

by the simple Euler method with $h = 0.1$, to get $y(1)$. Then repeat with $h = 0.2$ to get another estimate of $y(0.1)$. Extrapolate these results assuming errors are proportional to step size, and compare to the analytical result. (Analytical result is $y^2 = 1 + x^2$.)

9. Repeat Problem 1, but with the modified Euler method with $h = 0.025$, so that the solution is obtained after four steps. Comparing with Problem 7, about how much less effort is it to solve this problem to four decimals with the modified Euler method in comparison to the simple Euler method?

10. Find the solution to Problem 2 by the modified Euler method, using $h = 0.1$.

11. Solve $y' = \sin x + y, y(0) = 2$, by the modified Euler method to get y at $x = 0.1(0.1)0.5$.

12. A sky diver jumps from a plane, and during the time before he opens his parachute, the air resistance is proportional to the $3/2$ power of his velocity. If it is known that the maximum rate of fall under these conditions is 80 mph, determine his velocity during the first 2 seconds of fall using the modified Euler method with $\Delta t = 0.2$. Neglect horizontal drift and assume an initial velocity of zero.

Section 3

13. Solve Problem 1 by the Runge-Kutta method but with $h = 0.1$ so the solution is obtained in only one step. Carry five decimals, and compare the accuracy and amount of work required with this method against the simple and modified Euler techniques in Problems 7 and 9.

14. Solve Problem 2 by the Runge-Kutta method, using $h = 0.5$.

15. Determine y at $x = 0.2(0.2)0.6$ by the Runge-Kutta technique, given that

$$\frac{dy}{dx} = \frac{1}{x + y}, \quad y(0) = 2.$$

16. Use the Runge-Kutta method to find y at $t = 3$, given that

$$\frac{dy}{dt} = y^2 e^{-t}, \quad y(1) = 0.5.$$

Use $\Delta t = 0.5$.

17. Using the conditions of Problem 12, determine how long it takes for the jumper to reach 90% of his maximum velocity, by integrating the equation using the Runge-Kutta technique with $\Delta t = 0.5$ until the velocity exceeds this value, and then interpolating. Then use numerical integration on the velocity values to determine the distance he falls in attaining $0.9v_{max}$.

18. It is not easy to know the accuracy with which the function has been determined by either the Euler methods or the Runge-Kutta method. A possible way to measure accuracy is to repeat the problem with a smaller step size, and compare results. If the two computations agree to n decimal places, one then assumes the values are correct to that many places. Repeat Problem 17 with $\Delta t = 0.3$, which should give a global error about one-tenth as large, and by comparing results, determine the accuracy in Problem 17.

Section 4

19. For the differential equation,

$$\frac{dy}{dx} = y - x^2, \qquad y(0) = 1,$$

starting values are known: $y(0.2) = 1.2186$, $y(0.4) = 1.4682$, $y(0.6) = 1.7379$. Use the Milne method to advance the solution to $x = 1.2$. Carry four decimals and compare to the analytical solution.

20. For the differential equation $\dfrac{dy}{dx} = \dfrac{x}{y}$, the following values are given:

x	y
0	$\sqrt{1}$
1	$\sqrt{2}$
2	$\sqrt{5}$
3	$\sqrt{10}$

To how many decimal places will Milne's method give the value at $x = 4$? How many decimal places must be carried in the starting values of y to ensure this accuracy?

21. For the equation $y' = y \sin \pi x$, $y(0) = 1$, get starting values by the Taylor series for $x = 0.2(0.2)0.6$, and advance the solution to $x = 1.0$ by Milne's method.

22. Continue the results of Problem 15 to $x = 2.0$ by the method of Milne. If you find that the corrector formula reproduces the predictor values, double the value of h after sufficient values are available.

Section 5

23. Check that y_n as defined by Eq. (5.3) is a solution of the difference equation, Eq. (5.2).

24. Perform the long division to show that

$$Z_1 = 1 + Ah + O(h^2), \qquad Z_2 = -\left(1 - \frac{Ah}{3}\right) + O(h^2),$$

as given in Eq. (5.6).

Section 6

25. Express the differences in Eqs. (6.1) and (6.2) in terms of functional values to show that Eqs. (6.3) are equivalent.

26. Solve Problem 19 using the Adams-Moulton method.

27. Repeat Problem 21, except use the Adams-Moulton method.

28. For the equation

$$\frac{dy}{dx} = x^3 + y^2, \qquad y(0) = 0,$$

using $h = 0.2$, compute three new values by the Runge-Kutta method (four decimals). Then advance to $x = 1.4$ using the Adams-Moulton method. If you find the accuracy criterion is not met, use Eqs. (6.4) to interpolate additional values so that four-place accuracy is maintained.

29. Derive the interpolation formulas given in Eq. (6.4).

Section 7

30. Given the linear differential equation $dy/dx = y \sin x$. (a) What is the maximum value of h that ensures convergence of the Milne method when continuing applications of the corrector formula are made? (b) If an h one-tenth of this maximum value is used, how close must the predictor and corrector values be so that recorrections are not required? (c) In terms of the maximum h in part (a), what size of h is implied in the accuracy criterion, $D \cdot 10^n < 29$?

31. Repeat Problem 30 for the differential equation

$$\frac{dy}{dx} = x^3 + y^2, \qquad y(0) = 0,$$

in the neighborhood of the point $(1.0, 0.15)$.

32. Repeat Problem 30, except for the Adams-Moulton method.

33. Derive Eq. (7.5).

34. Derive convergence criteria similar to Eqs. (7.3) and (7.4) for the Euler predictor-corrector method. Why can one not derive an accuracy criterion similar to those for the methods of Milne and Adams-Moulton?

Section 8

35. Estimate the propagated error at each step when the equation

$$\frac{dy}{dx} = x + y, \qquad y(0) = 1,$$

is solved by the simple Euler method with $h = 0.02$, for $x = 0(0.02)0.1$. (The equation is solved by this method in Section 2.) Compare to the actual errors.

36. Follow the propagated error between $x = 1$ and $x = 1.6$ when the simple Euler method is used to solve

$$\frac{dy}{dx} = xy^2, \qquad y(1) = 1.$$

Take $h = 0.1$.   Compare to the actual errors at each step.   The analytical solution is $y = 2/(3 - x^2)$.

Section 9

37. The mathematical model of an electrical circuit is given by the equation

$$0.5 \frac{d^2Q}{dt^2} + 6 \frac{dQ}{dt} + 50Q = 24 \sin 10t,$$

with $Q = 0$ and $I = dQ/dt = 0$ at $t = 0$.   Express as a pair of first-order equations.

38. In the theory of beams it is shown that the radius of curvature at any point is proportional to the bending moment:

$$EI \frac{y''}{[1 + (y')^2]^{3/2}} = M(x),$$

where y is the deflection of the neutral axis.   In the usual approach, $(y')^2$ is neglected in comparison to unity, but if the beam has appreciable curvature, this is invalid. For the cantilever beam for which $y(0) = y'(0) = 0$, express the equation as a pair of simultaneous first-order equations.

39. The motion of the compound spring system as sketched is given by the solution of the pair of simultaneous equations

$$m_1 \frac{d^2y_1}{dt^2} = -k_1 y_1 - k_2(y_1 - y_2),$$

$$m_2 \frac{d^2y_2}{dt^2} = k_2(y_1 - y_2),$$

where y_1 and y_2 are the displacements from their equilibrium positions of the two masses.   The initial conditions are

$$y_1(0) = A, \qquad y_1'(0) = B,$$
$$y_2(0) = C, \qquad y_2'(0) = D.$$

Express as a set of first-order equations.

40. Solve the pair of simultaneous equations

$$dx/dt = xy + t, \qquad x(0) = 0, \qquad dy/dt = x - t, \qquad y(0) = 1,$$

by the modified Euler method for $t = 0.2(0.2)0.6$.

41. Express the third-order equation

$$y''' + ty'' - ty' - 2y = t, \qquad y(0) = y''(0) = 0, \qquad y'(0) = 1,$$

as a set of first-order equations and solve at $t = 0.1, 0.2$ by the Runge-Kutta method $(h = 0.1)$.

42. Advance the solution of Problem 40 to $x = 1.0$ $(h = 0.2)$ by the Milne method.

43. Find y at $x = 0.6$, given that

$$y'' = yy', \qquad y(0) = 1, \qquad y'(0) = -1.$$

Begin the solution by the Taylor series method, getting $y(0.1)$, $y(0.2)$, $y(0.3)$. Then advance to $x = 0.6$ employing the Adams-Moulton technique with $h = 0.1$ on the equivalent set of first-order equations.

Section 10

44. For a resonant spring system with a periodic forcing function, the differential equation is

$$\frac{d^2x}{dt^2} + 64x = 16 \cos 8t, \qquad x(0) = x'(0) = 0.$$

Determine the displacement at $t = 0.1(0.1)0.8$ by the method of Eq. (10.1) and also by any other method of this chapter. Compare to the analytical solution $t \sin 8t$.

Section 11

45. Use the first program given in the text to solve Problem 44.

46. Use the second program of the text to solve Problem 21. Determine the effect on $y(1.0)$ of varying step sizes by rerunning with $h = 0.2, 0.01, 0.05, 0.001$.

47. Write a program that will handle initial value problems up to order four by solving the equivalent system of first-order equations. Start the solution by the Runge-Kutta technique, and then advance the solution by the Adams-Moulton method. Provide for testing of the accuracy of corrector values (to N decimals) with a message being output if the criterion is not met, but let the program continue, using the calculated value.

7
solving
sets of
equations

A variety of engineering problems may be solved through network analysis, such as determining the current flowing in certain electrical networks, finding the stresses in a building frame or a bridge truss, computing the flow rates in a hydraulic system with interconnected branches, or estimating the concentrations of reactants subject to simultaneous chemical reactions. The mathematical problem in all these cases reduces to solving a set of equations simultaneously. We shall also find, when we study the numerical methods of solving boundary value problems and partial differential equations, that these also require the solution of sets of equations, often very large in number.

The solution of sets of equations is much more difficult when the equations are nonlinear. Fortunately the majority of applications involve only linear equations, although when such systems are large, careful choice of methods is required to avoid overly long procedures and to preserve maximum accuracy. In this chapter we shall first discuss some of the basic ways of solving sets of linear equations, and later take up the harder problem of nonlinear equations.

1. MATRIX NOTATION

Our discussion will be facilitated by some of the concepts and notation of matrix algebra. Only the more elementary ideas will be needed.

A *matrix* is a rectangular array of numbers in which not only the value of the number is important but also its position in the array. The size of the matrix is described by the number of its rows and columns. A matrix of n rows and m columns is said to be $n \times m$. The elements of the matrix are generally enclosed in brackets, and double subscripting is the common way of indexing the elements. Capital letters are used to refer to matrices. For example,

$$A = \begin{bmatrix} a_{11} a_{12} \cdots a_{1m} \\ a_{21} a_{22} \cdots a_{2m} \\ \vdots \\ a_{n1} a_{n2} \cdots a_{nm} \end{bmatrix} = [a_{ij}], \qquad i = 1, 2, \ldots, n, \qquad j = 1, 2, \ldots, m.$$

Two matrices of the same size may be added or subtracted. The sum of

$$A = [a_{ij}] \quad \text{and} \quad B = [b_{ij}]$$

is the matrix whose elements are the sum of the corresponding elements of A and B,

$$C = A + B = [a_{ij} + b_{ij}] = [c_{ij}].$$

Similarly we get the difference of two equal-sized matrices by subtracting corresponding elements. If two matrices are not equal in size, they cannot be added or subtracted. Two matrices are equal if and only if each element of one is the same as the corresponding element of the other. Obviously, equal matrices must be of the same size.

Multiplication of two matrices is defined as follows, when A is $n \times m$ and B is $m \times r$:

$$[a_{ij}][b_{ij}] = [c_{ij}] =$$

$$\begin{bmatrix} (a_{11}b_{11} + a_{12}b_{21} + \cdots + a_{1m}b_{m1}) \cdots (a_{11}b_{1r} + \cdots + a_{1m}b_{mr}) \\ (a_{21}b_{11} + a_{22}b_{21} + \cdots + a_{2m}b_{m1}) \cdots (a_{21}b_{1r} + \cdots + a_{2m}b_{mr}) \\ \vdots \\ (a_{n1}b_{11} + a_{n2}b_{21} + \cdots + a_{nm}b_{m1}) \cdots (a_{n1}b_{1r} + \cdots + a_{nm}b_{mr}) \end{bmatrix},$$

$$[c_{ij}] = \sum_{k=1}^{m} a_{ik}b_{kj}, \qquad i = 1, 2, \ldots, n, \quad j = 1, 2, \ldots, r.$$

It is simplest to select the proper elements if one counts across the rows of A with the left hand while counting down the columns of B with the right. Unless the number of columns of A equals the number of rows of B (so the counting comes out even), the matrices cannot be multiplied. Hence if A is $n \times m$, B must have m rows or else they are said to be "nonconformable for multiplication" and their product is undefined. In general $AB \neq BA$, so the order of factors must be preserved in matrix multiplication.

A matrix with only one column, $n \times 1$ in size, is termed a *column vector*, and one of only one row, $1 \times m$ in size, is called a *row vector*. Frequently the elements of vectors are only singly subscripted.

Certain square matrices have special properties. The diagonal elements are the line of elements a_{ii} from upper left to lower right of the matrix. When the diagonal elements are each equal to unity while all off-diagonal elements are zero, the matrix is said to be the *identity matrix of order n*. The usual symbol for such a matrix is I_n, and it has properties similar to unity. For example, the order 4 identity matrix is

$$\begin{bmatrix} 1 & 0 & 0 & 0 \\ 0 & 1 & 0 & 0 \\ 0 & 0 & 1 & 0 \\ 0 & 0 & 0 & 1 \end{bmatrix} = I_4.$$

The subscript is omitted when the order is clear from the context.

If all the elements above the diagonal are zero, the matrix is called *lower triangular;* it is called *upper triangular* when all the elements below the diagonal are zero. For example, these order-3 matrices are lower and upper triangular:

$$L = \begin{bmatrix} 1 & 0 & 0 \\ 4 & 6 & 0 \\ -2 & 1 & -4 \end{bmatrix},$$

$$U = \begin{bmatrix} 1 & -3 & 3 \\ 0 & -1 & 0 \\ 0 & 0 & 1 \end{bmatrix}.$$

Tridiagonal matrices are those which have nonzero elements only on the diagonal and in the positions adjacent to the diagonal; they will be of special importance in certain partial differential equations.   An example of a tridiagonal matrix is

$$\begin{bmatrix} -4 & 2 & 0 & 0 & 0 \\ 1 & -4 & 1 & 0 & 0 \\ 0 & 1 & -4 & 1 & 0 \\ 0 & 0 & 1 & -4 & 1 \\ 0 & 0 & 0 & 2 & -4 \end{bmatrix}.$$

The above definition of matrix multiplication permits us to write the set of linear equations,

$$a_{11}x_1 + a_{12}x_2 + \cdots + a_{1n}x_n = b_1,$$
$$a_{21}x_1 + a_{22}x_2 + \cdots + a_{2n}x_n = b_2,$$
$$\vdots$$
$$a_{n1}x_1 + a_{n2}x_2 + \cdots + a_{nn}x_n = b_n,$$

much more simply in matrix notation:

$$Ax = b,$$

where

$$A = \begin{bmatrix} a_{11}a_{12} \cdots a_{1n} \\ a_{21}a_{22} \cdots a_{2n} \\ \vdots \\ a_{n1}a_{n2} \cdots a_{nn} \end{bmatrix}, \quad x = \begin{bmatrix} x_1 \\ x_2 \\ \vdots \\ x_n \end{bmatrix}, \quad b = \begin{bmatrix} b_1 \\ b_2 \\ \vdots \\ b_n \end{bmatrix}.$$

2. ELIMINATION METHOD

The first method we shall study for the solution of a set of equations is just an enlargement of the familiar method of eliminating one unknown between a pair of simultaneous equations.   It is generally called Gaussian elimination and is the basic pattern of a large number of methods which can be classed as direct methods. (This is to distinguish them from indirect, or iterative, methods that we discuss later.)

Consider the simple example of three equations

$$3x_1 - x_2 + 2x_3 = 12,$$
$$x_1 + 2x_2 + 3x_3 = 11,$$
$$2x_1 - 2x_2 - x_3 = 2.$$

Multiplying the first equation by -1 and the second by 3 and adding will eliminate x_1.   Similarly, multiplying the first by -2 and the third by 3 and adding also eliminates x_1.   (We prefer to multiply by the negative values and add to avoid making mistakes when subtracting quantities of unlike sign.)   The result is

$$3x_1 - x_2 + 2x_3 = 12,$$
$$7x_2 + 7x_3 = 21,$$
$$-4x_2 - 7x_3 = -18.$$

We eliminate x_2 between the second and third equations by multiplying the second by -4 and the third by 7 and adding.   (Of course, just adding them as they stand would eliminate x_3, which is equally satisfactory, but we wish to keep our method systematic to lead up to an algorithm that can be readily programmed.)   After this operation we have the upper diagonal system

$$
\begin{aligned}
3x_1 - x_2 + 2x_3 &= 12, \\
7x_2 + 7x_3 &= 21, \\
-21x_3 &= -42.
\end{aligned}
$$

Obviously $x_3 = 2$ from the third equation, and back substitution gives $x_2 = 1$, $x_1 = 3$.

We now present the same problem, solved in exactly the same way, in matrix notation:

$$
\begin{bmatrix} 3 & -1 & 2 \\ 1 & 2 & 3 \\ 2 & -2 & -1 \end{bmatrix} \begin{bmatrix} x_1 \\ x_2 \\ x_3 \end{bmatrix} = \begin{bmatrix} 12 \\ 11 \\ 2 \end{bmatrix}.
$$

The arithmetic operations we have performed affect only the coefficients and the constant terms, of course, so we work with the matrix of coefficients "augmented" with the b-vector:

$$
A \vdots b = \begin{bmatrix} 3 & -1 & 2 & \vdots & 12 \\ 1 & 2 & 3 & \vdots & 11 \\ 2 & -2 & -1 & \vdots & 2 \end{bmatrix}.
$$

(The dotted line is usually omitted.)

We perform "elementary row transformations" to convert A to upper triangular:

$$
\begin{bmatrix} 3 & -1 & 2 & 12 \\ 1 & 2 & 3 & 11 \\ 2 & -2 & -1 & 2 \end{bmatrix} \begin{matrix} \\ 3R_2 + (-1)R_1 \rightarrow \\ 3R_3 + (-2)R_1 \rightarrow \end{matrix} \begin{bmatrix} 3 & -1 & 2 & 12 \\ 0 & 7 & 7 & 21 \\ 0 & -4 & -7 & -18 \end{bmatrix} 7R_3 + 4R_2 \rightarrow
$$

$$
\begin{bmatrix} 3 & -1 & 2 & 12 \\ 0 & 7 & 7 & 21 \\ 0 & 0 & -21 & -42 \end{bmatrix}. \qquad (2.1)
$$

The steps are to add 3 times the second row to -1 times the first row and 3 times the third row to -2 times the first row.   The next phase adds 7 times the third row to 4 times the second row.

We are now ready for back substitution.   Note that, except for notation and terminology, there is nothing new here.   We depend on our memory to know which numbers in the converted augmented matrix are coefficients and which are constant terms.

The back substitution step can be performed quite mechanically by eliminating the coefficients above the diagonal in (2.1).   Adding the third row of (2.1) to 3 times the second row, and adding twice the third row to 21 times the first row gives

$$
\begin{bmatrix} 63 & -21 & 0 & 168 \\ 0 & 21 & 0 & 21 \\ 0 & 0 & -21 & -42 \end{bmatrix}.
$$

We finish the elimination of off-diagonal elements by adding the second row to the first:

$$\begin{bmatrix} 63 & 0 & 0 & 189 \\ 0 & 21 & 0 & 21 \\ 0 & 0 & -21 & -42 \end{bmatrix}.$$

If we divide each row by the diagonal element, we get a form in which the elements of x, the vector whose components are the unknowns x_i, $i = 1, 2, \ldots, n$, are equal to the components of the transformed b-vector:

$$\begin{bmatrix} 1 & 0 & 0 & 3 \\ 0 & 1 & 0 & 1 \\ 0 & 0 & 1 & 2 \end{bmatrix}, \quad x = \begin{bmatrix} 3 \\ 1 \\ 2 \end{bmatrix}, \quad x_1 = 3, \quad x_2 = 1, \quad x_3 = 2.$$

Thinking of this procedure in terms of matrix operations, we transform the augmented coefficient matrix by elementary row operations until the identity matrix is created on the left. The x-vector then stands as the rightmost column.

Suppose a set of equations is given in which the number of unknowns is greater than the number of equations. Quite obviously, these cannot give a numerical solution for each unknown, but, if there are n equations and m variables, we can usually solve for n of the unknowns in terms of the others, which will be $m - n$ in number. Which of the unknowns are chosen to be solved for is arbitrary. Such a set of equations is said to be *underdetermined*.

If we are looking for values of the m variables which will satisfy the n equations in an underdetermined system, we can find an infinite set of these for each of the variables because $m - n$ of them can be specified at will.

Sometimes an n by n set of equations is also underdetermined. If two or more of the equations are identical, it is immediately obvious that there are really not n independent equations; the duplicate equation does not contain any new information. If one equation is a linear combination of the others, the system will be likewise underdetermined. One way of checking on this is to determine the ranks of the coefficient matrix and of the augmented matrix; they must be the same for a unique solution to exist.

When there are more equations than unknowns and the equations are linearly independent, the system is said to be *overdetermined*. Under these conditions, the solution of a subset of equations equal in number to the number of unknowns will not satisfy the remaining equations. We then say the equations are *inconsistent* and there is no solution.

Even when the number of equations and of unknowns is the same, there may be no solution. The following simple illustration shows inconsistency by inspection:

$$2x - 3y = 2,$$
$$2x - 3y = 4.$$

In larger systems such inconsistencies are not obvious. Fortunately, the set of equations that are derived for the mathematical models of physical problems has a unique solution, though errors in the coefficients or errors in the course of the solution may cause one of the pathological cases to occur.

3. GAUSS AND GAUSS-JORDAN METHODS

There are several objections to what we have done in the previous sections, and these require certain modifications. In a larger set of equations, and that is the situation we must prepare for, the multiplications will give very large and unwieldy numbers. We therefore will eliminate the first coefficient in the ith row by adding $-a_{i1}/a_{11}$ times the first equation to the ith equation. This is equivalent to making the leading coefficient 1 in the equation that retains that leading term. We proceed to eliminate the leading coefficients of the successive rows by again multiplying by the corresponding negative ratio of coefficients and adding.

Numerical mistakes plague us in calculations such as these, so, for hand calculation, it is worthwhile to do some extra work to control errors. A check column may be added to the augmented coefficient matrix. This number is the sum of the elements in its row, and is operated on arithmetically just like all the other elements. One checks the work by adding the new elements—the sum should agree with the check number.

We repeat the above example incorporating these ideas, carrying three decimal places in our work:

$$
\begin{array}{ccccc}
 & & \text{(Sum)} & & \\
\begin{bmatrix} 3 & -1 & 2 & 12 & 16 \\ 1 & 2 & 3 & 11 & 17 \\ 2 & -2 & -1 & 2 & 1 \end{bmatrix}
& \begin{array}{l} \\ R_2 + (-\tfrac{1}{3})R_1 \rightarrow \\ R_3 + (-\tfrac{2}{3})R_1 \rightarrow \end{array}
\end{array}
$$

$$
\begin{array}{ccccc}
 & & \text{(Sum)} & \text{(Check} \\
 & & & \text{sum)} \\
\begin{bmatrix} 3 & -1 & 2 & 12 & 16 \\ 0 & 2.333 & 2.334 & 7.004 & 11.672 \\ 0 & -1.334 & -2.332 & -5.992 & -9.656 \end{bmatrix}
& \begin{array}{l} \\ (11.671) \\ (-9.658) \end{array}
\end{array}
$$

$$
R_3 + \left(\frac{1.334}{2.333}\right)R_2 \rightarrow
\begin{bmatrix} 3 & -1 & 2 & 12 & 16 \\ 0 & 2.333 & 2.334 & 7.004 & 11.672 \\ 0 & 0 & -1.000 & -1.993 & -2.992 \end{bmatrix}
\quad (-2.993)
$$

(Check
(Sum) sum)

The method we have just illustrated is called *Gaussian elimination*. Back substitution gives $x_3 = 1.993$, $x_2 = 1.008$, $x_1 = 3.007$. The differences of these values from 2, 1, 3 are due to the effects of round-off, which is also the cause of discrepancies between the sum and the check sum numbers. Actually, in this example, the numbers were truncated after three decimals rather than rounded, but the errors are similar. One can imagine that in a large set of equations, the cumulative effects of round-off error could be extremely serious.

Carrying our work further, eliminating the coefficient elements above the diagonal is the equivalent of back substitution. Proceeding from the last step above, we have

$$R_1 + \left(\frac{-2}{-1.000}\right)R_3 \rightarrow$$

$$R_2 + \left(\frac{-2.334}{-1.000}\right)R_3 \rightarrow$$

				(Sum)	(Check sum)
3	−1	0	8.014	10.016	(10.014)
0	2.333	0	2.353	4.689	(4.686)
0	0	−1.000	−1.993	−2.992	

$$R_1 + \left(\frac{1}{2.333}\right)R_2 \rightarrow$$

				(Sum)	(Check sum)
3	0	0	9.021	12.022	(12.021)
0	2.333	0	2.353	4.689	
0	0	−1.000	−1.993	−2.992	

$$\frac{R_1}{3} \rightarrow$$

$$\frac{R_2}{2.333} \rightarrow$$

$$\frac{R_3}{-1.000} \rightarrow$$

				(Sum)
1	0	0	3.007	4.007
0	1	0	1.008	2.009
0	0	1	1.993	2.992

.

The fourth column is now the x-vector.

There are many variants on the Gaussian elimination scheme as we have presented it here. One may divide all the coefficients in the working row by its diagonal element (called the *pivot*) at the beginning of each step; this avoids the later divisions and gives 1's on the diagonal of the final matrix from the beginning. If one is working to a fixed number of decimal places, as with most desk calculators, or in fixed point or in floating point arithmetic on digital computers, it reduces round-off errors to interchange rows so that the largest number is on the diagonal. Similar interchanges of columns will also help control round-off error, but keeping track of the variables is then quite awkward. The last modification is termed "elimination with pivoting" or "with positioning." When only row interchanges are employed, it is "partial pivoting."

The Gauss-Jordan scheme eliminates the coefficients in the rows above the working row as well as in those below during the same step. This actually involves more arithmetical operations, but the back substitution is then avoided.

It is essential in these methods that the pivot element not be zero, for it is used as a divisor at some stage of the operation. If the set of equations has a unique solution (the set is then *consistent*), it will always be possible to find a nonzero coefficient in one of the rows below, and these rows should then be interchanged. If one is using elimination with positioning, the interchange to put the largest element as the pivot will automatically avoid zero divisors. In the exercises, the student will experience how pivoting can reduce round-off errors.

Some sets of equations are particularly sensitive to the values of the coefficients. For example, consider the pair of equations

$$x + y = 2.00,$$
$$x + 1.01y = 2.01,$$

which has the obvious solution $x = 1$, $y = 1$. However, if only small changes are made in the coefficients, such as to

$$x + y = 2.00,$$
$$x + 0.99y = 2.02,$$

we get a set of equations with the solutions $x = 4$, $y = -2$! This phenomenon is called *ill conditioning*.

One needs to be on guard for this situation; it can be predicted from a comparison of the elements of the coefficient matrix with the elements of the inverse of this matrix (see Section 5). In particular, one should be cautious about the accuracy of a solution to a set of simultaneous equations when the coefficients are only imperfectly known, if the system is ill conditioned.

Of present concern is that round-off errors can affect the values of ill conditioned systems because such errors cause a similar behavior as original errors in the coefficients. It is in such situations that pivoting or partial pivoting can pay big dividends. For the same reason, sophisticated computer programs will employ double precision arithmetic at judiciously chosen places in the algorithm. (Double precision throughout is wasteful of core storage and limits the size of the system that can be handled.) Ralston (1965) discusses these points.

4. CROUT REDUCTION

A modification of the elimination method, called Crout reduction (also named after another discoverer, Cholesky), is frequently used in computer programs. In it the matrix of coefficients A is transformed into the product of two matrices L and U, where L is a lower triangular and U is an upper triangular matrix with 1's on its diagonal. The rules for getting the elements of L and U are deduced from the fact that $LU = A$. We illustrate with the case of a 4×4 matrix:

$$\begin{bmatrix} l_{11} & 0 & 0 & 0 \\ l_{21} & l_{22} & 0 & 0 \\ l_{31} & l_{32} & l_{33} & 0 \\ l_{41} & l_{42} & l_{43} & l_{44} \end{bmatrix} \begin{bmatrix} 1 & u_{12} & u_{13} & u_{14} \\ 0 & 1 & u_{23} & u_{24} \\ 0 & 0 & 1 & u_{34} \\ 0 & 0 & 0 & 1 \end{bmatrix} = \begin{bmatrix} a_{11} & a_{12} & a_{13} & a_{14} \\ a_{21} & a_{22} & a_{23} & a_{24} \\ a_{31} & a_{32} & a_{33} & a_{34} \\ a_{41} & a_{42} & a_{43} & a_{44} \end{bmatrix}. \quad (4.1)$$

Multiplying the rows of L by the first column of U, we get $l_{11} = a_{11}$, $l_{21} = a_{21}$, $l_{31} = a_{31}$, $l_{41} = a_{41}$; the first column of L is the same as the first column of A.

We now multiply the first row of L by the columns of U:

$$l_{11}u_{12} = a_{12}, \qquad l_{11}u_{13} = a_{13}, \qquad l_{11}u_{14} = a_{14}, \qquad (4.2)$$

from which

$$u_{12} = \frac{a_{12}}{l_{11}}, \qquad u_{13} = \frac{a_{13}}{l_{11}}, \qquad u_{14} = \frac{a_{14}}{l_{11}}. \tag{4.3}$$

Thus the first row of U is determined.

In this method we alternate between getting a column of L and a row of U, so we next get the equations for the second column of L by multiplying the rows of L by the second column of U:

$$\begin{aligned} l_{21}u_{12} + l_{22} &= a_{22}, \\ l_{31}u_{12} + l_{32} &= a_{32}, \\ l_{41}u_{12} + l_{42} &= a_{42}, \end{aligned} \tag{4.4}$$

which gives

$$\begin{aligned} l_{22} &= a_{22} - l_{21}u_{12}, \\ l_{32} &= a_{32} - l_{31}u_{12}, \\ l_{42} &= a_{42} - l_{41}u_{12}. \end{aligned} \tag{4.5}$$

Proceeding in the same fashion, the equations we need are

$$u_{32} = \frac{a_{32} - l_{21}u_{13}}{l_{22}}, \qquad u_{24} = \frac{a_{24} - l_{21}u_{14}}{l_{22}},$$

$$l_{33} = a_{33} - l_{31}u_{13} - l_{32}u_{23}, \qquad l_{43} = a_{43} - l_{41}u_{13} - l_{42}u_{23}, \tag{4.6}$$

$$u_{34} = \frac{a_{34} - l_{31}u_{14} - l_{32}u_{24}}{l_{33}},$$

$$l_{44} = a_{44} - l_{41}u_{14} - l_{42}u_{24} - l_{43}u_{34}.$$

The general formula for getting elements can be written

$$l_{ij} = a_{ij} - \sum_{k=1}^{j-1} l_{ik}u_{kj},$$

$$u_{ij} = \frac{a_{ij} - \sum_{k=1}^{i-1} l_{ik}u_{kj}}{l_{ii}}.$$

The reason this method is popular in programs is that storage space may be economized. There is no need to store the 0's in either L or U, and the 1's on the diagonal of U can also be omitted. One can then store the essential elements of U where the 0's appear in the L array. Examination of Eqs. (4.1) through (4.6) shows that after any element of A, a_{ij}, is once used, it never again appears in the equations. Hence its place in the original $n \times n$ array A can be used to store an element of either L or U. In other words, the A array is transformed by the above equations and becomes

$$
\begin{bmatrix}
a_{11} & a_{12} & a_{13} & a_{14} \\
a_{21} & a_{22} & a_{23} & a_{24} \\
a_{31} & a_{32} & a_{33} & a_{34} \\
a_{41} & a_{42} & a_{43} & a_{44}
\end{bmatrix}
\rightarrow
\begin{bmatrix}
l_{11} & u_{12} & u_{13} & u_{14} \\
l_{21} & l_{22} & u_{23} & u_{24} \\
l_{31} & l_{32} & l_{33} & u_{34} \\
l_{41} & l_{42} & l_{43} & l_{44}
\end{bmatrix}.
$$

We apply this method to solving a set of linear equations by augmenting the A matrix with the vector b $(Ax = b)$, and proceeding to reduce its components according to the rule for u_{ij}, extending the range of the second index by one:

$$
\begin{bmatrix}
3 & -1 & 2 & 12 \\
1 & 2 & 3 & 11 \\
2 & -2 & -1 & 2
\end{bmatrix}
\qquad
\begin{bmatrix}
3 & -\frac{1}{3} & \frac{2}{3} & 4 \\
1 & \frac{7}{3} & 1 & 3 \\
2 & -\frac{4}{3} & -1 & 2
\end{bmatrix}
\begin{matrix} \leftarrow ② \\ \leftarrow ④ \\ \leftarrow ⑥ \end{matrix}.
\tag{4.7}
$$
$$
 \begin{matrix} \uparrow & \uparrow & \uparrow \\ ① & ③ & ⑤ \end{matrix}
$$

The circled numbers show the order in which the columns and rows of the new matrix are obtained.

The components of the x-vector are obtained from the relation $Ux = b'$, where b' is the rightmost column of the new matrix. We can write this out explicitly from Eq. (4.7) by remembering that U is upper triangular, and the 1's on its diagonal have been suppressed:

$$
U \vdots b =
\begin{bmatrix}
1 & -\frac{1}{3} & \frac{2}{3} & 4 \\
0 & 1 & 1 & 3 \\
0 & 0 & 1 & 2
\end{bmatrix}.
$$

We have, then, by straightforward back substitution,

$$
x_3 = 2,
$$
$$
x_2 = 3 - (1)(2) = 1,
$$
$$
x_1 = 4 - (\tfrac{2}{3})(2) - (-\tfrac{1}{3})(1) = 3.
$$

5. DETERMINANT OF A MATRIX AND MATRIX INVERSION

The student has perhaps wondered why there has been no reference heretofore to solution of linear equations by determinants (Cramer's rule). The reason for this is, except for systems of only two or three equations, the determinant method is too inefficient. For example, for a set of 10 simultaneous equations, about 70,000,000 multiplications and divisions are required if the usual method of expansion in terms of minors is used. A more efficient method of evaluating the determinants can reduce this to about 3000 multiplications, but even this is inefficient compared to Gaussian elimination which would require about 380. (The Gauss-Jordan scheme requires about 50% more arithmetic operations than Gauss with back substitution.)

In fact, the evaluation of a determinant can perhaps best be done by adapting the Gauss elimination procedure. Its utility derives from the fact that the determinant of a triangular matrix (either upper or lower triangular) is just the product of its diagonal elements. This is easily seen, in the case of an upper triangular matrix, by expansion by minors of the first column at each step. For example,

$$\begin{vmatrix} a_{11} & a_{12} & a_{13} & a_{14} \\ 0 & a_{22} & a_{23} & a_{24} \\ 0 & 0 & a_{22} & a_{34} \\ 0 & 0 & 0 & a_{44} \end{vmatrix} = a_{11}\begin{vmatrix} a_{22} & a_{23} & a_{24} \\ 0 & a_{33} & a_{34} \\ 0 & 0 & a_{44} \end{vmatrix} - 0 + 0 - 0$$

$$= a_{11}(a_{22}\begin{vmatrix} a_{33} & a_{34} \\ 0 & a_{44} \end{vmatrix} - 0 + 0)$$

$$= a_{11}a_{22}(a_{33}a_{44} - 0) = a_{11}a_{22}a_{33}a_{44}.$$

Since elementary row transformations do not change the value of a determinant, using the procedure of Gaussian elimination to convert to upper triangular is a simple way to evaluate determinants. If row interchanges are made to take advantage of positioning, the value of the determinant is multiplied by -1 for each such interchange. If a row is multiplied by a constant, the value of the determinant is also multiplied.

Example. Find the value of the determinant by using elementary row transformations to make it upper triangular.

$$\begin{vmatrix} 1 & 4 & -2 & 3 \\ 2 & 2 & 0 & 4 \\ 3 & 0 & -1 & 2 \\ 1 & 2 & 2 & -3 \end{vmatrix} = \begin{vmatrix} 1 & 4 & -2 & 3 \\ 0 & -6 & 4 & -2 \\ 0 & -12 & 5 & -7 \\ 0 & -2 & 4 & -6 \end{vmatrix} = \begin{vmatrix} 1 & 4 & -2 & 3 \\ 0 & -6 & 4 & -2 \\ 0 & 0 & -3 & -3 \\ 0 & 0 & \frac{8}{3} & -\frac{16}{3} \end{vmatrix}$$

$$= \begin{vmatrix} 1 & 4 & -2 & 3 \\ 0 & -6 & 4 & -2 \\ 0 & 0 & -3 & -3 \\ 0 & 0 & 0 & -8 \end{vmatrix} = (1)(-6)(-3)(-8) = -144.$$

If the product of two square matrices is the identity matrix, matrices are said to be inverses. If $AB = I$, we write $B = A^{-1}$, also $A = B^{-1}$. Inverses commute on multiplication, which is not true for matrices in general: $AB = BA = I$. Not all square matrices have an inverse. (Those that do not are called *singular*.) If a set of equations has a unique solution, the matrix of coefficients will be non-singular.

The inverse of a matrix can be defined in terms of the matrix of the minors of its determinant, but this is not a useful way to find an inverse. Gaussian elimination can be adapted to provide a practical way to invert a matrix. The procedure is to augment the given matrix with the identity matrix of the same order. One then reduces the original matrix to the identity matrix by elementary row transformations, performing the same operations on the augmentation columns. When the identity matrix stands as the left half of the augmented matrix, the inverse of the original stands as the right half.

Example. Find the inverse of

$$\begin{bmatrix} 1 & -1 & 2 \\ 3 & 0 & 1 \\ 1 & 0 & 2 \end{bmatrix}.$$

$$\begin{bmatrix} 1 & -1 & 2 & 1 & 0 & 0 \\ 3 & 0 & 1 & 0 & 1 & 0 \\ 1 & 0 & 2 & 0 & 0 & 1 \end{bmatrix} \rightarrow \begin{bmatrix} 1 & -1 & 2 & 1 & 0 & 0 \\ 0 & 3 & -5 & -3 & 1 & 0 \\ 0 & 1 & 0 & -1 & 0 & 1 \end{bmatrix}$$

$$\overset{(1)}{\rightarrow} \begin{bmatrix} 1 & -1 & 2 & 1 & 0 & 0 \\ 0 & 1 & 0 & -1 & 0 & 1 \\ 0 & 0 & -5 & 0 & 1 & -3 \end{bmatrix} \overset{(2)}{\rightarrow} \begin{bmatrix} 1 & -1 & 0 & 1 & \frac{2}{5} & -\frac{6}{5} \\ 0 & 1 & 0 & -1 & 0 & 1 \\ 0 & 0 & 1 & 0 & -\frac{1}{5} & \frac{3}{5} \end{bmatrix}$$

$$\rightarrow \begin{bmatrix} 1 & 0 & 0 & 0 & \frac{2}{5} & -\frac{1}{5} \\ 0 & 1 & 0 & -1 & 0 & 1 \\ 0 & 0 & 1 & 0 & -\frac{1}{5} & \frac{3}{5} \end{bmatrix}.$$

(1) Interchange the third and second rows before eliminating from the third row.
(2) Divide the third row by -5 before eliminating from the first row.

We confirm the fact that we have found the inverse by multiplication:

$$\begin{bmatrix} 1 & -1 & 2 \\ 3 & 0 & 1 \\ 1 & 0 & 2 \end{bmatrix} \begin{bmatrix} 0 & \frac{2}{5} & -\frac{1}{5} \\ -1 & 0 & 1 \\ 0 & -\frac{1}{5} & \frac{3}{5} \end{bmatrix} = \begin{bmatrix} 1 & 0 & 0 \\ 0 & 1 & 0 \\ 0 & 0 & 1 \end{bmatrix}.$$

Inverses can be used to solve a set of equations, as we readily see by the following matrix operations:

$$\begin{aligned} Ax &= b, \\ A^{-1}(Ax) &= A^{-1}b, \\ A^{-1}Ax = (A^{-1}A)x &= Ix = x = A^{-1}b. \end{aligned} \tag{5.1}$$

Equation (5.1) tells us that the unknowns $x_1, x_2, \ldots, x_n$ (these are the components of vector x) can be found by multiplying together the inverse of the coefficient matrix and the vector b, whose components are the constant terms of the equations. Solving equations by finding the inverse matrix is less efficient than by Gauss elimination or Crout reduction, and should be used only if the inverse is required for other purposes, or where a series of sets of simultaneous equations are to be solved which differ only in their constant terms.

6. GAUSS-SEIDEL ITERATION

As opposed to the direct method of solving a set of linear equations by elimination, we now discuss iterative methods. In certain cases, these methods are preferred over the direct methods—when the coefficient matrix is sparse (has many zeros) they may be more rapid.* They may be more economical in core storage

---

* When the occurrence of zeros follows some easy pattern, elimination methods can take advantage of this. See the program for a tridiagonal system at the end of the chapter.

requirements of a computer. For hand computation they have the distinct advantage that they are self-correcting if an error is made; they may sometimes be used to reduce round-off error in the solutions computed by direct methods. They can also be applied to sets of nonlinear equations.

We illustrate the method by a simple example:

$$8x_1 + x_2 - x_3 = 8,$$
$$2x_1 + x_2 + 9x_3 = 12, \qquad\qquad (6.1)$$
$$x_1 - 7x_2 + 2x_3 = -4.$$

The solution is $x_1 = 1$, $x_2 = 1$, $x_3 = 1$. We begin our iterative scheme by solving each equation for one of the variables, choosing, when possible, to solve for the variable with largest coefficient:

$$x_1 = 1 - 0.125x_2 + 0.125x_3 \quad \text{(from first equation)},$$
$$x_2 = 0.571 + 0.143x_1 + 0.286x_3 \quad \text{(from third equation)}, \qquad (6.2)$$
$$x_3 = 1.333 - 0.222x_1 - 0.111x_2 \quad \text{(from second equation)}.$$

We begin with some initial approximation to the value of the variables. (Each can be taken as equal to zero if no better initial estimates are at hand.) Substituting these into the right-hand sides of the set of equations generates new approximations which are closer to the true value. These new values are substituted into the right-hand sides to generate a second approximation, and the process is repeated until successive values of each of the variables are sufficiently alike. For the set of equations given above we get:

Successive Estimates of Solution (Jacobi Method)

	First	Second	Third	Fourth	Fifth	Sixth	Seventh	Eighth
x_1	0	1.000	1.095	0.995	0.993	1.002	1.001	1.000
x_2	0	0.571	1.095	1.026	0.990	0.998	1.001	1.000
x_3	0	1.333	1.048	0.969	1.000	1.004	1.001	1.000

Note that this method is exactly the same as the method of iteration for a single equation that was discussed in Chapter 1, but applied to a set of equations.

The procedure as we have described it is known as the *Jacobi method*, also called "the method of simultaneous displacements" because each of the equations is simultaneously changed by using the most recent set of x-values.

Actually, the x-values of the next trial are not all calculated "simultaneously." In the above, we calculated the second estimate of x_1 before we did the x_2, and new values of both x_1 and x_2 were available before we improved the value of x_3. In nearly all cases the new values are better than the old, and should be used in preference to the poorer values.* When this is done, the method is known by the name *Gauss-Seidel*. In this method our first step is to rearrange the set of equa-

---

* There may be times when this obvious move is not beneficial, due to cancellation of errors, but in general this is good strategy in numerical analysis.

tions by solving each equation for one of the variables in terms of the others, exactly as we have done above in the Jacobi method. One then proceeds to improve each x-value in turn, using always the most recent approximations to the values of the other variables. The rate of convergence is more rapid, as shown by reworking the same example as above (Eqs. 6.2, rearranged form of 6.1):

Successive Estimates of Solution (Gauss-Seidel Method)

	First	Second	Third	Fourth	Fifth	Sixth
x_1	0	1.000	1.041	0.997	1.001	1.000
x_2	0	0.714	1.014	0.996	1.000	1.000
x_3	0	1.032	0.990	1.002	1.000	1.000

For hand computation of a set of equations by the Gauss-Seidel method, an arrangement of the equations in a vertical array can help speed the work and avoid errors. This is illustrated by Fig. 6.1, where the coefficients of Eqs. (6.2) are listed in columns, one column for each equation. The coefficient of 1 for x_1 in the first equation is put in parentheses to remind us that x_1 stands on the left side—similarly for the coefficients of x_2 and x_3 in the other equations.

We begin the computations in the first column by listing the product of the initial estimates of x_2 and x_3 (here zero) times their coefficients and adding these to the constant term in the column (all this is working with Eq. I). This sum, which is circled, is the improved estimate of x_1. We now work across the row, listing the product of this value (1.000) with the x_1 coefficients in each equation. The column is completed by tabulating the initial x_3 value times its coefficient in the second column (the second equation). The improved value of x_2 is obtained by summing the contributions due to x_1 and x_3 in the second equation, as shown by numbers in row II; the result is circled. A row is now completed (we anticipate our x_2 term in Eq. I) and numbers summed to give a new x_3.

The iteration method will not converge for all sets of equations, nor for all possible rearrangements of the equations. When the equations can be ordered so that each diagonal entry is larger in magnitude than the sum of the magnitudes of the other coefficient in that row, the iteration will converge for any starting values. This is easy to visualize, because all the equations can be put in the form

$$x_i = c_i - \frac{a_1}{a_i} x_1 - \frac{a_2}{a_i} x_2 - \cdots.$$

The error in the next value of x_i will be the sum of the errors in all other x's multiplied by the coefficients, and if the sum of the magnitudes of the coefficients is less than unity, the error will decrease as the iteration proceeds. The above convergence condition is a sufficient condition only; i.e., if the condition holds, the system always converges, but sometimes the system converges even if the condition is violated.

Coefficient of	Eq. I	Eq. II	Eq. III
x_1	(1)	0.143	−0.222
x_2	−0.125	(1)	−0.111
x_3	0.125	0.286	(1)
Constant	1.000	0.571	1.333

	Eq. I	Eq. II	Eq. III
(x_2)	0		
(x_3)	0	0	
(x_1)	⟨1.000⟩	0.143	0.222
(x_2)	−0.088	⟨0.714⟩	−0.079
(x_3)	0.129	0.295	⟨1.032⟩
	⟨1.041⟩	0.148	−0.231
	−0.126	⟨1.014⟩	−0.112
	0.123	0.283	⟨0.990⟩
	⟨0.997⟩	0.142	−0.221
	−0.124	⟨0.996⟩	−0.110
	0.125	0.286	⟨1.002⟩
	⟨1.001⟩	0.143	−0.222
	−0.125	⟨1.000⟩	−0.111
	0.125	0.286	⟨1.000⟩
	⟨1.000⟩	0.143	−0.222
	−0.125	⟨1.000⟩	−0.111
			⟨1.000⟩

Fig. 6.1. Vertical array for Gauss-Seidel method

7. RELAXATION METHOD

There is an iteration method which is more rapidly convergent than Gauss-Seidel and which can be used for hand calculations to advantage. It is unfortunately not well adapted to computer application. The method is due to a British engineer, Richard Southwell, and has been applied to a wide variety of problems. [Allen (1954) is an excellent reference.]

If we consider the Gauss-Seidel scheme, we realize that the order in which the equations are used is important. We should improve that x which is most in error, since, in the rearranged form, that variable does not appear on the right, and hence its own error will not affect the next iterate. By using that equation, then, we introduce lesser errors into the computation of the next iterate. The *method of relaxation* is a scheme which permits one to select the best equation to be used for maximum rate of convergence.

	x_1	x_2	x_3	Initial Residuals
Eq. I	-1	-0.125	0.125	1.000
Eq. II	0.143	-1	0.286	0.571
Eq. III	-0.222	-0.111	-1	1.333

x_1	R_{I}	x_2	R_{II}	x_3	R_{III}
0	~~1000~~	0	~~571~~	0	~~1333~~
	+167		+381		
	~~1167~~		~~952~~	1333	~~0~~
			+167		−259
1167	~~0~~		~~1119~~		~~−259~~
	−140				−124
	~~−140~~	1119	~~0~~		~~−383~~
	−48		−109		
	~~−189~~		~~−109~~	−383	~~0~~
			−27		+42
−189	~~0~~		~~−136~~		~~42~~
	+17				+15
	~~17~~	−136	~~0~~		~~57~~
	+ 7		+16		
	~~24~~		~~16~~	57	~~0~~
			+ 3		− 5
24	~~0~~		~~19~~		~~−5~~
					− 2
	~~−2~~	19	~~0~~		~~−7~~
	− 1		− 2		
	~~−3~~		~~−2~~	−7	~~0~~
					+ 1
−3	~~0~~		~~−2~~		~~1~~
					0
	~~0~~	−2	~~0~~		~~1~~
	0		0	1	0
999		1000		1001	
Check residuals:	1		−1		0

Fig. 7.1. Solving a set of linear equations by relaxation

We illustrate the method by the same example as solved in Section 6. The original equations are

$$
\begin{aligned}
8x_1 + x_2 - x_3 &= 8, \\
2x_1 + x_2 + 9x_3 &= 12, \\
x_1 - 7x_2 + 2x_3 &= -4.
\end{aligned}
\tag{7.1}
$$

We again begin by a rearrangement of the equations, but different from that for Gauss-Seidel or Jacobi methods. We transpose all the terms to one side, and then divide by the negative of the largest coefficient. Equations (7.1) become

$$
\begin{aligned}
-x_1 - 0.125x_2 + 0.125x_3 + 1 \qquad\ &= 0, \\
-0.222x_1 - 0.111x_2 - \qquad x_3 + 1.333 &= 0, \\
0.143x_1 - \qquad x_2 + 0.286x_3 + 0.571 &= 0.
\end{aligned}
\tag{7.2}
$$

If we begin with some initial set of values, and substitute in Eqs. (7.2), the equations will not be satisfied (unless, by chance, we have stumbled onto the solution); the left sides will not be zero, but some other value which we call the *residual* and denote by R_i. It is also convenient to reorder the equation so the -1 coefficients are on the diagonal. Equations (7.2) become, with these rearrangements

$$
\begin{aligned}
-x_1 - 0.125x_2 + 0.125x_3 + 1 \qquad\ &= R_1, \\
0.143x_1 - \qquad x_2 + 0.286x_3 + 0.571 &= R_2, \\
-0.222x_1 - 0.111x_2 - \qquad x_3 + 1.333 &= R_3.
\end{aligned}
$$

For example, with $x_1 = 0$, $x_2 = 0$, $x_3 = 0$, we have $R_1 = 1$, $R_2 = 0.571$, $R_3 = 1.333$. The largest residual in magnitude, R_3, tells us that the third equation is most in error and should be improved first. The method gets its name "relaxation" from the fact that we make a change in x_3 to relax R_3 (the greatest residual) so as to make it zero. Observing the coefficients of the various equations, we see that increasing the value of x_3 by one, say, will decrease R_3 by one, will increase R_1 by 0.125, and increase R_2 by 0.286. To change R_3 from its initial value of 1.333 to zero, we should increase x_3 by that same amount.

We then select the new residual of greatest magnitude, and relax it to zero. We continue until all residuals are zero and when this is true, the values of the x's will be at the exact solution. In implementing this method, there are some modifications that make the work easier. We illustrate in Fig. 7.1.

We make three double columns, one for each variable and for the residual of the equation in which that variable appears with -1 coefficient. The initial x values and the initial residuals are entered. It is convenient to work entirely with integers by multiplying the initial x-values and residuals by 1000, and then to scale down the solution by dividing by 1000 at the end of the computations. We avoid fractions; if a fractional change in a variable is needed to relax to zero we only relax to near zero.

In Fig. 7.1, we set down the increments to the x's but record the cumulative effect on the residuals. (The old values of the residuals are crossed out when replaced by a new value.) When the residuals are zero, we add the various increments to the initial value to get the final value. In this example, round-off errors cause an error of one in the third decimal.

It is important to make a final check by recomputing residuals at the end of the calculation to check for mistakes in arithmetic. The method is not readily programmed because searching on the computer for the largest residual is slow. It can be done rapidly by scanning the residuals in a hand calculation, however.

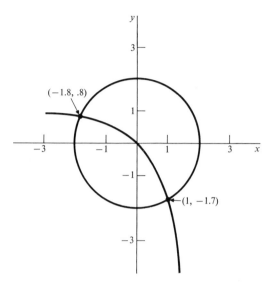

Fig. 8.1

8. SETS OF NONLINEAR EQUATIONS

As mentioned previously, the problem of finding the solution of a set of nonlinear equations is much more difficult than for linear equations. (In fact, some sets have no real solutions.) Consider the example of a pair of nonlinear equations:

$$x^2 + y^2 = 4,$$
$$e^x + y = 1. \tag{8.1}$$

Graphically, the solution to this system is represented by the intersections of the circle $x^2 + y^2 = 4$ with the curve $y = 1 - e^x$. Fig. 8.1 shows that these are near $(-1.8, 0.8)$ and $(1, -1.7)$. We can use the method of linear iteration to improve these approximations. Just as in Section 4, Chapter 1, we rearrange both equations to a form of the pattern $x = f(x,y)$, $y = g(x,y)$, and use the method of linear iteration on each equation in turn. Under proper conditions, these will converge. For example, if we rearrange Eqs. (8.1) in the form

$$x = \pm\sqrt{4 - y^2}, \quad (- \text{ sign for leftmost root}),$$
$$y = 1 - e^x,$$

we get the following successive values, beginning with $y_1 = 0.8$ in the first equation:

x-values: −1.83   −1.815   −1.8163   −1.8162

y-values: 0.8   0.84   0.8372   0.8374   0.8374

When we begin at $y = -1.7$ to find the root to the right of the origin, we get

x: 1.05   0.743   1.669   imaginary value

y: −1.7   −1.857   −1.102   −4.307

The equations diverge! (Beginning with $x = 1.0$ in the second equation is no help. This also diverges.) However, with a different rearrangement of the

original equations, such as

$$x = \ln(1 - y),$$
$$y = \pm\sqrt{4 - x^2}, \qquad (-\text{ sign for rightmost root}), \qquad (8.2)$$

we get

x: 0.993 1.006 1.0038 1.0042 1.0042

y: -1.7 -1.736 -1.7286 -1.7299 -1.7296

The pair of rearranged equations in (8.2) converges.

Some of the difficulties with sets of nonlinear equations are apparent from this simple example. If there are more than two equations in the system, finding a convergent form of the equations is increasingly difficult. A criterion for convergence (sufficiency condition only) is as follows:

The set of equations $x = f(x,y,z, \cdot \cdot \cdot)$, $y = g(x,y,z, \cdot \cdot \cdot)$, $z = h(x,y,z, \cdot \cdot \cdot)$, $\cdot \cdot \cdot$ *will converge if, in an interval about the root,*

$$|f_x| + |f_y| + |f_z| + \cdots < 1,$$
$$|g_x| + |g_y| + |g_z| + \cdots < 1,$$
$$|h_x| + |h_y| + |h_z| + \cdots < 1,$$
$$\cdot \quad \cdot \quad \cdot \quad \cdot \quad \cdot \quad \cdot \quad \cdot \quad < 1.$$

Computing all the partial derivatives and knowing where the root is are major problems. Getting starting values for the multidimensioned system is also correspondingly difficult.

Newton's method can also be applied to systems. We begin with the forms

$$f(x,y) = 0,$$
$$g(x,y) = 0.$$

Let $x = r$, $y = s$ be a root, and expand both functions as a Taylor series about the point (x_1, y_1) in terms of $(r - x_1)$, $(s - y_1)$, where (x_1, y_1) is a point near the root:

$$f(r,s) = 0 = f(x_1,y_1) + f_x(x_1,y_1)(r - x_1) + f_y(x_1,y_1)(s - y_1) + \cdots, \quad (8.3)$$
$$g(r,s) = 0 = g(x_1,y_1) + g_x(x_1,y_1)(r - x_1) + g_y(x_1,y_1)(s - y_1) + \cdots.$$

If (x_1,y_1) is near enough to (r,s), we can truncate after the first derivative terms and solve for the unknowns $(r - x_1)$, $(s - y_1)$. We exhibit the solution in determinant form in Eq. (8.4). Because we have truncated the series, we get only an approximation:

$$r - x_1 \doteq \frac{\begin{vmatrix} -f & f_y \\ -g & g_y \end{vmatrix}}{\begin{vmatrix} f_x & f_y \\ g_x & g_y \end{vmatrix}} \qquad (8.4)$$

$$s - y_1 \doteq \frac{\begin{vmatrix} f_x & -f \\ g_x & -g \end{vmatrix}}{\text{Denominator}}$$

[Note: All functions are evaluated at (x_1,y_1).]

The right-hand sides of (8.4) are increments to add to x_1, y_1 to get an improved estimate of the root (r,s). The extension to more than two simultaneous equations is straightforward in principle, but the effort soon becomes tremendous.

We illustrate by repeating the previous example:

$$f(x,y) = 4 - x^2 - y^2 = 0,$$
$$g(x,y) = 1 - e^x - y = 0.$$

The partials are

$$f_x = -2x, \qquad f_y = -2y,$$
$$g_x = -e^x, \qquad g_y = -1.$$

Beginning at $x_1 = 1$, $y_1 = -1.7$, we get

$$r = 1 + \frac{\begin{vmatrix} -0.110 & 3.4 \\ -0.0183 & -1 \end{vmatrix}}{\begin{vmatrix} -2 & 3.4 \\ -2.7183 & -1 \end{vmatrix}} = 1 + \frac{0.1722}{11.2422} = 1.0153,$$

$$s = -1.7 + \frac{\begin{vmatrix} -2 & -0.110 \\ -2.7183 & -0.0183 \end{vmatrix}}{11.2422} = -1.7 + \frac{-0.2624}{11.2422} = -1.7233.$$

The next repetition agrees with the true value of the root within three in the fourth decimal place.

9. COMPUTER PROGRAMS FOR SETS OF EQUATIONS

Three examples of computer programs utilizing the methods of this chapter are exhibited. The first two employ the Gaussian elimination method—Program 1 for a general $n \times n$ array of coefficients and Program 2 for a tridiagonal system. The latter embodies an ingeneous scheme to compress the matrix so that less storage is used by it for 500 equations than is required by the former for only 40. Program 3 uses iteration to solve a set of equations which must have the diagonal coefficients dominant.

The first Gaussian elimination program proceeds quite conventionally, utilizing the straightforward method of Section 3. Nothing fancy is attempted, such as pivoting or partial pivoting, although a division by zero is avoided by testing each pivot to see if it is zero and interchanging with the next lower row that has a nonzero element in the pivot column. After the coefficient matrix has been made upper triangular, back substitution is performed to give a pure diagonal matrix. The values of the variables are then computed by a simple division. Notice how round-off affects the results of the example used to test the program. Each component of the x-vector should be 1.

The tridiagonal matrix program has a very simple algorithm. The coefficients are represented by the symbol D for diagonal elements, A for elements above the diagonal, B for elements below the diagonal, and C for the constant vector. Be-

Program 1

```
ZZJOB 5                           CSC   001    GERALD, C. F.
ZZFORX5
*LIST PRINTER
C  PROGRAM TO SOLVE A SET OF LINEAR EQUATIONS BY GAUSSIAN ELIMINATION.
C  SIZE OF CORE MAY LIMIT NUMBER OF EQUATIONS THAT CAN BE SOLVED.
C  THIS PROGRAM DESIGNED FOR UP TO 40 EQUATIONS.
       DIMENSION A(40,41)
C  READ IN NUMBER OF EQUATIONS, N, AND COEFFICIENTS.
C  COEFFICIENTS ARE PUNCHED 8 PER CARD, EXCEPT EACH EQUATION BEGINS
C     ON A NEW CARD.
       READ 100, N
       NP1 = N + 1
       DO 10 I = 1,N
   10 READ 101, (A(I,J), J = 1,NP1)
C  PRINT HEADING
       PRINT 200, N
C  ELIMINATE COEFFICIENTS BELOW DIAGONAL.
       DO 20 I = 2,N
       DO 20 J = I,N
C  TEST IF PIVOT ELEMENT IS ZERO. IF SO, SWITCH ROWS.
       IF (A(I-1, I-1)) 1,2,1
    2 IM1 = I - 1
       DO 21 M = I,N
       IF (A(M,IM1)) 3,21,3
    3 DO 22 MM = IM1,NP1
       SAVE = A(M,MM)
       A(M,MM) = A(IM1,MM)
   22 A(IM1,MM) = SAVE
   21 CONTINUE
       PRINT 199
    1 R = A(J,I-1) / A(I-1,I-1)
       DO 20 K = I,NP1
   20 A(J,K) = A(J,K) - R*A(I-1,K)
C  BACK SUBSTITUTE BY ELIMINATION OF COEFFICIENTS ABOVE DIAGONAL.
       DO 30 I = 2,N
       K = N - I + 2
       R = A(K,NP1) / A(K,K)
       DO 30 J = I,N
       L = N - J + 1
   30 A(L,NP1) = A(L,NP1) - R*A(L,K)
C  VALUE OF VARIABLES IS COLUMN OF CONSTANTS DIVIDED BY CORRESPONDING NUMBER
C     ON THE DIAGONAL.  COMPUTE AND PRINT.
       DO 40 I = 1,N
       X = A(I,NP1) / A(I,I)
   40 PRINT 201, I, X
       CALL EXIT
  100 FORMAT (I2)
  101 FORMAT (8F10.0)
  199 FORMAT (72H COEFFICIENT MATRIX IS SINGULAR. NO UNIQUE SOLUTION TO
      1SET OF EQUATIONS. )
  200 FORMAT (/ 20H SOLUTION TO SET OF , I3,
      1 34H EQUATIONS BY GAUSSIAN ELIMINATION // )
  201 FORMAT (3H X( , I3, 4H) = , E14.8)
       END
04
1.0         0.96        0.84        0.64        3.44
0.96        0.9214      0.4406      0.2222      2.5442
0.84        0.4406      1.0         0.3444      2.6250
0.64        0.2222      0.3444      1.0         2.2066
ZZZZ
```

```
SOLUTION TO SET OF    4 EQUATIONS BY GAUSSIAN ELIMINATION


X(  1) =  .10002153E+01
X(  2) =  .99980000E+00
X(  3) =  .99990438E+00
X(  4) =  .10000893E+01
```

Program 2

```
ZZJOB 5                              CSC   001      GERALD. C. F.
ZZFORX5
*LIST PRINTER
C   PROGRAM TO SOLVE TRIDIAGONAL SYSTEM BY ELIMINATION.
C   STORAGE IS MINIMIZED BY COMPRESSING COEFFICIENTS INTO AN N BY 4 ARRAY.
C   PROGRAM WILL HANDLE UP TO 500 EQUATIONS.   D = DIAGONAL ELEMENTS.
C   A = ABOVE DIAGONAL, B = BELOW DIAGONAL, C = CONSTANT VECTOR.
C   THE C VECTOR IS FINALLY USED TO STORE X COMPONENTS FOR OUTPUT.
        DIMENSION A(500), D(500), B(500), C(500)
C   READ IN NUMBER OF EQUATIONS, N, AND COEFFICIENTS.
C   THE THREE COEFFICIENTS PLUS THE CONSTANT TERM OF EACH EQUATION
C   ARE ON ONE CARD.
        READ 100, N, (B(I), D(I), A(I), C(I), I = 1,N)
C   PRINT HEADING
        PRINT 200, N
C   COMPUTE THE NEW MATRIX. VALUES OF VARIABLES WILL BE STORED IN C ARRAY.
        DO 10 I = 2,N
        R = B(I)/D(I-1)
        D(I) = D(I) - R*A(I-1)
   10   C(I) = C(I) - R*C(I-1)
C   BACK SUBSTITUTION.
        C(N) = C(N)/D(N)
        DO 20 I = 2,N
        J = N - I + 1
   20   C(J) = (C(J) - A(J)*C(J+1))/D(J)
C   PRINT OUT VALUES OF VARIABLES
        PRINT 201, (I, C(I), I = 1,N)
        CALL EXIT
  100 FORMAT (I3 / (4F10.0))
  200 FORMAT (/ 34H SOLUTION TO TRIDIAGONAL SYSTEM OF , I4,
     1 26H EQUATIONS BY ELIMINATION //)
  201 FORMAT (3H X( , I4, 5H ) = , E14.7)
        END
005
0.0          4.0          -1.0          0.4
-1.0         4.0          -1.0          0.8
-1.0         4.0          -1.0          1.2
-1.0         4.0          -1.0          1.6
-2.0         4.0          0.0           1.6
ZZZZ

SOLUTION TO TRIDIAGONAL SYSTEM OF    5 EQUATIONS BY ELIMINATION

X(   1 ) =   1.9889502E-01
X(   2 ) =   3.9558009E-01
X(   3 ) =   5.8342537E-01
X(   4 ) =   7.3812149E-01
X(   5 ) =   7.6906072E-01
```

cause of the easy pattern of coefficients, both triangularization and back substitution are very quick:

$$\begin{bmatrix} d_1 & a_1 & 0 & 0 & \cdots & 0 \\ b_2 & d_2 & a_2 & 0 & \cdots & 0 \\ 0 & b_3 & d_3 & a_3 & \cdots & 0 \\ \vdots & & & & & \\ 0 & \cdot & \cdot & 0 & 0 & b_n & d_n \end{bmatrix} \begin{bmatrix} x_1 \\ x_2 \\ x_3 \\ \\ x_n \end{bmatrix} = \begin{bmatrix} c_1, \\ c_2, \\ c_3, \\ \\ c_n. \end{bmatrix}$$

Triangularization:

$$\text{new } d_i = d_i - \frac{b_i}{d_{i-1}} a_{i-1}, \qquad i = 2, 3, \ldots, n,$$

$$\text{new } c_i = c_i - \frac{b_i}{d_{i-1}} c_{i-1}, \qquad i = 2, 3, \ldots, n.$$

Program 3

```
ZZJOB 5                        CSC  001    GERALD. C. F.
ZZFORX5
*LIST PRINTER
C  PROGRAM TO SOLVE A SET OF EQUATIONS BY GAUSS SEIDEL ITERATION.
C  PROGRAM HANDLES UP TO 20 EQUATIONS.
C  EQUATIONS ARE TO BE ORDERED SUCH THAT REARRANGING BY SOLVING FOR VARIABLE
C    ON THE DIAGONAL WILL BE CONVERGENT.
      DIMENSION A(20.21). X(20)
C  READ IN NUMBER OF EQUATIONS. N. LIMIT TO ITERATIONS. MAX. AND TOLERANCE. TOL.
      READ 100. N. MAX. TOL
C  READ COEFFICIENTS. PUNCH 8 PER CARD. BEGINNING EACH EQUATION ON A NEW CARD.
      NP1 = N + 1
      DO 10 I = 1.N
   10 READ 101. (A(I.J). J = 1.NP1)
C  READ IN STARTING VALUES OF X. PUNCH 8 PER CARD.
      READ 101. (X(I). I = 1.N)
C  PRINT HEADING.
      PRINT 200
C  PERFORM ITERATIONS. STORE MAXIMUM CHANGE IN X VALUES FOR TESTING.
      DO 30 I = 1.MAX
      DIFF = 0.0
      DO 20 J = 1.N
      SAVE = X(J)
      X(J) = A(J.NP1)
      DO 21 K = 1.N
      IF (J - K) 22.21.22
   22 X(J) = X(J) - A(J.K)*X(K)
   21 CONTINUE
      X(J) = X(J) /A(J.J)
      IF (DIFF - ABSF(X(J) - SAVE)) 23.20.20
   23 DIFF = ABSF(X(J) - SAVE)
   20 CONTINUE
      IF (TOL - DIFF) 30.40.40
   30 CONTINUE
C  TOLERANCE WAS NOT MET. PRINT NOTE AND LAST VALUES.
      PRINT 201. TOL. MAX
      GO TO 50
   40 PRINT 202. I
   50 PRINT 203. (J. X(J). J = 1.N)
      CALL EXIT
  100 FORMAT (2I3. E14.0)
  101 FORMAT (8F10.0)
  200 FORMAT (/50H SOLUTION TO SET OF LINEAR EQUATIONS BY ITERATION  //)
  201 FORMAT(14H TOLERANCE OF . E8.0. 15H NOT MET AFTER .I4.
     1 12H ITERATIONS. / 29H LAST VALUES CALCULATED WERE /)
  202 FORMAT (16H X VALUES. AFTER. I4. 16H ITERATIONS ARE /)
  203 FORMAT (3H X( . I3. 4H) = . E14.7)
      END
003050          1.E-7
8.0        1.0        -1.0        8.0
1.0       -7.0         2.0       -4.0
2.0        1.0         9.0       12.0

ZZZZ
SOLUTION TO SET OF LINEAR EQUATIONS BY ITERATION

X VALUES. AFTER  11 ITERATIONS ARE

X(  1) =   1.0000000E+00
X(  2) =   1.0000000E+00
X(  3) =   1.0000000E+00
```

Back substitution:
$$x_j = \frac{c_j - a_j x_{j+1}}{d_j}, \qquad j = n - 1, \ldots, 1.$$

Program 3, illustrating iteration, begins with a set of starting values of the variable read in from cards. In the test example, these were all zero and were obtained by reading a blank card. The standard Gauss-Seidel method of Section

6 is used.   The iteration is continued until the largest magnitude by which any x-component changes is less than TOL, unless the limit MAX to the number of iterations is reached.   To determine the maximum change, each old x-component is saved until after the new value is computed, so that the change in value can be compared with the change in previously computed components which is largest in magnitude, and, if greater, replaces that one which has been stored in DIFF.

PROBLEMS

Section 1

1. Given matrices A, B, and C:

$$A = \begin{bmatrix} 1 & 2 & -1 & 3 \\ 0 & 4 & -2 & 1 \\ 3 & -1 & 1 & 1 \end{bmatrix}, \quad B = \begin{bmatrix} 0 & 1 & 0 \\ 1 & 2 & -1 \\ 1 & -1 & 3 \\ 2 & -2 & 1 \end{bmatrix}, \quad C = \begin{bmatrix} 0 & 1 & 4 & 2 \\ -1 & -2 & 3 & 1 \\ 0 & 1 & 2 & -1 \end{bmatrix}.$$

   (a) Find $A + B$, $A + C$, $A - 2C$.

   (b) Find AB, BA, AC.

   (c) What is the size of ABC?

   (d) Show that $(AB)C = A(BC)$.

2. Given the square matrix M:

$$M = \begin{bmatrix} 1 & 1 & -1 \\ -1 & 0 & 1 \\ 1 & -1 & -1 \end{bmatrix}.$$

   (a) Find M^2, M^3.

   (b) Show that $MI = IM = M$, where I is the order 3 unitary matrix.

   (c) A square matrix can always be expressed as a sum of an upper triangular and a lower triangular matrix.   Find a U and an L such that $U + L = M$.

3. Write the set of equations in matrix notation:

$$a_1 x + b_1 y + c_1 z = d_1,$$
$$a_2 x + b_2 y + c_2 z = d_2,$$
$$a_3 x + b_3 y + c_3 z = d_3.$$

4. Write as a set of equations:

$$\begin{bmatrix} 1 & 3 & 1 & -1 \\ 2 & 0 & 1 & 1 \\ 0 & -1 & 4 & 1 \\ 0 & 1 & 1 & -5 \end{bmatrix} \begin{bmatrix} x_1 \\ x_2 \\ x_3 \\ x_4 \end{bmatrix} = \begin{bmatrix} 3 \\ 1 \\ 6 \\ 16 \end{bmatrix}.$$

Section 2

5. Solve by elimination:

$$\begin{bmatrix} 5 & -1 & 0 \\ -1 & 5 & -1 \\ 0 & -1 & 5 \end{bmatrix} \begin{bmatrix} x_1 \\ x_2 \\ x_3 \end{bmatrix} = \begin{bmatrix} 9 \\ 4 \\ -6 \end{bmatrix}.$$

6. Solve by elimination:

$$\begin{bmatrix} 5 & -1 & 1 \\ 2 & 4 & 0 \\ 1 & 1 & 5 \end{bmatrix} x = \begin{bmatrix} 10 \\ 12 \\ -1 \end{bmatrix}.$$

7. Solve the set of equations in Problem 4.

8. Show that the following set of equations does not have a solution:

$$\begin{aligned} 3x_1 + 2x_2 - x_3 - 4x_4 &= 10, \\ x_1 - x_2 + 3x_3 - x_4 &= -4, \\ 2x_1 + x_2 - 3x_3 &= 16, \\ - x_2 + 8x_3 - 5x_4 &= 3. \end{aligned}$$

9. If the constant vector in Problem 8 has 2, 3, 1, 3 as components, show that the equations have an infinite set of solutions.

Section 3

10. Solve Problem 5 by Gaussian elimination. Provide a check column to control numerical errors. Do the back substitutions by eliminating the elements above the diagonal through elementary row operations.

11. Solve Problem 6 by Gaussian elimination. Carry four decimals in your work.

12. Reorder the equations in Problem 6 by putting the first equation last, and then solve by Gaussian elimination (four decimals). During the back substitution steps observe how the errors increase due to the larger coefficients in comparison to Problem 11.

13. (a) Solve the system

$$\begin{aligned} 2.51x_1 + 1.48x_2 + 4.53x_3 &= 0.05, \\ 1.48x_1 + 0.93x_2 - 1.30x_3 &= 1.03, \\ 2.68x_1 + 3.04x_2 - 1.48x_3 &= -0.53, \end{aligned}$$

by Gaussian elimination, carrying four decimals. Do not interchange rows. Notice that one divides by a small coefficient in reducing the third equation.

(b) Solve the system again, using elimination with pivoting. Notice now that the small coefficient is no longer a divisor.

(c) Substitute each set of answers in the original equations and observe that the left- and right-hand sides match better with the (b) answers. The solution when seven places is carried to eliminate round-off is $x_1 = 1.45310$, $x_2 = -1.58919$, $x_3 = -0.27489$.

Section 4

14. Use Crout reduction to solve Problem 6.

15. Use Crout reduction to solve the set of equations in Problem 4.

16. Solve the system in Problem 13(a) by Crout reduction. Note that the small divisor again appears and causes round-off error to be severe. Pivoting can be used with

the Crout scheme; one must interchange corresponding rows in both the working matrix (LU) and in the original matrix (A).

17. Solve the tridiagonal system

$$\begin{bmatrix} 4 & -1 & 0 & 0 & 0 \\ -1 & 4 & -1 & 0 & 0 \\ 0 & -1 & 4 & -1 & 0 \\ 0 & 0 & -1 & 4 & -1 \\ 0 & 0 & 0 & -1 & 4 \end{bmatrix} x = \begin{bmatrix} 100 \\ 200 \\ 200 \\ 200 \\ 100 \end{bmatrix}$$

by Crout reduction. Notice how readily the operations proceed for this particular case. The coefficient matrix occurs in the solution of partial differential equations as discussed in Chapter 10.

Section 5

18. Find the determinant of the matrix

$$\begin{bmatrix} 0 & 1 & -1 \\ 3 & 1 & -4 \\ 2 & 1 & 1 \end{bmatrix}$$

by row operations to make it (a) upper triangular, (b) lower triangular.

19. Find the determinant of the matrix

$$\begin{bmatrix} 1 & 4 & -2 & 1 \\ -1 & 2 & -1 & 1 \\ 3 & 3 & 0 & 4 \\ 4 & -4 & 2 & 3 \end{bmatrix}.$$

20. Invert the coefficient matrix in Problem 5, and then use the inverse to generate the solution. Note that a symmetric matrix has a symmetric inverse.

21. If the constant vector in Problem 5 is changed to one with components (13, 11, −22), what now is the solution? Observe that the inverse obtained in Problem 22 gives the answer readily.

22. Attempt to find the inverse of the coefficient matrix in Problem 8. Note that a singular matrix has no inverse.

23. Invert the coefficient matrix in Problem 17. Work with decimals to four places.

Section 6

24. Solve Problem 5 by the Jacobi method, beginning with the initial vector (0,0,0).

25. Solve Problem 5 by Gauss-Seidel iteration, beginning with the vector (0,0,0).

26. Solve Problem 6 by Gauss-Seidel iteration, beginning with the approximate solution (2,2,−1).

27. In Problem 12, answers were obtained as 2.5550, 1.7225, −1.0555, which are slightly in error due to round-off. Using these as starting values for Gauss-Seidel iteration

of Problem 6, make just one iteration and compare to correct values of 2.5555, 1.7222, -1.0555.

28. In Problem 13(a), inaccurate values were obtained because round-off error was magnified with a small divisor. Use these values as an initial approximation and improve by the Gauss-Seidel method. Make three iterations and observe that while the solution appears to be converging, it is rather slow. This is because the system cannot be arranged so that the diagonal coefficients are strongly dominant.

29. Solve Problem 17 by Gauss-Seidel iteration.

Section 7

30. Beginning with (0,0,0), use relaxation to solve the system

$$6x_1 - 3x_2 + x_3 = 11,$$
$$2x_1 + x_2 - 8x_3 = -15,$$
$$x_1 - 7x_2 + x_3 = 10.$$

31. Solve Problem 5 by relaxation.

32. Solve Problem 6 by relaxation. Begin with the approximate solution $(2,2,-1)$.

33. Repeat Problem 28, except use relaxation. Is the convergence more rapid?

34. Relaxation is especially well adapted to problems like Problem 17. Solve by the relaxation method starting with the vector (45,80,90,80,45) which one obtains by inspection.

Section 8

35. The system

$$xy = 1,$$
$$x^2 + y^2 = 4,$$

has four solutions, all symmetrical about the line $y = x$. Use the method of iteration to locate them.

36. Find the two intersections nearest the origin of the curves $x^2 + x - y^2 = 1$ and $y - \sin x^2 = 0$. Use the method of linear iteration.

37. In Problem 35, it is easy to eliminate y between the two equations. Finding roots of the resulting equation then gives the x-values of the intersections, for which we may use a variety of methods. Solve the problem by this technique, using Newton's method to improve your initial estimates.

38. Solve the system

$$x^2 + y^2 + z^2 = 9,$$
$$xyz = 1,$$
$$x + y - z^2 = 0,$$

by linear iteration to obtain the solution near (2.5, 0.2, 1.6).

39. Solve Problem 35 by Newton's method for a system. Begin with the approximate solution (0.5,2).

40. Solve Problem 36 by Newton's method.

41. Solve by using Newton's method:

$$x^3 + 3y^2 = 21,$$
$$x^2 + 2y + 2 = 0.$$

Make sketches of the graphs to locate approximate values of the intersections.

42. For a set of three nonlinear equations, all of the form $f_i(x,y,z) = 0$, each function can be expanded in a Taylor series about the point (x_1,y_1,z_1), analogously to Eqs. (8.3). If the point (x_1,y_1,z_1) is near enough to a root, all the terms beyond the first partials can be neglected, and each series may be set equal to zero. From this linear set one can solve for the increments to add to x_1,y_1, and z_1 to more nearly approximate the root. Perform the indicated derivation and use it to solve Problem 38.

8
boundary
value
problems

In differential equations of order greater than one, two or more values must be known to evaluate the constants in the particular function that satisfies the differential equation. In the problems discussed in Chapter 6, these several values were all specified at the same value of the independent variable, generally at the start or *initial value*. For that reason, such problems are termed initial value problems.

For an important class of problems, the several values of the function or its derivatives are not all known at the same point, but rather at two different values of the independent variable. Because these values of the independent variable are usually the end points, or boundaries, of some domain of interest, problems with this type of conditions are classed as *boundary value problems*. Determining the deflection of a simply supported beam is a typical example, where the conditions specified are the deflections and second derivatives of the elastic curve at the supports. Heat flow problems also fall in this class when the temperatures or temperature gradients are given at two points. A special case of boundary value problem is involved in vibration problems.

1. THE "SHOOTING METHOD"

Suppose we wish to solve the second-order boundary value problem

$$\frac{d^2y}{dt^2} + t\frac{dy}{dt} + y = 3t^2 + 2, \qquad y(0) = 0, \quad y(1) = 1. \tag{1.1}$$

Note that if $y'(0)$ were given in addition to $y(0)$, this would be an initial value problem. There is a way to adapt our previous methods to this problem, as illustrated in Fig. 1.1. We know the value of y at $t = 0$ and at $t = 1$, as given by the dots. The curve that represents y between these two points is desired. We know that some such curve, such as the dotted line, exists; its slope and curvature are interrelated to y and t by the differential equation (1.1).

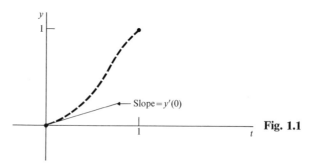

Fig. 1.1

If we assume the slope of the curve at $t = 0$, say $y'(0) = 0.4$, we could solve the equation as an initial value problem using this assumed value. The test of our assumption is whether we calculate y at $t = 1$ to match the known value, $y(1) = 1$. In Table 1.1 we show the results of this computation employing the

Table 1.1. Solving a Second-Order Equation by the Shooting Method

$$y'' = 3t^2 + 2 - ty' - y$$

Time	Distance	Velocity
Assume $y'(0) = 0.4$		
0.00	0.0000000	0.4000000
0.10	0.0498015	0.5960297
0.20	0.1188168	0.7842759
0.30	0.2062906	0.9651998
0.40	0.3115277	1.1395400
0.50	0.4339183	1.3082701
0.60	0.5729558	1.4725452
0.70	0.7282622	1.6336325
0.80	0.8995850	1.7928500
0.90	1.0868026	1.9514990
1.00	1.2899194	2.1108035
Now assume $y'(0) = 0.1$		
0.00	0.0000000	0.1000000
0.10	0.0199500	0.2990075
0.20	0.0597032	0.4960692
0.30	0.1190709	0.6913005
0.40	0.1978793	0.8848861
0.50	0.2959762	1.0770694
0.60	0.4132358	1.2681384
0.70	0.5495626	1.4584104
0.80	0.7048935	1.6482149
0.90	0.8791982	1.8378772
1.00	1.0724776	2.0277033

Linearly interpolate: new slope $= 0.1 - \dfrac{0.4 - 0.1}{1.2899 - 1.0725}(0.0725) = 0.000$

Time	Distance	Velocity
Using $y'(0) = 0.0$		
0.00	0.0000000	0.0000000
0.10	0.0100000	0.2000000
0.20	0.0400000	0.4000000
0.30	0.0900000	0.6000000
0.40	0.1600000	0.8000000
0.50	0.2500000	1.0000000
0.60	0.3600000	1.2000000
0.70	0.4900000	1.4000000
0.80	0.6400000	1.6000000
0.90	0.8100000	1.8000000
1.00	1.0000000	2.0000000

computer program of Chapter 6, Section 11, modified for this equation, using the Euler method with $\Delta t = 0.1$. Since the result of this gives $y(1) = 1.289\ldots$, and not the desired $y(1) = 1$, we assume another value for $y'(0)$ and repeat. Since the calculated value of $y(1)$ is too high we assume a smaller value for the slope, say $y'(0) = 0.1$. The second computation gives $y(1) = 1.072\ldots$, which is better but still too high. After these two trials, we linearly interpolate (here extrapolate) for a third trial. At $y'(0) = 0.0$, we get the correct value of $y(1)$. In most problems* more attempts are needed to get the correct solution.

The method we have illustrated is called the *shooting method* because it resembles an artillery problem. One sets the elevation of the gun and fires a preliminary round at the target. After successive shots have straddled the target, one zeros in on it by using intermediate values of the gun's elevation. This corresponds to our using assumed values of the initial slope and interpolating based on how close we come to $y(1)$.

This method is often quite laborious. Especially with problems of third and higher order, the necessity to assume two or more conditions at the starting point (and match with the same number of conditions at the end) is slow and tedious. We shall ordinarily use the method of central difference approximations of the next section.

2. SOLUTION THROUGH A SET OF EQUATIONS

We have seen in Chapter 4 how the derivatives of a function can be approximated by finite difference quotients. In Chapter 5 they were derived by an alternative method. If we replace the derivatives in a differential equation by such expressions, we convert it to a difference equation whose solution is an approximation to the solution of the differential equation. This method is sometimes preferred over the shooting method discussed above. Consider the same example as before:

$$\frac{d^2y}{dt^2} + t\frac{dy}{dt} + y = 3t^2 + 2, \tag{2.1}$$

$$y(0) = 0, \quad y(1) = 1.$$

Central difference approximations to derivatives are more accurate than forward or backward approximations ($O(h^2)$ versus $O(h)$), so we replace the derivatives with

$$\left.\frac{dy}{dt}\right|_{t=t_i} = \frac{y_{i+1} - y_{i-1}}{2h} + O(h^2),$$

$$\left.\frac{d^2y}{dt^2}\right|_{t=t_i} = \frac{y_{i+1} - 2y_i + y_{i-1}}{h^2} + O(h^2).$$

---

* In only special situations is the functional value at the end of the interval proportional to the starting value of the slope, but such interpolation gives improved values for $y'(x_0)$.

The quantity h is the constant difference in t-values. Substituting these equivalences into Eq. (2.1), and rearranging, we get

$$\frac{y_{i+1} - 2y_i + y_{i-1}}{h^2} + t_i \left(\frac{y_{i+1} - y_{i-1}}{2h}\right) + y_i = 3t_i^2 + 2, \qquad (2.2)$$

$$(2 - ht_i)y_{i-1} + (2h^2 - 4)y_i + (2 + ht_i)y_{i+1} = 6h^2t_i^2 + 4h^2.$$

In Eqs. (2.2), we have replaced y by y_i and t by t_i, since these values correspond to the point at which the difference quotients represent the derivatives. Our problem now reduces to solving Eqs. (2.2) at points in the interval from $t = 0$ to $t = 1$. Let us subdivide the interval into a number of equal subintervals. For example, if $h = \Delta t = \frac{1}{4}$, the points $t_1 = 0$, $t_2 = \frac{1}{4}$, $t_3 = \frac{1}{2}$, $t_4 = \frac{3}{4}$, $t_5 = 1$ subdivide the interval into four subintervals.

We write the difference equation in (2.2) for each of these values of t at which y is unknown, giving

$t = t_2 = \frac{1}{4}$: $(2 - (\frac{1}{4})(\frac{1}{4}))y_1 + (2(\frac{1}{4})^2 - 4)y_2 + (2 + (\frac{1}{4})(\frac{1}{4}))y_3 = 6(\frac{1}{4})^2(\frac{1}{4})^2 + 4(\frac{1}{4})^2,$

$t = t_3 = \frac{1}{2}$: $(2 - (\frac{1}{2})(\frac{1}{4}))y_2 + (2(\frac{1}{4})^2 - 4)y_3 + (2 + (\frac{1}{2})(\frac{1}{4}))y_4 = 6(\frac{1}{4})^2(\frac{1}{2})^2 + 4(\frac{1}{4})^2,$

$t = t_4 = \frac{3}{4}$: $(2 - (\frac{3}{4})(\frac{1}{4}))y_3 + (2(\frac{1}{4})^2 - 4)y_4 + (2 + (\frac{3}{4})(\frac{1}{4}))y_5 = 6(\frac{1}{4})^2(\frac{3}{4})^2 + 4(\frac{1}{4})^2.$

We know $y_1 = 0$, $y_5 = 1$; hence the difference equation is not written corresponding to $x = 0$ or $x = 1$. Substituting the values $y_1 = 0$, $y_5 = 1$ and simplifying, we have, in matrix form,

$$\begin{bmatrix} -3.875 & 2.062 & 0 \\ 1.875 & -3.875 & 2.125 \\ 0 & 1.812 & -3.875 \end{bmatrix} \begin{bmatrix} y_2 \\ y_3 \\ y_4 \end{bmatrix} = \begin{bmatrix} 0.273 \\ 0.344 \\ -1.727 \end{bmatrix}.$$

Note that the coefficient matrix is tridiagonal, so that the solution comes readily by elimination: $y_2 = 0.062$, $y_3 = 0.250$, $y_4 = 0.562$. These match to three decimals the analytical solution to the problem, $y = t^2$. The reason that we agree with the analytical solution even though the difference equation is an approximation to the differential equation of error $O(h^2)$ is that with the exact solution a second-degree polynomial, the third derivative in the error term (compare lozenges in Chapter 4) is zero. In general we do not get exact answers as the next example shows.

Example. Solve $d^2y/dx^2 = y$, $y(1) = 1.175$, $y(3) = 10.018$. (The analytical solution is $y = \sinh x$, to which we can compare our estimates of the function.)

While it is not strictly necessary, it is common to normalize the function to the interval $(0,1)$.* This we can do by the change of variable $x = (b - a)t + a$,

---

* Perhaps the most important reason is to make a computer program more general. Also, if we find the inverse to the coefficient matrix in order to solve the forthcoming set of equations, it is then of more general applicability.

where (a,b) is the original interval to be normalized.   Letting $x = 2t + 1$,

$$\frac{dy}{dx} = \frac{dy}{dt}\frac{dt}{dx} = \frac{1}{2}\frac{dy}{dt},$$

$$\frac{d^2y}{dx^2} = \frac{d}{dt}\left(\frac{1}{2}\frac{dy}{dt}\right)\frac{dt}{dx} = \frac{1}{2^2}\frac{d^2y}{dt^2},$$

the problem becomes

$$\frac{d^2y}{dt^2} = 4y, \qquad y(0) = 1.175, \quad y(1) = 10.018.$$

Replacing the second derivative by the central difference approximation, converting the differential equation to a difference equation, we have

$$\frac{y_{i+1} - 2y_i + y_{i-1}}{h^2} = 4y_i.$$

Subdividing the interval $(0,1)$ into three parts, $h = 0.25$, and writing the difference equation at the four internal points where y is unknown, we have the set of equations

$$\begin{bmatrix} -2.25 & 1 & 0 \\ 1 & -2.25 & 1 \\ 0 & 1 & -2.25 \end{bmatrix} \begin{bmatrix} y_2 \\ y_3 \\ y_4 \end{bmatrix} = \begin{bmatrix} -1.175 \\ 0 \\ -10.018 \end{bmatrix}. \qquad (2.3)$$

In (2.3) we have used $y_1 = 1.175$, $y_5 = 10.018$.   Again, a tridiagonal matrix of coefficients occurs.   The solution to this set of equations is:

t	x	Calculated y	$\sinh x$	Error
0	1.0	(1.175)	(1.175)	
0.25	1.5	2.147	2.129	−0.018
0.50	2.0	3.655	3.627	−0.028
0.75	2.5	6.077	6.050	−0.027
1.0	3.0	(10.018)	(10.018)	

If we desire greater accuracy than given above, we can, of course, improve them by using a smaller value of h_i, say $h = 0.1$.   If this is done, the number of equations becomes larger, since there are now nine interior points at which y is unknown.

An easier way to improve the midpoint value, at $x = 2$, is to recalculate y with h *larger*, getting a less accurate value, but extrapolating with the assumption that our $O(h^2)$ errors can be taken as proportional to h^2.   Hence, with only two subdivisions, $\Delta x = 1$, and

$$\frac{y(1) - 2y(2) + y(3)}{(1)^2} = y(2).$$

(Here we have elected not to normalize the interval.)

The only unknown is $y(2)$. We get $y(2) = 3.731$. An improved value of $y(2)$ is then

$$3.655 + \tfrac{1}{3}(3.655 - 3.731) = 3.630.$$

This is in error by only 3 in the third decimal place.

3. DERIVATIVE BOUNDARY CONDITIONS

The conditions which the solution of a differential equation must satisfy need not necessarily be just the value of the function. In many applied problems, some derivative of the function may be known at the boundaries of an interval. In the more general case a linear combination of the function and its derivatives is specified. Our above procedure needs modification for this type of boundary conditions. We illustrate by an example:

$$\frac{d^2y}{dx^2} = y,$$

$$y'(1) = 1.175,$$
$$y'(3) = 10.018.$$

(This problem has the same differential equation as the example of Section 2 but with the values specified for the derivative at $x = 1$ and $x = 3$, it now has the analytical solution $y = \cosh x$.)

We begin just as before. We change variables to make the interval $(0,1)$ by letting $x = 2t + 1$:

$$\frac{d^2y}{dt^2} = 4y,$$

$$y'(0) = 1.175,$$
$$y'(1) = 10.018.$$

We now replace the derivative by a central difference approximation and with the difference equation at each point where y is unknown. With $h = 0.25$,

$$
\begin{aligned}
t = 0: & \quad y_l - 2.25y_1 + y_2 = 0, \\
t = 0.25: & \quad y_1 - 2.25y_2 + y_3 = 0, \\
t = 0.50: & \quad y_2 - 2.25y_3 + y_4 = 0, \\
t = 0.75: & \quad y_3 - 2.25y_4 + y_5 = 0, \\
t = 1.00: & \quad y_4 - 2.25y_5 + y_r = 0.
\end{aligned}
\tag{3.1}
$$

Two more equations are required than in Section 2 because y is unknown at $t = 0$ and $t = 1$ as well as at the interior points. These two equations involve the values y_l, y_r, points one space to the left and to the right of the interval $(0,1)$. We assume that the domain of the differential equation can be so extended.

Our problem is now that Eqs. (3.1) contain seven unknowns, and we have only five equations. The boundary conditions, however, have not yet been involved. Let us express these also as difference quotients, preferring central difference approximations of $O(h^2)$ error as used in replacing derivatives in the original equation:

$$\frac{dy}{dx}\Big|_{x=1} = \frac{dy}{dt}\frac{dt}{dx}\Big|_{t=0} \doteq \left(\frac{1}{2}\right)\frac{y_2 - y_l}{2h} = 1.175, \qquad y_l = y_2 - (4)(0.25)(1.175),$$

$$\frac{dy}{dx}\Big|_{x=3} = \frac{dy}{dt}\frac{dt}{dx}\Big|_{t=1} \doteq \left(\frac{1}{2}\right)\frac{y_r - y_4}{2h} = 10.018, \qquad y_r = y_4 + (4)(0.25)(10.018).$$

(3.2)

The relations of (3.2), when substituted into (3.1), reduce the number of unknowns to five, and we solve the equations in the usual way.   The solution is:

t	x	y	cosh x	Error
0	1.0	1.552	1.543	−0.009
0.25	1.5	2.334	2.352	0.018
0.50	2.0	3.699	3.762	0.063
0.75	2.5	5.988	6.132	0.144
1.00	3.0	9.776	10.067	0.291

We observe here that the errors are much greater than for the previous example, being very large at $x = 3.0$.   The explanation is that our approximation for the derivative is poor at large x-values.   While of $O(h^2)$, the third derivative appearing in the error term is large in magnitude.   In the previous example, knowing the function at the end points eliminated the errors there.

Repeating the calculation with $h = 0.5$ and extrapolation should improve the estimate.   When this is done, we find

x	$h = 0.5$	Extrapolated	cosh x
1.0	$y = 1.575$	1.545	1.543
2.0	$y = 3.537$	3.753	3.762
3.0	$y = 9.037$	10.022	10.067

The results are significantly improved.

We could of course use better finite difference approximations to the derivatives, not only for the boundary values, but for the equation itself.   The disadvantage of doing this is that the system of equations is then not tridiagonal, with attendant greater difficulty of solution.

If the more general form of boundary condition applies, $ay + by' = c$, equations similar to (3.2) result, and after the boundary conditions have been approximated and the exterior y-values eliminated from the set of equations, the solution of the problem proceeds as before.

Equations of order higher than second will involve approximations for the third or higher derivatives.   Central difference formulas will involve points more than h away from the point where the derivative is being approximated, and may be unsymmetrical in the case of the odd derivatives.   Probably the method of

Chapter 5 is the easiest way to derive the necessary formulas. Again, nontridiagonal systems result. Fortunately, most of the important physical problems are simulated by equations of order two.

4. CHARACTERISTIC VALUE PROBLEMS

Problems in the fields of elasticity and vibration (including applications of the wave equations of modern physics) fall into a special class of boundary value problems known as characteristic value problems. Certain problems in statistics also reduce to such problems. We discuss here only the most elementary forms of characteristic value problems.

Consider the homogeneous* second-order equation with homogeneous boundary conditions:

$$\frac{d^2y}{dt^2} + k^2y = 0, \qquad y(0) = 0, \quad y(1) = 0, \tag{4.1}$$

where k^2 is a parameter. We first solve this equation nonnumerically to show that there is a solution for only certain particular, or "characteristic," values of the parameter. The general solution is $y = a \sin kx + b \cos kx$, which can easily be verified by substituting into the differential equation; the solution contains the two arbitrary constants a and b, because the differential equation is second order. The constants a and b are to be determined to make the general solution agree with the boundary conditions.

At $x = 0$, $y = 0 = a \sin (0) + b \cos (0) = b$. Then b must be zero. At $x = 1$, $y = 0 = a \sin (k)$; we may either have $a = 0$ or $\sin (k) = 0$ to satisfy this condition. The former leads to $y(x) = 0$, which we call the *trivial solution*. This function, y everywhere zero, will always be a solution to any homogeneous differential equation with homogeneous boundary conditions. It is usually of no interest. To get a nontrivial solution, we must choose the other alternative, $\sin (k) = 0$, which is satisfied only for certain values of k, the characteristic† values of the system. The solution to Eq. (4.1) must then be

$$k = n\pi, \qquad n = 1, 2, \ldots,$$
$$y = a \sin n\pi x. \tag{4.2}$$

Note that the arbitrary constant a can have any value, so that the solution is determined only to within a multiplicative constant. In Fig. 4.1 we sketch several of the solutions as given by Eq. (4.2).

The information of interest in characteristic value problems are the eigenvalues for the system. If we are dealing with a vibration problem, these give the natural frequencies of the system, which are especially important because, with external loads applied at near these frequencies, resonance will cause an amplification of

---

* Homogeneous here means all the terms are alike in being functions of y or its derivatives.
† These are also called "eigenvalues," derived from the German term Eigenwerte.

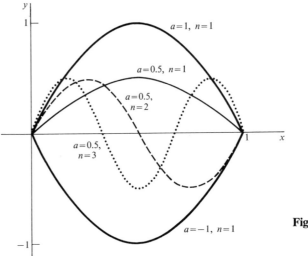

Fig. 4.1

motion so that failure is likely.   In the field of elasticity, there is an eigenfunction corresponding to each eigenvalue, and these determine the possible shapes of the elastic curve when the system is in equilibrium.   Often the smallest nonzero value of the parameter is especially important; this gives the fundamental frequency of the system.

To illustrate the use of numerical methods, we will solve Eq. (4.1) again.   Our attack is the general one of replacing the differential equation by a difference equation, and writing this at all points into which the x-interval has been subdivided and where y is unknown.*   Replacing the derivative in (4.1) by a central difference approximation, we have

$$\frac{y_{i+1} - 2y_i + y_{i-1}}{h^2} + k^2 y_i = 0. \tag{4.3}$$

Letting $h = 0.2$, and writing out Eq. (4.3) at each of the four interior points, we get

$$\begin{aligned}
y_1 - (2 - 0.04k^2)y_2 + y_3 &= 0, \\
y_2 - (2 - 0.04k^2)y_3 + y_4 &= 0, \\
y_3 - (2 - 0.04k^2)y_4 + y_5 &= 0, \\
y_4 - (2 - 0.04k^2)y_5 + y_6 &= 0.
\end{aligned} \tag{4.4}$$

---

* There is an alternative trial and error approach, analogous to the shooting method, but it is generally more work.

The boundary conditions give $y_1 = y_6 = 0$. Making this substitution and writing in matrix form, after multiplying all equations by -1 yields

$$\begin{bmatrix} 2 - 0.04k^2 & -1 & 0 & 0 \\ -1 & 2 - 0.04k^2 & -1 & 0 \\ 0 & -1 & 2 - 0.04k^2 & -1 \\ 0 & 0 & -1 & 2 - 0.04k^2 \end{bmatrix} \begin{bmatrix} y_2 \\ y_3 \\ y_4 \\ y_5 \end{bmatrix} = 0. \qquad (4.5)$$

Note that we can write this as the matrix equation $(A - \lambda I)y = 0$, where

$$A = \begin{bmatrix} 2 & -1 & 0 & 0 \\ -1 & 2 & -1 & 0 \\ 0 & -1 & 2 & -1 \\ 0 & 0 & -1 & 2 \end{bmatrix}, \qquad \lambda = 0.04k^2,$$

and I is the identity matrix of order 4.

We shall consider the problem in this form in the next section.

The solution to our characteristic value problem, Eq. (4.1), reduces to solving the set of equations in (4.4) or (4.5).

Such a set of homogeneous linear equations has a nontrivial solution only if the determinant of the coefficient matrix is equal to zero. Hence

$$\begin{bmatrix} 2 - 0.04k^2 & -1 & 0 & 0 \\ -1 & 2 - 0.04k^2 & -1 & 0 \\ 0 & -1 & 2 - 0.04k^2 & -1 \\ 0 & 0 & -1 & 2 - 0.04k^2 \end{bmatrix} = 0.$$

Expanding this determinant will give an eighth-degree polynomial in k; the roots of this polynomial will be approximations to the characteristic values of the system. Letting $2 - 0.04k^2 = A$ makes the expansion simpler, and we get

$$A^4 - 3A^2 + 1 = 0.$$

The roots of this biquadratic are $A^2 = (3 \pm \sqrt{5})/2$, or $A = 1.618, -1.618, 0.618, -0.618$.

We will get an estimate of the principal eigenvalues from $2 - 0.04k^2 = 1.618$, giving $k = \pm 3.09$ (compare to $\pm \pi$). The next eigenvalues are obtained from

$$\begin{aligned} 2 - 0.04k^2 &= 0.618, & \text{giving} \quad k &= \pm 5.88 \text{ (compare to } \pm 2\pi), \\ 2 - 0.04k^2 &= -0.618, & \text{giving} \quad k &= \pm 8.09 \text{ (compare to } \pm 3\pi), \\ 2 - 0.04k^2 &= -1.618, & \text{giving} \quad k &= \pm 9.51 \text{ (compare to } \pm 4\pi). \end{aligned}$$

Our estimates of the characteristic values get progressively worse. Fortunately the smallest values of k are of principal interest in most applied problems. To improve accuracy, we will need to write our difference equation using smaller values of h. The work soon gets inordinately long, and we look for a simpler method to find the eigenvalues of a matrix. This is the subject of the next section.

5. EIGENVALUES OF A MATRIX BY ITERATION

We have seen that particular values of the parameter occurring in a characteristic value problem are of special interest, and the numerical solution involves finding the eigenvalues of the coefficient matrix of the set of difference equations. To do this, we need to find the values of λ that satisfy the matrix equation

$$Ax = \lambda x$$

where A is a square matrix. If A is $n \times n$ and nonsingular, there are n different vectors x which, when multiplied by the matrix A, give a scalar multiple of x. These vectors are called eigenvectors or characteristic vectors. The multipliers λ, which are n in number, are termed the eigenvalues or characteristic values of the matrix A. As we have seen in the previous section, one method is to solve the equivalent matrix equation

$$(A - \lambda I)x = 0$$

by determining the values of λ which make the determinant $|A - \lambda I|$ equal to zero. This involves expanding the determinant to give a polynomial in λ, the characteristic equation, whose roots must be evaluated. Solving for the roots of the characteristic equation is laborious. However, a straightforward iterative procedure will give the dominant eigenvalue. Consider a simple 2×2 system:

$$\begin{bmatrix} 3 & 0 \\ 1 & -1 \end{bmatrix} \begin{bmatrix} x_1 \\ x_2 \end{bmatrix} = \lambda \begin{bmatrix} x_1 \\ x_2 \end{bmatrix},$$

$$\begin{bmatrix} 3 - \lambda & 0 \\ 1 & -1 - \lambda \end{bmatrix} \begin{bmatrix} x_1 \\ x_2 \end{bmatrix} = 0. \tag{5.1}$$

Finding values of λ and the vectors x that satisfy Eq. (5.1) is our problem.

In this simple case the characteristic equation is easy to solve:

$$(3 - \lambda)(-1 - \lambda) - 0 = 0, \qquad \lambda^2 - 2\lambda - 3 = (\lambda - 3)(\lambda + 1) = 0.$$

The eigenvalues are $\lambda_1 = 3$, $\lambda_2 = -1$. We can find the eigenvectors corresponding to these from the two sets of equations

$$\begin{bmatrix} 3 & 0 \\ 1 & -1 \end{bmatrix} \begin{bmatrix} x_1 \\ x_2 \end{bmatrix} = 3 \begin{bmatrix} x_1 \\ x_2 \end{bmatrix}, \qquad \begin{cases} 3x_1 = 3x_1 \\ x_1 - x_2 = 3x_2 \end{cases}, \qquad \begin{cases} x_1 \text{ arbitrary} \\ x_2 = \frac{1}{4}x_1 \end{cases}$$

$$\begin{bmatrix} 3 & 0 \\ 1 & -1 \end{bmatrix} \begin{bmatrix} x_1 \\ x_2 \end{bmatrix} = -1 \begin{bmatrix} x_1 \\ x_2 \end{bmatrix}, \qquad \begin{cases} 3x_1 = -x_1 \\ x_1 - x_2 = -x_2 \end{cases}, \qquad \begin{cases} x_1 = 0 \\ x_2 \text{ arbitrary.} \end{cases}$$

We see that any nonzero value can be taken for x_1 in the first vector and for x_2 in the second. Our eigenvectors are then any multiple of

$$\begin{bmatrix} 4 \\ 1 \end{bmatrix} \quad \text{and} \quad \begin{bmatrix} 0 \\ 1 \end{bmatrix}.$$

We now illustrate the iterative procedure of determining the eigenvalue of largest magnitude, $\lambda = 3$. We start with an arbitrary vector and multiply it by the matrix repeatedly. Using $\begin{bmatrix} 1 \\ 1 \end{bmatrix}$, we have

$$\begin{bmatrix} 3 & 0 \\ 1 & -1 \end{bmatrix} \begin{bmatrix} 1 \\ 1 \end{bmatrix} = \begin{bmatrix} 3 \\ 0 \end{bmatrix} = 3 \begin{bmatrix} 1 \\ 0 \end{bmatrix},$$

$$\begin{bmatrix} 3 & 0 \\ 1 & -1 \end{bmatrix} \begin{bmatrix} 3 \\ 0 \end{bmatrix} = \begin{bmatrix} 9 \\ 3 \end{bmatrix} = 3 \begin{bmatrix} 3 \\ 1 \end{bmatrix},$$

$$\begin{bmatrix} 3 & 0 \\ 1 & -1 \end{bmatrix} \begin{bmatrix} 9 \\ 3 \end{bmatrix} = \begin{bmatrix} 27 \\ 6 \end{bmatrix} = 6 \begin{bmatrix} 4.5 \\ 1 \end{bmatrix},$$

$$\begin{bmatrix} 3 & 0 \\ 1 & -1 \end{bmatrix} \begin{bmatrix} 27 \\ 6 \end{bmatrix} = \begin{bmatrix} 81 \\ 21 \end{bmatrix} = 21 \begin{bmatrix} 3.86 \\ 1 \end{bmatrix},$$

$$\begin{bmatrix} 3 & 0 \\ 1 & -1 \end{bmatrix} \begin{bmatrix} 81 \\ 21 \end{bmatrix} = \begin{bmatrix} 243 \\ 60 \end{bmatrix} = 60 \begin{bmatrix} 4.05 \\ 1 \end{bmatrix},$$

$$\begin{bmatrix} 3 & 0 \\ 1 & -1 \end{bmatrix} \begin{bmatrix} 243 \\ 60 \end{bmatrix} = \begin{bmatrix} 729 \\ 183 \end{bmatrix} = 183 \begin{bmatrix} 3.98 \\ 1 \end{bmatrix},$$

$$\begin{bmatrix} 3 & 0 \\ 1 & -1 \end{bmatrix} \begin{bmatrix} 729 \\ 183 \end{bmatrix} = \begin{bmatrix} 2187 \\ 546 \end{bmatrix} = 546 \begin{bmatrix} 4.01 \\ 1 \end{bmatrix}.$$

We see that the successive values of the product vector approach closer and closer to multiples of the eigenvector $\begin{bmatrix} 4 \\ 1 \end{bmatrix}$. The ratio of corresponding components of two successive products approaches the largest eigenvalue. In this case

$$\lambda = \tfrac{2187}{729} = 3, \qquad \lambda \doteq \tfrac{546}{183} = 2.98.$$

This method is slow to converge if the magnitudes of the largest eigenvalues are nearly the same. If the two eigenvalues of largest magnitude are equal but of opposite sign, we need to adapt the method. We illustrate with

$$\begin{bmatrix} 3 & 0 \\ 1 & -3 \end{bmatrix} \qquad \text{which has} \qquad \lambda_1 = 3, \quad \lambda_2 = -3.$$

Beginning with $\begin{bmatrix} 1 \\ 1 \end{bmatrix}$ again

$$\begin{bmatrix} 1 \\ 1 \end{bmatrix} \rightarrow \begin{bmatrix} 3 \\ -2 \end{bmatrix} \rightarrow \begin{bmatrix} 9 \\ 9 \end{bmatrix} \rightarrow \begin{bmatrix} 27 \\ -18 \end{bmatrix} \rightarrow \begin{bmatrix} 81 \\ 81 \end{bmatrix};$$

ratios are 9, 9.

Here the vector oscillates and does not approach the eigenvector, and the ratios of components of successive products also oscillates. However, the ratio of components of every other product approaches 9, which is λ_1^2. In this case we reach this ratio from the beginning.

With more realistically sized matrices, the iteration will take many steps and requires a computer for its practical execution. The method is known as the *power method* and will converge only when the largest eigenvalue in modulus is uniquely determined.

Program 1

```
ZZJOB 5                              CSC   001     GERALD, C. F.
ZZFORX5
*LIST PRINTER
C   PROGRAM TO FIND THE DOMINANT EIGENVALUE BY ITERATION.
C   PROGRAM WILL HANDLE UP TO 10 BY 10 MATRIX.
      DIMENSION A(10,10), X(10), Y(10), Z(10)
C   READ IN SIZE OF MATRIX. THEN READ IN MATRIX A AND STARTING VECTOR X.
C   COMPONENTS OF A ARE PUNCHED ROW-WISE, 8 PER CARD. X ALSO 8 PER CARD.
      READ 100, N
      DO 5 I = 1,N
    5 READ 101, (A(I,J), J = 1,N)
      READ 101, (X(I), I = 1,N)
C   COMPUTE SUCCESSIVE PRODUCTS, REPEATING 50 TIMES.
C   STORE LAST THREE PRODUCTS IN X, Y, Z ARRAYS.
      DO 20 I = 1,48
      DO 10 J = 1,N
      Y(J) = 0.0
      DO 10 K = 1,N
   10 Y(J) = Y(J) + A(J,K) * X(K)
C   NORMALIZE THE PRODUCT VECTOR AND STORE BACK IN X.
      DO 20 L = 1,N
   20 X(L) = Y(L) / Y(1)
C   COMPUTE THE LAST TWO PRODUCTS. THESE NOT NORMALIZED.
      DO 40 J = 1,N
      Y(J) = 0.0
      DO 40 K = 1,N
   40 Y(J) = Y(J) + A(J,K) * X(K)
      DO 30 J = 1,N
      Z(J) = 0.0
      DO 30 K = 1,N
   30 Z(J) = Z(J) + A(J,K) * Y(K)
C   PRINT OUT LAST THREE PRODUCTS.
      PRINT 102,(X(I),Y(I),Z(I),I=1,N)
      CALL EXIT
  100 FORMAT (I2)
  101 FORMAT (8F10.0)
  102 FORMAT(52H RESULTS OF ITERATIVE METHOD FOR EIGENS OF A MATRIX /
     1 44H 48TH PRODUCT    49TH PRODUCT   50TH PRODUCT /
     2(1H , E14.8, 2E15.8))
      END
04
-5.509882  1.870086   0.008814
0.287865  -11.811654 5.711900   0.058717
0.049099   4.308033  -12.970687 0.229326
0.006235   0.269851   1.397369  -17.596207
10.        0.0        0.0        0.0
ZZZZ

    RESULTS OF ITERATIVE METHOD FOR EIGENS OF A MATRIX
    48TH PRODUCT    49TH PRODUCT    50TH PRODUCT
    .10000000E+01  -.17667620E+02   .31224563E+03
   -.65334549E+01   .11548563E+03  -.20419698E+04
    .68509739E+01  -.12127564E+03   .21474245E+04
   -.18822878E+02   .33902776E+03  -.61040156E+04
```

6. PROGRAMMING CONSIDERATIONS

In Section 1, the shooting method of solving boundary value problems can profit from programs which solve second and higher order initial value problems. Examples of these have been given in Chapter 6; they would be of more universal value if written to accommodate equations of order higher than the second, probably by reducing to a system of simultaneous first-order equations. They would be of special utility if written as subroutines which are called by a main program that inputs the initial estimated values of the assumed initial conditions, and perhaps automatically interpolates to progressively improve the estimates.

Similarly, the programs presented in Chapter 7, especially the program that solves sets of equations whose coefficient matrix is tridiagonal, are adapted to solving boundary value problems by the difference equation method of Sections 2 and 3 of this chapter.

A program to find the dominant eigenvalue of a matrix by the power method is given on page 184 (Program 1). The program does not attempt to measure when the normalized vector products stabilize but rather employs the power method an arbitrary 50 times. The 48th, 49th, and 50th products are printed. The dominant eigenvalue will be the square root of the ratio of corresponding components of the 50th and 48th vectors. In the example given, $\lambda \doteq -\sqrt{312.2} = -17.66$. The validity of this value is indicated by how near to this same quantity the ratio of the various corresponding components are. The 49th product is of value to show whether there is a second eigenvalue of the same magnitude but of opposite sign, or to indicate the sign of the dominant eigenvalue when its magnitude is uniquely determined. In the example shown, the sign of the dominant eigenvalue is negative and its magnitude is uniquely determined. The 48th product, or better yet, the normalized 50th product, gives an estimate of the eigenvector corresponding to the dominant eigenvalue.

PROBLEMS

Section 1

1. Solve the boundary value problem

$$y'' + xy' - 3y = 4.2x,$$
$$y(0) = 0, \quad y(1) = 1.9$$

   by the shooting method, assuming values of the initial slope, which is near unity. Use $h = 0.25$ and use the modified Euler method, carrying three decimals. Compare to the analytical solution $y = x^3 + 0.9x$.

2. Use the shooting method to solve

$$y'' - yy' = e^x,$$
$$y(0) = 1, \quad y(1) = -1.$$

   Use the computer program of Chapter 6 or another one that you have written to solve the problem with assumed values of the initial slope.

3. Write a computer program which, after making calculations with two initial guesses of the slope, automatically interpolates to get an improved estimate of the slope for the next calculation. The program should terminate when the end condition is reached within a given tolerance.

4. The shooting method can work backwards through the interval as well as forward. Solve Problem 1 by moving backward from $x = 1$ (Δx is then negative), and use an assumed value for the slope at $x = 1$.

Section 2

5. Given the boundary problem

$$\frac{d^2y}{d\theta^2} + y = 0, \qquad y(0) = 0, \quad y(\pi/2) = 1.$$

(a) Normalize to the interval [0,1] by an appropriate change of variable.

(b) Solve the normalized problem by the finite difference method, replacing the derivative by a difference quotient of error $O(h^2)$, and then solving the set of equations. Use $h = 0.25$.

(c) Solve the original equation (without normalizing) by finite differences. Take $\Delta\theta = \pi/8$.

(d) Compare your solutions to the analytical solution $y = \sin\theta$.

(e) Repeat (a) with $h = 0.5$, and extrapolate the value at the midpoint.

6. Solve the boundary value problem

$$y'' + xy' - xy = 2x, \qquad y(0) = 1, \quad y(1) = 0.$$

Take $h = 0.2$

7. Solve Problem 1 by replacing the derivatives with finite difference quotients.

8. Solve Problem 2 by rewriting the differential equation as a set of difference equations.

Section 3

9. Repeat Problem 5, except for the boundary value problem with derivative conditions:

$$\frac{d^2y}{d\theta^2} + y = 0, \qquad y'(0) = 0, \quad y'(\pi/2) = 1.$$

In part (d), compare to $y = -\cos\theta$.

10. Solve

$$\frac{d^2y}{d\theta^2} + y = 0, \qquad y'(0) + y(0) = 1, \quad y'\left(\frac{\pi}{2}\right) + y\left(\frac{\pi}{2}\right) = 0.$$

Subdivide the interval into four equal parts.

11. The most general form of conditions that can be specified for a second-order boundary value problem is a linear combination of the function and its derivative at both ends of the interval. Set up the equations to solve

$$y'' = xy' + x^2y = x^3, \qquad \begin{cases} y(0) + y'(0) - y(1) - y'(1) = 1, \\ y(0) - y'(0) - y(1) + y'(1) = 2. \end{cases}$$

Use $h = \frac{1}{3}$.

12. Solve the third-order boundary value problem

$$y''' - y' = e^x, \qquad y(0) = 0, \quad y(1) = 1, \quad y'(1) = 0.$$

Use $h = 0.2$. Approximate the third derivative in terms of the average third central difference

$$y_0''' = \frac{f_2 - 2f_1 + 2f_{-1} - f_{-2}}{2h^3} + O(h^2)$$

at $x = 0.4, 0.6, 0.8$. Using the derivative condition will eliminate the assumed function value at $x = 1.2$, but we are short one equation. Obtain this by writing the equation at $x = 0.2$ using the unsymmetrical approximation for y''':

$$y_0''' = \frac{-f_3 + 6f_2 - 12f_1 + 10f_0 - 3f_{-1}}{2h^3} + O(h^2).$$

13. Derive the unsymmetrical approximation for y_0''' given in Problem 12. The method of undetermined coefficients (Chapter 5) may be used, or one may differentiate an interpolation polynomial three times and then set $S = 0$. In the latter, be sure the polynomial fits at the correct points.

14. Solve the fourth-order problem

$$y^{iv} - y''' + y = x^2, \quad y(0) = 0, \quad y'(0) = 0, \quad y(1) = 2, \quad y'(1) = 0.$$

Use symmetrical expressions for the derivatives.

15. Solve the differential equation in Problem 14 subject to the conditions

$$y(0) = y'(0) = y''(0) = 0, \quad y(1) = 2.$$

Unsymmetrical expressions will be required.

Section 4

16. Consider the characteristic value problem with k restricted to real values:

$$\frac{d^2y}{dx^2} - k^2y = 0, \quad y(0) = 0, \quad y(1) = 0.$$

(a) Show analytically that there is no solution except the trivial solution $y = 0$.

(b) Show, by setting up the set of difference equations corresponding to the differential equation with $h = 0.2$, that there are no real values of k for which a solution to the difference equations exists.

17. Solve the characteristic value problem by the method of finite differences:

$$y'' + 2y' + k^2y = 0, \quad y(0) = y(1) = 0.$$

Compare the principal eigenvalue to $k = \pm\sqrt{1 + \pi^2} = \pm 3.297$.

(a) Use $h = \frac{1}{2}$.

(b) Use $h = \frac{1}{3}$.

(c) Use $h = \frac{1}{4}$.

(d) Assuming errors are proportional to h^2, extrapolate from a and c, b and c to improved values for the principal eigenvalue.

18. Find the principal eigenvalue of

$$y'' + k^2x^2y = 0, \quad y(0) = y(1) = 0.$$

19. Using the principal value, $k = 3.297$, in Problem 17, find y as a function of x over the interval $[0,1]$. This function is the corresponding eigenfunction.

20. The second eigenvalue in Problem 17 is $k = \sqrt{1 + 4\pi^2} = 6.3623$. Find the corresponding eigenfunction.

Section 5

21. Find the larger eigenvalue and the corresponding eigenvector by the iteration method:

  (a) $\begin{bmatrix} 1 & 0 \\ 1 & 1 \end{bmatrix}$

  (b) $\begin{bmatrix} 1 & 2 \\ 3 & 4 \end{bmatrix}$

  (c) $\begin{bmatrix} 1 & 1 \\ 1 & -1 \end{bmatrix}$ (the two eigenvalues are equal but of opposite sign)

22. Find the dominant eigens:
$$\begin{bmatrix} 1 & 0 & 2 \\ 0 & 1 & 3 \\ 1 & 1 & 1 \end{bmatrix}.$$

23. Use iteration to find the eigenvalues of the matrices for Problem 17(a), (b), (c).

9
numerical
solution of
elliptic partial
differential
equations

Physical phenomena that are a function of more than one independent variable are conveniently represented by a partial differential equation, by which term we mean an equation involving partial derivatives. Most scientific problems have mathematical models which are second-order equations with the highest order of derivative being the second. If u is a function of the two independent variables x and y, there are three second-order partial derivatives

$$\frac{\partial^2 u}{\partial x^2}, \quad \frac{\partial^2 u}{\partial x\,\partial y}, \quad \frac{\partial^2 u}{\partial y^2}.$$

(We shall treat only of functions for which the order of differentiation is unimportant, so $\partial^2 u/\partial y\,\partial x = \partial^2 u/\partial x\,\partial y$.) Depending on the values of the coefficients of the second derivative terms, partial differential equations are classified as elliptic, parabolic, or hyperbolic. The most important distinctions to us are the kinds of problems and the nature of boundary conditions which lead to one type of equation or another. Steady-state potential distribution problems fall in the class of elliptic equations, which we discuss first.

1. EQUATION FOR STEADY-STATE HEAT FLOW

We derive the relationship for temperature, u, as a function of the two space variables, x and y, for the equilibrium temperature distribution on a flat plate (Fig. 1.1). Consider the element of the plate whose surface area is $dx\,dy$. We assume that heat flows only in the x- and y-directions and not in the perpendicular direction. If the plate is very thin, or if the upper and lower surfaces are both well insulated, the physical situation will agree with our assumption. Let t be the thickness of the plate.

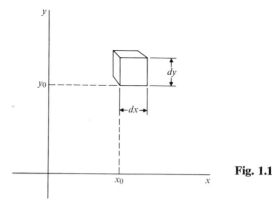

Fig. 1.1

Heat flows at a rate proportional to the cross-sectional area, to the temperature gradient ($\partial u/\partial x$ or $\partial u/\partial y$), and to the thermal conductivity k, which we shall assume constant at all points. The flow of heat is from high to low temperature, of course, meaning opposite to the direction of increasing gradient. We use a minus sign in the equation to account for this:

Rate of heat flow into element at $x = x_0$:

$$-k(t\ dy)\frac{\partial u}{\partial x}.$$

The gradient at $x_0 + dx$ is the gradient at x_0 plus the increment in gradient over the distance dx:

Gradient at $x_0 + dx$:

$$\frac{\partial u}{\partial x} + \frac{\partial}{\partial x}\left(\frac{\partial u}{\partial x}\right)dx,$$

Rate of heat flow out of element at $x = x_0 + dx$:

$$-k(t\ dy)\left[\frac{\partial u}{\partial x} + \frac{\partial}{\partial x}\left(\frac{\partial u}{\partial x}\right)dx\right],$$

Net rate of heat flow into element in x-direction:

$$-k(t\ dy)\left[\frac{\partial u}{\partial x} - \left(\frac{\partial u}{\partial x} + \frac{\partial}{\partial x}\left(\frac{\partial u}{\partial x}\right)dx\right)\right] = -kt\ dx\ dy\ \frac{\partial^2 u}{\partial x^2}.$$

Similarly, in the y-direction we have the following.

Net rate of heat flow into element in y-direction:

$$-k(t\ dx)\left[\frac{\partial u}{\partial y} - \left(\frac{\partial u}{\partial y} + \frac{\partial}{\partial y}\left(\frac{\partial u}{\partial y}\right)dy\right)\right] = -kt\ dx\ dy\ \frac{\partial^2 u}{\partial y^2}.$$

If there is equilibrium as to temperature distribution, i.e., steady state, there will be no net flow into the element; hence

$$-kt\ dx\ dy\left(\frac{\partial^2 u}{\partial x^2} + \frac{\partial^2 u}{\partial y^2}\right) = 0,$$

$$\frac{\partial^2 u}{\partial x^2} + \frac{\partial^2 u}{\partial y^2} = \nabla^2 u = 0.$$

(1.1)

The operator

$$\nabla^2 = \left(\frac{\partial^2}{\partial x^2} + \frac{\partial^2}{\partial y^2}\right)$$

is called the *Laplacian*, and Eq. (1.1) is called *Laplace's equation*. For three-dimensional heat flow problems, we would have analogously,

$$\nabla^2 u = \left(\frac{\partial^2}{\partial x^2} + \frac{\partial^2}{\partial y^2} + \frac{\partial^2}{\partial z^2}\right)u = 0.$$

(1.2)

Equation (1.2), which has been derived with reference to heat flow, applies as well to steady-state diffusion problems (where u is now concentration of material) and to electrical potential distribution (where u is electromotive force); in fact

Laplace's equation holds for the steady-state distribution of the potential of any quantity where the rate of flow is proportional to the gradient, and where the proportionality constant does not vary with position or the value of u. In our examples, we shall generally use terminology corresponding to heat flow as being more closely related to the average student's everyday experience.

In the above, we have tacitly assumed that no heat was being generated at points in the plate, as by electric heaters imbedded in it, nor any removal, as by cooling coils. In the presence of such "sources" or "sinks," we would not equate the net flow into the element to zero, as in Eq. (1.1), but the net flow into the element of volume would equal the net rate of heat removal from the element, if steady state applies. Assuming this removal rate to be a function of the location of the element in the xy-plane, $f(x,y)$, we would have

$$\nabla^2 u = f(x,y). \tag{1.3}$$

This equation is called *Poisson's equation*. Our numerical methods of solving elliptic differential equations apply equally well to both Laplace's and Poisson's equations. Analytical methods find the latter type considerably more difficult. These two equations include most of the physical applications of elliptic partial differential equations.

Equation (1.2) or Eq. (1.3) describes how u varies within the interior of a closed region. The function u is determined on the boundaries of the region by boundary conditions. The boundary conditions may specify the value of u at all points on the boundary, or some combination of the potential and the normal derivative.

2. REPRESENTATION AS A DIFFERENCE EQUATION

Our scheme for solving all kinds of partial differential equations will be to replace the derivatives by difference quotients, converting the equation to a difference equation. We then write the difference equation corresponding to each point at the intersections (nodes) of a gridwork that subdivides the region of interest at which the function values are unknown. Solving these equations simultaneously gives values for the function at each node which approximate the true values. We begin with the two-dimensional case.

Approximating derivatives by difference quotients was the subject matter of Chapters 4 and 5. We rederive the few relations we need independently.

Let $h = \Delta x = $ equal spacing of gridwork in the x-direction. We assume that the function $f(x)$ has a continuous fourth derivative. By Taylor series,

$$f(x_n + h) = f(x_n) + f'(x_n)h + \frac{f''(x_n)}{2}h^2 + \frac{f'''(x_n)}{6}h^3 + \frac{f^{iv}(\xi_1)}{24}h^4,$$

$$x_n < \xi_1 < x_n + h,$$

$$f(x_n - h) = f(x_n) - f'(x_n)h + \frac{f''(x_n)}{2}h^2 - \frac{f'''(x_n)}{6}h^3 + \frac{f^{iv}(\xi_2)}{24}h^4,$$

$$x_n - h < \xi_2 < x_n.$$

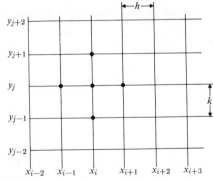

Fig. 2.1

It follows that

$$\frac{f(x_n + h) - 2f(x_n) + f(x_n - h)}{h^2} = f''(x_n) + \frac{f^{iv}(\xi)}{12} h^2, \qquad x_n - h < \xi < x_n + h.$$

A subscript notation is convenient:

$$\frac{f_{n+1} - 2f_n + f_{n-1}}{h^2} = f_n'' + O(h^2). \tag{2.1}$$

In Eq. (2.1) the subscripts on f indicate the x-values at which it is evaluated. The order relation $O(h^2)$ signifies that the error approaches proportionality to h^2 as $h \to 0$.

Similarly, the first derivative is approximated,

$$\frac{f(x_n + h) - f(x_n - h)}{2h} = f'(x_n) + \frac{f'''(\xi)}{6} h^2, \qquad x_n - h < \xi < x_n + h,$$

$$\frac{f_{n+1} - f_{n-1}}{2h} = f_n' + O(h^2). \tag{2.2}$$

When f is a function of both x and y, we get the second partial derivative with respect to x, $\partial^2 u/\partial x^2$, by holding y constant and evaluating the function at three points where x equals x_n, $x_n + h$, and $x_n - h$. The partial derivative $\partial^2 u/\partial y^2$ is similarly computed, holding x constant. We require that fourth derivatives with respect to both variables exist.

To solve the Laplace equation on a region in the xy-plane, we subdivide the region with equispaced lines parallel to the x- and y-axes. Let $\Delta x = h$ and $\Delta y = k$. Consider a portion of the region near (x_i, y_j). (see Fig. 2.1):

$$\frac{\partial^2 u}{\partial x^2} + \frac{\partial^2 u}{\partial y^2} = 0.$$

Replacing the derivatives by difference quotients which approximate the derivatives at the point (x_i, y_j), we get

$$\nabla^2 u(x_i, y_j) = \frac{u(x_{i+1}, y_j) - 2u(x_i, y_j) + u(x_{i-1}, y_i)}{h^2}$$

$$+ \frac{u(x_i, y_{j+1}) - 2u(x_i, y_j) + u(x_i, y_{j-1})}{k^2} = 0.$$

Again, it is convenient to let double subscripts on u indicate the x- and y-values:

$$\nabla^2 u_{i,j} = \frac{u_{i+1,j} - 2u_{i,j} + u_{i-1,j}}{h^2} + \frac{u_{i,j+1} - 2u_{i,j} + u_{i,j-1}}{k^2} = 0.$$

It is common to take $h = k$, resulting in considerable simplification, whence

$$\nabla^2 u_{i,j} = \frac{1}{h^2} [u_{i+1,j} + u_{i-1,j} + u_{i,j+1} + u_{i,j-1} - 4u_{i,j}] = 0. \tag{2.3}$$

Note that five points are involved in the relationship of Eq. (2.3), points to the right, left, above, and below the central point (x_i, y_j). It is convenient to represent the relationship pictorially, where the linear combination of u's is represented symbolically. Equation (2.3) becomes

$$\nabla^2 u_{i,j} = \frac{1}{h^2} \left\{ \begin{matrix} & 1 & \\ 1 & -4 & 1 \\ & 1 & \end{matrix} \right\} u_{i,j} = 0.$$

The representation of the Laplacian operator by the pictorial operator is fundamental to our work in elliptic partial differential equations. The approximation has $O(h^2)$ error, provided that u is sufficiently smooth.

3. LAPLACE'S EQUATION ON A SQUARE REGION

Let us solve for the steady-state temperatures on a square plate whose side is 1 foot long. The plate is made of steel ($k = 0.16$ cal/sec · cm^2 · °C/cm) and the top and right-hand edges are held at 100°C, while the other edges are at 0°C. Subdivide the plate into 3-inch squares, so there are nine nodes or intersections on the interior of the region (Fig. 3.1). (We would normally subdivide more finely to get better approximations of the temperatures, but the present example is illustrative of the technique.)

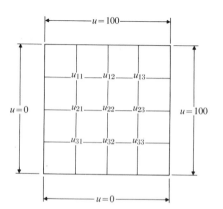

Fig. 3.1

Write the equation

$$u_{i,j} = \frac{1}{h^2}\left\{ \begin{array}{ccc} & 1 & \\ 1 & -4 & 1 \\ & 1 & \end{array} \right\} u_{i,j} = 0, \tag{3.1}$$

with $i = 1, 2, 3$, $j = 1, 2, 3$, a total of nine equations:*

$$\begin{aligned}
0 \;+\; 100 + u_{12} + u_{21} - 4u_{11} &= 0, \\
u_{11} + 100 + u_{13} + u_{22} - 4u_{12} &= 0, \\
u_{12} + 100 + 100 + u_{23} - 4u_{13} &= 0, \\
0 \;+\; u_{11} + u_{22} + u_{31} - 4u_{21} &= 0, \\
u_{21} + u_{12} + u_{23} + u_{32} - 4u_{22} &= 0, \\
u_{22} + u_{13} + 100 + u_{33} - 4u_{23} &= 0, \\
0 \;+\; u_{21} + u_{32} + 0 \;-\; 4u_{31} &= 0, \\
u_{31} + u_{22} + u_{33} + 0 \;-\; 4u_{32} &= 0, \\
u_{32} + u_{23} + 100 + 0 \;-\; 4u_{33} &= 0.
\end{aligned} \tag{3.2}$$

Note that neither the thermal properties of the plate nor its size enter into the computations. The equation is independent of scale. Also note that the ambiguous corner temperatures (where 100° and 0° conditions intersect) also do not enter.

The problem now reduces to solving this set of simultaneous linear equations. There are exactly enough equations to solve for the nine unknowns. We may use any of the methods of Chapter 7.

If we plan to use the elimination method, it is to our advantage to observe that symmetry gives $u_{12} = u_{23}$, $u_{11} = u_{33}$, $u_{21} = u_{32}$, so we can reduce the number of equations to six. The augmented coefficient matrix of Eqs. (3.2) becomes

(u_{11})	(u_{12})	(u_{13})	(u_{21})	(u_{22})	(u_{31})	
-4	1		1			-100
1	-4	1		1		-100
	2	-4				-200
1			-4	1	1	0
	2		2	-4		0
			2		-4	0

$$ \tag{3.3}$$

Solving by Gaussian elimination gives the following to two decimals:

$$\begin{aligned}
u_{11} = u_{33} &= 50.00, \\
u_{12} = u_{23} &= 71.43, \\
u_{13} &= 85.72, \\
u_{21} = u_{32} &= 28.57, \\
u_{22} &= 50.00, \\
u_{31} &= 14.29.
\end{aligned}$$

---

* It is usually best to adopt a consistent pattern of writing the $u_{i,j}$ to avoid mistakes. The author prefers a clockwise pattern starting at 9 o'clock.

As we carry out the solution we find the equations are such as to give pivoting without rearrangement.

The relatively large number of zero coefficients suggests that iterative methods are well suited. The Gauss-Seidel method would select the diagonal variables as the ones to be solved for, and since the coefficients of the diagonal variables are dominant, we know it will be convergent. The rearranged equations are

$$
\begin{aligned}
u_{11} &= \tfrac{1}{4}(100 + u_{12} + u_{21}), \\
u_{12} &= \tfrac{1}{4}(100 + u_{11} + u_{13} + u_{22}), \\
u_{13} &= \tfrac{1}{4}(200 + 2u_{12}), \\
u_{21} &= \tfrac{1}{4}(0 + u_{11} + u_{22} + u_{31}), \\
u_{22} &= \tfrac{1}{4}(0 + 2u_{12} + 2u_{21}), \\
u_{31} &= \tfrac{1}{4}(0 + 2u_{21}).
\end{aligned}
\tag{3.4}
$$

Iterative solution in the form of Eqs. (3.4) is termed *Liebmann's method.* The pictorial operator relation, Eq. (3.1), gives the set of equations directly, without going through the steps of (3.2) and (3.3):

$$
\frac{1}{h^2}\left\{\begin{matrix} & 1 & \\ 1 & -4 & 1 \\ & 1 & \end{matrix}\right\} u_{i,j} = \frac{1}{h^2}\left\{\begin{matrix} & 1 & \\ 1 & 0 & 1 \\ & 1 & \end{matrix}\right\} u_{i,j} - 4u_{i,j} = 0.
$$

Multiplying by h^2 and rearranging, we have

$$
u_{i,j} = \tfrac{1}{4}\left\{\begin{matrix} & 1 & \\ 1 & 0 & 1 \\ & 1 & \end{matrix}\right\} u_{i,j}.
\tag{3.5}
$$

Equations (3.4) result by letting i and j vary over all interior points with unique temperatures.

In solving by iteration, if we can discover good estimates to start the solution, we will converge rapidly. One scheme is to subdivide the region very coarsely at first. Imagine there to be only one interior point, u_{22}; we immediately write $u_{22} = 0.25(0 + 100 + 100 + 0) = 50$. (By coincidence we get the same value as with nine points.) With this value for u_{22} we can "eyeball" the rest: $u_{21} = u_{32} \doteq 25$, $u_{12} = u_{23} \doteq 75$, $u_{11} = u_{33} \doteq 50$, $u_{13} \doteq 87$, $u_{31} \doteq 12$.

The easiest way to perform the computation by hand is to lay out the figures on a diagram of the region so the position on the paper indicates at which point u is being evaluated. Figure 3.2 shows the computation from the above starting values. We revise each estimate of u by replacing the current estimate with the arithmetic average of its four closest neighbors, according to Eq. (3.5). Any consistent pattern is satisfactory. (Though some patterns will converge more rapidly, it is nearly impossible to predict which are best.) In Fig. 3.2, we have adopted a pattern from left to right in each row, moving from top to bottom among the rows.

It is of advantage to utilize symmetry by changing each symmetrical point when a change to its counterpart is recalculated. This was done in the example.

```
           100          100          100
  0         50           75           87          100
            50           71.75        85.88
            49.94        71.42        85.71
            49.97        71.40        85.70
            49.98        71.41        85.70
            49.99        71.42        85.71

  0         25           50           75          100
            28           49.88        71.75
            28.46        49.94        71.42
            28.53        49.96        71.40
            28.55        49.98        71.41
            28.56        49.99        71.42

            12           25           50
  0         14           28           50          100
            14.23        28.46        49.94
            14.26        28.53        49.97
            14.28        28.55        49.98
            14.28        28.56        49.99

             0            0            0
```

Fig. 3.2. Liebmann's method.

Because our starting values were close, three-significant-figure accuracy comes quickly (after about three circuits have been made), but convergence in the second decimal place comes with some agony—two or three more rounds of calculation will be required beyond the point of Fig. 3.2 to complete the computations.

The Liebmann method is well suited for computer programming (see the example at the end of the chapter).   For hand computation, as we have just seen, it converges only slowly, and we might hope for a more rapid technique.

4. THE METHOD OF RELAXATION

The method of relaxation is a more rapid method than Liebmann's, although it is not as suitable for computer application.   To establish the basis for this scheme, we look again at Eq. (3.1).   If we start with initial estimates of $u_{i,j}$, the linear combination of Eq. (3.1) will not normally add to zero, but there will be a nonzero total, termed the *residual*, whose magnitude is an indication of the errors in u at the five points being considered, especially for the central point, since its coefficient is -4.   We use this fact to establish where in the region our initial estimates are most faulty, and improve these first, rather than to use a fixed pattern as in Liebmann's method.   Thereby we get accelerated convergence.

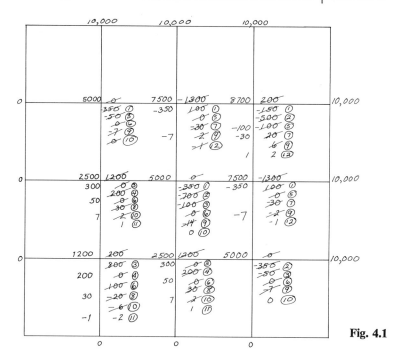

Fig. 4.1

We start the calculation by laying out the initial estimates for the function on a sketch of the region. Residuals are calculated at each point according to

$$\left\{\begin{matrix} & 1 & \\ 1 & -4 & 1 \\ & 1 & \end{matrix}\right\} u_{i,j} = R_{i,j}.$$

We now change the value of $u_{i,j}$ at the point whose residual is greatest (point of probable greatest error). This causes corresponding changes (relaxations) in the residuals at the point in question as well as at all interior points which are adjacent. A change of Δu at a given point increases its own residual by $-4\Delta u$; the adjacent residuals are increased by Δu. (If the residual at a point is -60, a change of -15 in the value of u will relax this residual to zero, and add -15 to the adjacent residuals.) One proceeds next to the point which now has the greatest residual, and continues in like manner until the residuals are sufficiently small.

Figure 4.1 illustrates the method. The values of $u_{i,j}$ are to the left, the $R_{i,j}$ to the right. It is common to multiply all values of the function and the residuals by a power of 10 so as to be able to use integer arithmetic. Fractions are then ignored. Residuals are carried as cumulative totals, but only the increments of u are written. After all residuals have been relaxed to a magnitude of two or less, the u-values are accurate to the nearest whole number. One sums the original u estimates with the increments to get the final result.

In Fig. 4.1, circled numbers indicate the effect of each step. After steps 1 and 2, we observe that two symmetrical points combine to have a doubled effect on the residual of a point which is a neighbor to both. In steps 3, 6, 9, 10, and 13, we have made changes to both of the symmetrical pairs at once.

The art of relaxation has been highly developed.* In our example we relaxed each residual to near zero only—relaxing to exactly zero is pointless since future changes of the neighboring points will introduce new residuals. The skilled operator knows when to over-relax and when to under-relax in anticipation of future changes. Relaxation schemes to handle several points simultaneously (group and block relaxation) have been devised.

This method is much less important now than it was a few years ago because it does not lend itself to computer programming. The reason for this is that selecting the largest (or near largest) residual, while easy for the human by scanning his figures, is slow on the computer. The time lost in the search routine generally offsets the time saved by more rapid convergence. Consequently, elliptic partial differential equations are normally programmed using Liebmann's method. For hand computation, the relaxation scheme is useful. One may wish to use it to get fair starting values for input to a computer program, for example.

5. THE POISSON EQUATION

The methods of the previous section are readily applied to Poisson's equation. We illustrate with an analysis of torsion in a rectangular bar subject to twisting. The torsion function ϕ satisfies the Poisson equation:

$$\nabla^2\phi + 2 = 0, \qquad \phi = 0 \text{ on boundary.} \tag{5.1}$$

The tangential stresses are proportional to the partial derivatives of ϕ for a twisted prismatic bar of constant cross section. Let us find ϕ over the cross section of a rectangular bar 6 by 8 inches in size.

Subdivide the cross section into 2-inch squares, so that there are six interior points as in Fig. 5.1. In terms of difference quotients, Eq. (5.1) becomes

$$\frac{1}{h^2}\left\{ \begin{matrix} & 1 & \\ 1 & -4 & 1 \\ & 1 & \end{matrix} \right\} \phi_{i,j} + 2 = 0,$$

or $\tag{5.2}$

$$\frac{1}{4}\left\{ \begin{matrix} & 1 & \\ 1 & -4 & 1 \\ & 1 & \end{matrix} \right\} \phi_{i,j} + 2 = 0.$$

(The function ϕ is dimensional; with our choice of h, ϕ will have square inches as units.)

---

* See, for example, Allen (1954).

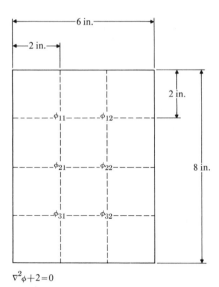

Fig. 5.1

$\nabla^2\phi + 2 = 0$

The set of equations, when (5.2) is applied at each interior point, is

$$
\begin{aligned}
0 \quad + 0 \quad + \phi_{12} + \phi_{21} - 4\phi_{11} + 8 &= 0, \\
\phi_{11} + 0 \quad + 0 \quad + \phi_{22} - 4\phi_{12} + 8 &= 0, \\
0 \quad + \phi_{11} + \phi_{22} + \phi_{31} - 4\phi_{21} + 8 &= 0, \\
\phi_{21} + \phi_{12} + 0 \quad + \phi_{32} - 4\phi_{22} + 8 &= 0, \\
0 \quad + \phi_{21} + \phi_{32} + 0 \quad - 4\phi_{31} + 8 &= 0, \\
\phi_{31} + \phi_{22} + 0 \quad + 0 \quad - 4\phi_{32} + 8 &= 0.
\end{aligned}
\tag{5.3}
$$

Symmetry considerations show that $\phi_{11} = \phi_{12} = \phi_{31} = \phi_{32}$ and $\phi_{21} = \phi_{22}$, so only two unknowns are left in (5.3) after substitutions:

$$
\begin{aligned}
\phi_{21} - 3\phi_{11} + 8 &= 0, \\
2\phi_{11} - 3\phi_{21} + 8 &= 0.
\end{aligned}
$$

Obviously these would be solved by elimination; $\phi_{11} = 4.56$, $\phi_{21} = 5.72$.

Let us now solve the problem with a 1-inch-square mesh; we can use the values just calculated to help us guess at starting values. We shall get different values at the same interior locations as in Fig. 5.1, because we will reduce our $O(h^2)$ error. Liebmann's method will be, with $h = 1$,

$$
\phi_{i,j} = \tfrac{1}{4}(\phi_{i+1,j} + \phi_{i-1,j} + \phi_{i,j+1} + \phi_{i,j-1} + 2).
$$

Starting with reasonable initial estimates we iterate until successive values agree.

In Fig. 5.2 we show the results of applying relaxation. Note that there is symmetry about both the vertical axis and the horizontal axis. This causes the points on these axes to be affected by two symmetrical points changing in unison. As is normal with such symmetry, we work with only a quarter of the figure.

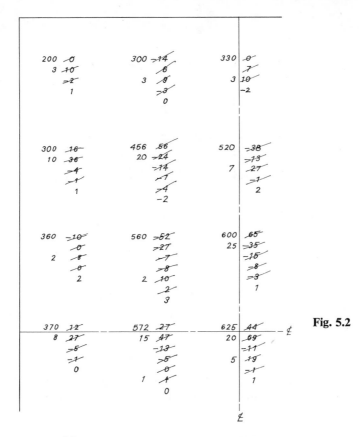

Fig. 5.2

Residuals are computed by

$$\frac{1}{12}\left\{\begin{matrix} & 1 & \\ 1 & -4 & 1 \\ & 1 & \end{matrix}\right\}\phi_{i,j} + 2 = R_{i,j}.$$

In Fig. 5.2 we have multiplied all values of ϕ and R by 100, anticipating two-decimal accuracy.

Note that after the initial residuals are computed, the relaxation pattern is identical to that for Laplace's equation.

It is important to recompute the residuals at the end of the calculations to check the work. It is easy to make arithmetic errors, or to overlook changing a residual, with this method.

6. DERIVATIVE BOUNDARY CONDITIONS

In Section 3, we solved for the steady-state temperatures at interior points of a square plate, with the boundary temperatures being specified. In many problems, instead of knowing the boundary temperatures, we know the temperature gradi-

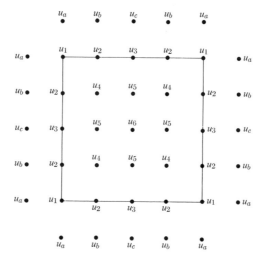

Fig. 6.1

ent in the direction normal to the boundary, as for example when heat is being lost from the surface by radiation and conduction. Consider a square plate within which heat is being uniformly generated at each point at a rate of

$$Q \text{ cal/cm}^3 \cdot \text{sec.}$$

The plate is of steel for which $k = 0.16$ cal/sec $\cdot$ cm^2 $\cdot$ °C/cm, and is 8 by 8 cm, 1 cm thick. For this situation Poisson's equation holds in the form

$$\nabla^2 u = -\frac{Q}{k} = \frac{1}{h^2} \left\{ \begin{matrix} & 1 & \\ 1 & -4 & 1 \\ & 1 & \end{matrix} \right\} u_{i,j}. \tag{6.1}$$

Suppose that $Q = 10$ for our example. The top and bottom faces are insulated so no heat is lost, and heat is lost from the edges of the plate so that

$$\frac{\partial u}{\partial x} = \frac{\partial u}{\partial y} = 1.5 \frac{°C}{cm},$$

taking the proper outward normal derivative for each edge.

In Fig. 6.1 we sketch the plate, with a gridwork to give nine interior points with $h = 2$ cm. In this problem, the edge temperatures are also unknown, so there is a total of 25 points at which u is to be determined. Because of fourfold symmetry, the set of equations can be written in terms of six quantities; these are indicated on the diagram. Note that only one octant of the region is required to show the six different variables.

We now write Eq. (6.1) at each unknown point, which is our general procedure in all elliptic partial differential equations, except that at the boundary points we cannot form the five-point combination—there are not enough points to form the star. We get around this by the device of extending our network to a row of exterior points. We utilize these fictitious exterior points to include in the set of equations those for which the central point is on the edge of the square:

$$\frac{1}{2^2}\left(u_a + u_a + u_2 + u_2 - 4u_1\right) = -\frac{10}{0.16},$$

$$\frac{1}{2^2}\left(u_1 + u_b + u_3 + u_4 - 4u_2\right) = -\frac{10}{0.16},$$

$$\frac{1}{2^2}\left(u_2 + u_c + u_2 + u_5 - 4u_3\right) = -\frac{10}{0.16},$$ (6.2)

$$\frac{1}{2^2}\left(u_2 + u_2 + u_5 + u_5 - 4u_4\right) = -\frac{10}{0.16},$$

$$\frac{1}{2^2}\left(u_4 + u_3 + u_4 + u_6 - 4u_5\right) = -\frac{10}{0.16},$$

$$\frac{1}{2^2}\left(u_5 + u_5 + u_5 + u_5 - 4u_6\right) = -\frac{10}{0.16}.$$

It would appear that we have not helped ourselves by the fictitious points outside the square. We have enabled six equations to be written, but three new unknowns have been introduced. However, we have not yet utilized the boundary conditions. Write the derivatives as difference quotients as discussed in Section 1:

$$\left(\frac{\partial u}{\partial y}\right)_1 = \frac{u_2 - u_a}{2(2)} = 1.5, \qquad u_a = u_2 - 6,$$

$$\left(\frac{\partial u}{\partial y}\right)_2 = \frac{u_4 - u_b}{2(2)} = 1.5, \qquad u_b = u_4 - 6,$$ (6.3)

$$\left(\frac{\partial u}{\partial y}\right)_3 = \frac{u_5 - u_c}{2(2)} = 1.5, \qquad u_c = u_5 - 6.$$

We write these approximations for the gradient condition choosing the proper order of points so the outward normal derivative is positive. Using a central difference approximation of error $O(h^2)$ makes these compatible with our other difference quotient approximations.

We solve the set of equations in (6.2) after eliminating the fictitious points using (6.3). Elimination, iteration (Liebmann's method), or relaxation may be used. The solution of the equations is left as an exercise.

We can apply this same method to the more general boundary condition,

$$au - b\frac{\partial u}{\partial x} = c,$$

a, b, and c are constants, in an obvious application of the relationships of (6.3).

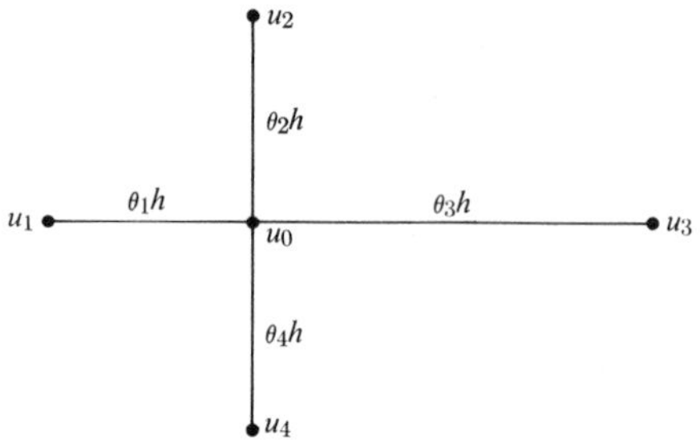

Fig. 7.1

7. IRREGULAR REGIONS

When the boundary of the region is not such that the network can be drawn to have the boundary coincide with the nodes of the mesh, we must proceed differently at points near the boundary. Consider the general case of a group of five points whose spacing is nonuniform, arranged in an unequal armed star. We represent each distance by $\theta_i h$, where θ_i is the fraction of the standard spacing h, which the particular distance represents (Fig. 7.1). Along the line from u_1 to u_0 to u_3, we may approximate the first derivatives:

$$\left(\frac{\partial u}{\partial x}\right)_{1-0} \doteq \frac{u_0 - u_1}{\theta_1 h}; \quad \left(\frac{\partial u}{\partial x}\right)_{0-3} \doteq \frac{u_3 - u_0}{\theta_3 h}.$$

Since

$$\frac{\partial^2 u}{\partial x^2} = \frac{\partial}{\partial x}\left(\frac{\partial u}{\partial x}\right),$$

we have

$$\frac{\partial^2 u}{\partial x^2} \doteq \frac{(u_3 - u_0)/\theta_3 h - (u_0 - u_1)/\theta_1 h}{\frac{1}{2}(\theta_1 + \theta_3)h} = \frac{2}{h^2}\left[\frac{u_1 - u_0}{\theta_1(\theta_1 + \theta_3)} + \frac{u_3 - u_0}{\theta_3(\theta_1 + \theta_3)}\right]. \quad (7.1)$$

Similarly,

$$\frac{\partial^2 u}{\partial y^2} \doteq \frac{2}{h^2}\left[\frac{u_2 - u_0}{\theta_2(\theta_2 + \theta_4)} + \frac{u_4 - u_0}{\theta_4(\theta_2 + \theta_4)}\right]. \quad (7.2)$$

The expressions in (7.1) and (7.2) have errors $O(h)$, which introduces larger errors in the computations than for points which are arranged in an equal armed star. Combining, we get

$$\nabla^2 u = \frac{\partial^2 u}{\partial x^2} + \frac{\partial^2 u}{\partial y^2} = \frac{2}{h^2}\left[\frac{u_1}{\theta_1(\theta_1 + \theta_3)} + \frac{u_2}{\theta_2(\theta_2 + \theta_4)}\right.$$

$$\left. + \frac{u_3}{\theta_3(\theta_1 + \theta_3)} + \frac{u_4}{\theta_4(\theta_2 + \theta_4)} - \left(\frac{1}{\theta_1\theta_3} + \frac{1}{\theta_2\theta_4}\right)u_0\right]. \quad (7.3)$$

We use the operator of Eq. (7.3) for points adjacent to boundary points when the boundary points do not coincide with the mesh, instead of our standard op-

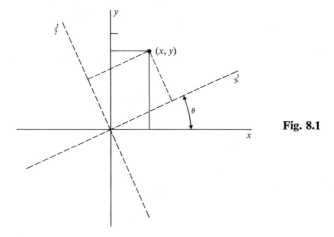

Fig. 8.1

erator.   If the boundary conditions involve normal derivatives, great complications arise, especially for curved boundaries.   Fox (1962) gives relationships for
this situation.

If our mesh of points is chosen very fine (as we would do to get high accuracy),
there is an even simpler way to handle irregular boundaries.   One uses the closest
mesh point as the boundary point, thus in effect perturbing the actual region to
one that coincides with the network and then the standard operator of Eq. (6.1)
applies everywhere.   This introduces some error, of course, but often its effect
is no worse than the O(h) operator of (7.3).   It is also usually easier to program a
computer solution using this perturbated region technique.

8. LAPLACIAN OPERATOR IN NONRECTANGULAR NETWORKS

It is occasionally useful to have a finite difference approximation to the Laplacian
for an equispaced triangular network of points.   To derive this, we need the formulas for a rotation of axes from analytical geometry.   The point (x,y), written
in terms of its coordinates with respect to the pair of $x'y'$-axes which are rotated
an angle θ from the xy-system is (see Fig. 8.1):

$$
\begin{aligned}
x &= x' \cos \theta - y' \sin \theta, \\
y &= x' \sin \theta + y' \cos \theta,
\end{aligned}
\tag{8.1}
$$

We first compute $\partial u / \partial x'$:

$$
\frac{\partial u}{\partial x'} = \frac{\partial u}{\partial x}\frac{\partial x}{\partial x'} + \frac{\partial u}{\partial y}\frac{\partial y}{\partial x'} = \frac{\partial u}{\partial x} \cos \theta + \frac{\partial u}{\partial y} \sin \theta = u_x \cos \theta + u_y \sin \theta.
$$

Then we get

$$
\frac{\partial^2 u}{\partial x'^2} = \cos \theta \, (u_{xx} \cos \theta + u_{xy} \sin \theta) + \sin \theta \, (u_{yx} \cos \theta + u_{yy} \sin \theta)
$$

$$
= u_{xx} \cos^2 \theta + 2u_{xy} \sin \theta \cos \theta + u_{yy} \sin^2 \theta.
$$

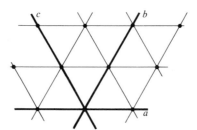

Fig. 8.2

For an equispaced triangular network, the connections between nodes make angles of 0°, 60°, and 120° with the horizontal (Fig. 8.2). Call these the *a*-, *b*-, and *c*-axes.

For $\theta = 0°$,

$$\frac{\partial^2 u}{\partial a^2} = u_{xx},$$

For $\theta = 60°$,

$$\frac{\partial^2 u}{\partial b^2} = \frac{1}{4}u_{xx} + \frac{\sqrt{3}}{2}u_{xy} + \frac{3}{4}u_{yy},$$

For $\theta = 120°$,

$$\frac{\partial^2 u}{\partial c^2} = \frac{1}{4}u_{xx} - \frac{\sqrt{3}}{2}u_{xy} + \frac{3}{4}u_{yy}.$$

Adding, we obtain

$$\frac{\partial^2 u}{\partial a^2} + \frac{\partial^2 u}{\partial b^2} + \frac{\partial^2 u}{\partial c^2} = \tfrac{3}{2}(u_{xx} + 0 + u_{yy})$$

$$= \tfrac{3}{2}\nabla^2 u.$$

Laplace's equation, $\nabla^2 u = 0$, can be represented by

$$\frac{2}{3}\left(\frac{\partial^2 u}{\partial a^2} + \frac{\partial^2 u}{\partial b^2} + \frac{\partial^2 u}{\partial c^2}\right) = 0.$$

Using finite difference quotients to approximate the partial derivatives gives a pictorial operator for a triangular network:

$$\nabla^2 u = \frac{2}{3h^2}\left\{\begin{matrix} & 1 & & 1 & \\ 1 & & -6 & & 1 \\ & 1 & & 1 & \end{matrix}\right\} u.$$

Note that, for Laplace's equation, we have that the potential at every point is the arithmetic average of the potentials at its six equidistant neighbors. We observe this rule of averages even in the three-dimensional situation below.

For circular regions, one may derive a finite difference approximation to the Laplacian in polar coordinates. Consider the group of points which are the

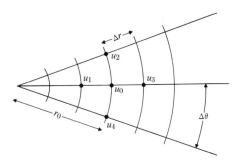

Fig. 8.3

nodes of a polar coordinate network (Fig. 8.3):

$$\nabla^2 u = \frac{\partial^2 u}{\partial r^2} + \frac{1}{r} \frac{\partial u}{\partial r} + \frac{1}{r^2} \frac{\partial^2 u}{\partial \theta^2} = \frac{u_3 - 2u_0 + u_1}{(\Delta r)^2} + \frac{1}{r_0} \frac{u_3 - u_1}{2\Delta r} + \frac{1}{r_0^2} \frac{u_2 - 2u_0 + u_4}{(\Delta \theta)^2}.$$

No "standard" operator can be written for this, so solving the set of equations by iteration when the problem is in polar coordinates is awkward.

9. ACCELERATED CONVERGENCE OF LIEBMANN'S METHOD

In Sections 4 and 5, we discussed both the Liebmann method and the relaxation method. The latter has the advantage of more rapid convergence, while the former is more adapted to program writing because the iterations are performed in a systematic manner. We can combine a part of the advantage of relaxation within the systematic iteration scheme. First we observe that Liebmann's method really uses the same kind of computation as does relaxation.

Liebmann's equation is, for Laplace's equation,

$$u_{i,j}^{n+1} = \tfrac{1}{4}(u_{i+1,j}^n + u_{i-1,j}^n + u_{i,j+1}^n + u_{i,j-1}^n), \tag{9.1}$$

where the superscripts show the order of the iteration, i.e., the $(n + 1)$ iterate utilizes the nth set of values around the point (i,j). The relaxation residue is

$$R_{i,j}^n = (u_{i+1,j}^n + u_{i-1,j}^n + u_{i,j+1}^n + u_{i,j-1}^n) - 4u_{i,j}^n,$$

or

$$\tfrac{1}{4}R_{i,j}^n = \tfrac{1}{4}(u_{i+1,j}^n + u_{i-1,j}^n + u_{i,j+1}^n + u_{i,j-1}^n) - u_{i,j}^n.$$

We can rewrite (9.1) as

$$u_{i,j}^{n+1} = u_{i,j}^n + \tfrac{1}{4}R_{i,j}^n. \tag{9.2}$$

Equation (9.2) shows that (a) we can consider Liebmann's process as a relaxation process wherein the residuals are brought exactly to zero, and (b) the pattern is systematic rather than attacking the largest residual first.

It can be shown that the rate of convergence of the iteration represented by Eq. (9.2) can be accelerated if, instead of the factor $\tfrac{1}{4}$ applied to the residual, we use a larger factor. In effect, this is over-relaxation; hence this modification of the Liebmann method is called "S.O.R." (successive over-relaxation) or "extrapolated

Liebmann." To do this we use, instead of (9.2), the equation

$$u_{i,j}^{n+1} = u_{i,j}^n + \alpha R_{i,j}^n$$

$$= u_{i,j}^n + \alpha(u_{i+1,j}^n + u_{i-1,j}^n + u_{i,j+1}^n + u_{i,j-1}^n - 4u_{i,j}^n)$$

$$= \alpha(u_{i+1,j}^n + u_{i-1,j}^n + u_{i,j+1}^n + u_{i,j-1}^n - \left(4 - \frac{1}{\alpha}\right)u_{i,j}^n).$$

The optimum value of α, for a rectangular region* which has p and q mesh divisions on its sides, is the smaller root of the quadratic

$$\left(\cos\frac{\pi}{p} + \cos\frac{\pi}{q}\right)^2 \alpha^2 - 4\alpha + 1 = 0. \qquad (9.3)$$

The value of α from Eq. (9.3) will lie between $\frac{1}{4}$ and $\frac{1}{2}$.

In the concluding section of this chapter, the results of a computer program using S.O.R. is compared to the ordinary Liebmann calculation in solving the Poisson equation over a rectangular region. The number of iterations to converge is cut almost in half.

10. THE LAPLACIAN OPERATOR IN THREE DIMENSIONS

Writing a difference equation to approximate the three-dimensional Laplacian is straightforward. We use triple subscripts to indicate the spatial position of points, and take the mesh distance the same in each direction:

$$\nabla^2 u = \frac{\partial^2 u}{\partial x^2} + \frac{\partial^2 u}{\partial y^2} + \frac{\partial^2 u}{\partial z^2} = \frac{u_{i+1,j,k} - 2u_{i,j,k} + u_{i-1,j,k}}{h^2}$$

$$+ \frac{u_{i,j+1,k} - 2u_{i,j,k} + u_{i,j-1,k}}{h^2} + \frac{u_{i,j,k+1} - 2u_{i,j,k} + u_{i,j,k-1}}{h^2}$$

$$= \frac{1}{h^2} \left\{ \qquad \qquad \right\} u_{i,j}.$$

We see again, for Laplace's equation $\nabla^2 u = 0$, that the potential u is the arithmetic average of its six nearest neighboring values. The set of equations for this case is more extensive and more tedious to solve, but in principle the methods are unchanged. In the iterative methods, keeping track of the successive values of u is awkward. An isometric projection of the points is generally recommended over the use of superimposed sheets of paper. In a computer, triple subscripting solves the problem, but large storage requirements are imposed by realistic problems.

---

* For nonrectangular regions, no simple rule for the optimum has been discovered. Varga (1959) discusses how α can be estimated from the earlier iterates.

11. A COMPUTER PROGRAM TO SOLVE POISSON EQUATIONS

Program 1 solves the Poisson equation $\nabla^2 u = K$, using Liebmann's method modified by a variable relaxation factor α, as discussed in Section 8. The boundary values and number of mesh spaces on each edge of the rectangular region are input parameters.

For a 4 by 8 region, $\nabla^2 u = -2$ with $u = 0$ on the boundary, 17 iterations were required to converge to a tolerance of 0.001 when $\alpha = 0.25$, starting with initial estimates of $u = 0$ everywhere. (The tolerance was compared to the maximum change in any u-value between successive iterations.)

Program 1

```
ZZJOB 5                      CSC  001     GERALD, C. F.
ZZFORX5
*LIST PRINTER
C    PROGRAM FOR SOLVING POISSON EQUATION, DEL SQUARE U = CONSTANT
C    READ IN PARAMETERS AND BOUNDARY VALUES, PUNCH HEADING
         DIMENSION U(13,13)
         READ 99, NWIDE, NHIGH, XK, TOL, H, ALPHA
         PRINT 100, XK, ALPHA
         NWPI = NWIDE + 1
         NHPI = NHIGH + 1
         DO 10 I = 1,NHPI, NHIGH
   10 READ 101,(U(I,J),J=1,NWPI)
         DO 11 J = 1, NWPI, NWIDE
   11 READ 101,(U(I,J),I=2,NHIGH )
C    INITIALIZE INTERIOR POINTS
         DO 12 I = 2,NHIGH
         DO 12 J = 2,NWIDE
   12 U(I,J) =   0.0
C    COMPUTE VALUES BY LIEBMANNS METHOD. END WHEN MAX CHANGE IS LESS THAN
C       THE TOLERANCE.
         DO 14 K = 1,50
         DEVMAX = 0.0
         DO 13 I = 2,NHIGH
         DO 13 J = 2,NWIDE
         SAVE = U(I,J)
         U(I,J) = ALPHA*(U(I+1,J) + U(I-1,J) + U(I,J+1) + U(I,J-1)
   1    - XK*H**2 - (4. - 1./ALPHA)*U(I,J))
         DEV = ABSF(U(I,J)-SAVE)
         IF (DEV - DEVMAX) 13,13,15
   15 DEVMAX = DEV
   13 CONTINUE
         IF (TOL - DEVMAX) 14,16,16
   14 CONTINUE
         PRINT 102
         GO TO 17
   16 PRINT 104, K
   17 DO 30 I = 1,NHPI
   30 PRINT 103, I, (U(I,J), J=1,NWPI)
         CALL EXIT
   99 FORMAT (2I2/4F10.0)
  100 FORMAT (/70H SOLUTION TO POISSON EQUATION - DEL SQUARE U = K - BY
     1LIEBMANN METHOD. / 9H FOR K = , F5.2, 29H, USING RELAXATION FACTOR
     2 OF , F5.3)
  101 FORMAT (8F10.0)
  102FORMAT (//51H TOLERANCE TO END COMPUTATION NOT MET IN 50 CYCLES./
     132H LAST ROUND OF CALCULATIONS IS -)
  103 FORMAT (4HOROW, I2/(1H ,12F10.4))
  104 FORMAT (// I3, 21H ITERATIONS WERE USED )
         END
0408
-2.0       0.001      0.5        0.25
```

(Continued)

Program 1 Cont.

```
SOLUTION TO POISSON EQUATION - DEL SQUARE U = K - BY LIEBMANN METHOD.
FOR K = -2.00, USING RELAXATION FACTOR OF   .250

17 ITERATIONS WERE USED

ROW 1
   0.0000      0.0000      0.0000      0.0000      0.0000

ROW 2
   0.0000       .3972       .5129       .3975      0.0000

ROW 3
   0.0000       .5770       .7579       .5774      0.0000

ROW 4
   0.0000       .6543       .8654       .6546      0.0000

ROW 5
   0.0000       .6759       .8956       .6762      0.0000

ROW 6
   0.0000       .6546       .8658       .6549      0.0000

ROW 7
   0.0000       .5776       .7586       .5778      0.0000

ROW 8
   0.0000       .3977       .5135       .3978      0.0000

ROW 9
   0.0000      0.0000      0.0000      0.0000      0.0000


SOLUTION TO POISSON EQUATION - DEL SQUARE U = K - BY LIEBMANN METHOD.
FOR K = -2.00, USING RELAXATION FACTOR OF   .316

10 ITERATIONS WERE USED

ROW 1
   0.0000      0.0000      0.0000      0.0000      0.0000

ROW 2
   0.0000       .3978       .5137       .3979      0.0000

ROW 3
   0.0000       .5780       .7591       .5781      0.0000

ROW 4
   0.0000       .6553       .8666       .6554      0.0000

ROW 5
   0.0000       .6768       .8968       .6769      0.0000

ROW 6
   0.0000       .6554       .8667       .6554      0.0000

ROW 7
   0.0000       .5781       .7592       .5781      0.0000

ROW 8
   0.0000       .3979       .5138       .3979      0.0000

ROW 9
   0.0000      0.0000      0.0000      0.0000      0.0000
```

When α was made 0.3167, its calculated optimum value, the number of iterations was reduced to 10 to meet the same tolerance.

PROBLEMS

Section 1

1. Newton's law states that the force of attraction between two masses is

$$f = g\frac{mM}{r^2},$$

where g is the gravitational constant, and m and M are the two masses a distance r apart. If one mass, M, is a unit mass, the force on it is

$$f = g\frac{m}{r^2}.$$

Note that for the function

$$p = -g\frac{m}{r}$$

its derivative, $\partial p/\partial r$, equals the force on the unit mass:

$$\frac{\partial p}{\partial r} = g\frac{m}{r^2} = f,$$

where p is called the potential function. If we know how the potential function varies throughout a given force field, we can calculate the force acting on a unit mass at any point in the field.

Suppose we have a distribution of masses in a planar area S, such that $\rho(a,b)$ is the concentration of mass (mass per unit area) at any point (a,b) within S. The potential, a function of the location (x,y) within the region S, is then

$$p(x,y) = -\iint\limits_{S} \frac{g\rho}{r} \, dA,$$

where $r = \sqrt{(x - a)^2 + (y - b)^2}$. Show that the potential function obeys Laplace's equation.

Section 2

2. Show using the Taylor series method that:

$$\text{(a) } f_n' = \frac{f_{n+1} - f_n}{h} + O(h)$$

$$\text{(b) } f_n'' = \frac{f_{n+2} - 2f_{n+1} + f_n}{h^2} + O(h)$$

We, hence, prefer to use Eqs. (2.1) and (2.2) which have errors $O(h^2)$.

3. The mixed second derivative $\partial^2 u/(\partial x\, \partial y)$ can be considered as

$$\frac{\partial}{\partial x}\left(\frac{\partial u}{\partial y}\right) = \frac{\partial^2 u}{\partial x\, \partial y} = \frac{\partial}{\partial y}\left(\frac{\partial u}{\partial x}\right).$$

Show that in terms of finite difference quotients with $\Delta x = \Delta y$, this derivative can be approximated by the pictorial operator

$$\frac{1}{4h^2} \begin{Bmatrix} -1 & & 1 \\ & & \\ 1 & & -1 \end{Bmatrix} u_{i,j} + O(h^2).$$

4. If

$$\frac{d^2u}{dx^2} = \frac{-u_{i+2} + 16u_{i+1} - 30u_i + 16u_{i-1} - u_{i-2}}{12h^2} + O(h^4),$$

find the fourth-order operator for the Laplacian. Assume that the function u has a continuous sixth derivative.

5. Suppose we have a rectangular plate with top and bottom edges held at the same constant temperature, while the right-hand edge is held at 100°, and the left edge at 0°. If heat flows in only two directions, it is obvious that the temperatures vary linearly in the x-direction, and are constant along vertical lines.

   (a) Show that such a temperature distribution satisfies Eq. (2.3).

   (b) Show that this temperature distribution also obeys the relationship derived in Problem 4 with $\Delta^2 u = 0$. What about points adjacent to the edges of the plate?

Section 3, 4

6. Set up equations analogous to Eqs. (3.2) for the example problem of Section 3, but with a grid spacing of 4 inches.

   (a) Solve the set of equations by elimination.

   (b) Solve the equations by Liebmann's method.

   (c) Solve the equations by the relaxation method.

7. Solve for the steady-state temperatures in a rectangular plate, 12 by 15 inches, if one 15-inch edge is held at 100° while the other edges are all held at 20°. The material is aluminum. Take $h = k = 3$ inches, and consider heat to flow only in the lateral directions. Sketch in the approximate location of the 50° isothermal curve.

8. Solve for the temperatures in the plate of Problem 7 when the edge temperatures are held as shown. Use the relaxation method.

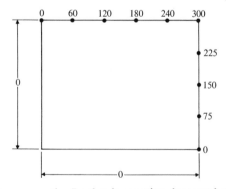

9. The region for which we can solve Laplace's equation does not have to be rectangular. We can apply the methods of Section 3 for any region just so long as the meshes of

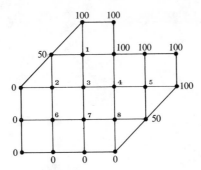

our network coincide with the boundary.   Solve for the steady state potentials at the eight interior points shown.

Section 5

10. Find the torsion function ϕ for a square bar.

    (a) Subdivide the cross section into nine equal squares, so there are four interior points.   Because of symmetry, all four values of ϕ are the same.

    (b) Then subdivide the cross section into 36 equal squares so there are 25 interior points.   Use the results of (a) to estimate starting values for iteration.

11. Solve for the torsion in a hollow square bar, 5 inches in outside dimension, and with walls 2 inches thick (so that the inside hole is 1 inch on a side).   On the inner surface as well as on the outer, $\phi = 0$.

12. Solve for the torsion in a prismatic bar of a cross section similar to the region of Problem 9.

13. Solve $\nabla^2 u = f(x,y)$ on the square region bounded by $x = 0$, $x = 1$, $y = 0$, $y = 1$ with $f(x,y) = xy$.   Use $h = \frac{1}{3}$.   Take $\mu = 0$ on the boundary.

Section 6

14. Solve the set of equations in Eqs. (6.2), using (6.3) to eliminate u_a, u_b, and u_c.

    (a) Use elimination.

    (b) Use Liebmann's method.

    (c) Use the relaxation method.

15. Solve the example problem of Section 6 except that three edges of the plate (in addition to the top and bottom faces) are insulated.   The plate then loses heat from only one edge at a rate such that the gradient there is 1.5°C/cm.

16. Solve the example problem of Section 6, but with boundary conditions $\partial u/\partial x = 0.2(u - 20)$.

17. Solve for the steady-state temperatures of the region in Problem 9 except that the plate is insulated at each exterior point marked zero.   The edge temperatures are maintained at the temperatures given at the other exterior points.

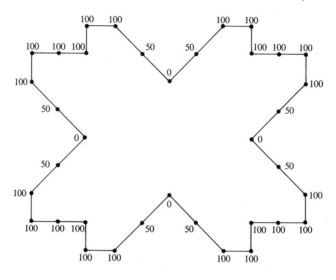

18. Consider a region which is obtained from Problem 9 by reflecting the figure across both of the edges marked zero. Show that the steady potentials at points in the original region are the same as for Problem 17.

Section 7

19. Find the potential distribution in a region whose shape is a 3-4-5 right triangle. The potential is maintained at zero except on the hypotenuse where it is 50. Use a square mesh with $h = 1$.

20. A coaxial cable has a circular outer conductor 10 cm in diameter, and a square inner conductor 3 cm on a side. (The square conductor is concentric.) The outer conductor is at zero volts while the inner is at 100 volts. Find the potential between the conductors. (Note that only one octant needs to be calculated due to symmetry.) Use a grid of 1-cm. squares.

21. If a hollow shaft has an outer circular cross section of diameter 10 in. and a concentric square inner cross section which is 3 inches on a side, what is the value of the torsion function at the nodes of a 1-inch grid?

Section 8

22. Using the Laplacian operator for an equispaced triangular network, determine the potential in the given region if $\nabla^2 p = 0$.

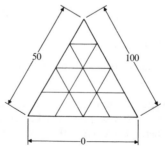

23. Repeat Problem 21, except take two edges at 100, the third at 0.

24. Find the solution to the Poisson equation over an equilateral triangle 5 inches on a side if

$$\nabla^2 u = k, \; u = 0 \text{ on boundary,}$$

and values of k are as shown.

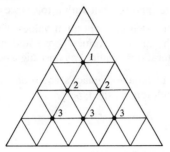

25. Use the Laplacian in polar coordinates to set up the set of difference equations to solve

$$\nabla^2 u = \theta.$$

on a semicircular region whose radius is 4.   Take $\Delta r = 1$ and $\Delta \theta = \pi/6$.   Boundary conditions are $u = 10$ on the straight edge and $u = 0$ on the curved boundary.

Section 9

26. Using Eq. (9.3), compute the optimum relaxation factor for a rectangular plate, 12 by 15 inches   (a) with $h = 3$ inches, (b) with $h = 1$ inch, (c) with $h = 0.1$ inch.

27. A rectangular plate is 12 by 15 inches and has both 15-inch edges held at 100° while the other edges are at 0°.

   (a) Beginning with initial estimates of temperature at 0°, how many iterations are required by the unaccelerated Liebmann method for the calculated interior temperatures to each agree within 1° to the previous iteration?   Use $h = 3$ inches and observe that symmetry considerations permit one to work in only one quadrant.

   (b) Using the optimum acceleration factor (see Problem 26), how much more rapidly does convergence occur, using the same criterion as for part (a)?

   (c) Vary the order of iteration, and repeat parts (a) and (b).

28. Repeat Problem 27 except now one of the 12-inch edges is imperfectly insulated so $u = \partial u/\partial x$ on that face (the gradient is measured in the outward direction).

Section 10

29. A cube is 3 cm along its edge.   Two opposite faces are held at 100°, the other four faces at 0°.   Find the internal temperatures at the nodes of a 1-cm network.

30. Repeat Problem 29 except that two adjacent faces are held at 100° while the rest are held at 0°.

31. Repeat Problem 29 except that one of the 0° faces has been insulated.

Section 11

32. (a) Use the computer program of the text to solve Problem 7 with a grid spacing of 3 inches, 1 inch, and 0.1 inch, all with $\alpha = 0.25$.   Set tolerance at 0.1.

    (b) Redo part (a) but with optimum values of α, calculated by Eq. (9.3).   Compare the number of iterations required.

    (c) Repeat parts (a) and (b), but with a tolerance of 0.001.

    (d) Verify that the calculated optimum values of α are indeed optimum by repeating part (b) with various α-values above and below the calculated optimum. Plot the number of iterations required against the α-value.

33. Revise the program to handle Laplace's equation in rectangular regions, but permitting boundary conditions of the type

$$au + b\frac{\partial u}{\partial x} = c,$$

where $\partial u/\partial x$ is the outward normal.

34. Write a program to handle the Poisson equation in rectangular regions where the Laplacian is equal to a function of position (x,y) in the region, $\nabla^2 u = f(x,y)$.

10
parabolic
partial
differential
equations

The problem of unsteady-state flow of heat is a physical situation that can be represented by a parabolic partial differential equation. The simplest situation is for flow of heat in one direction. Imagine a rod, uniform in cross section and in its perimeter so that heat flows only longitudinally. Consider a differential portion of the rod, dx in length with cross-sectional area A.

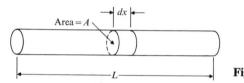

Fig. 0.1

We let u represent the temperature at any point in the rod, whose distance from the left end is x. Heat is flowing from left to right under the influence of the temperature gradient $\partial u / \partial x$. Make a balance of the rate of heat flow into and out of the element. Use k for thermal conductivity, $\text{cal}/\text{g} \cdot \text{cm}^2 \cdot {}^{\circ}\text{C}/\text{cm}$, which we assume is constant,

Rate of flow of heat in: $-kA \dfrac{\partial u}{\partial x}$,

Rate of flow of heat out: $-kA \left(\dfrac{\partial u}{\partial x} + \dfrac{\partial}{\partial x}\left(\dfrac{\partial u}{\partial x}\right) dx \right)$.

The difference between the rate of flow in and the rate of flow out is the rate at which heat is being stored in the element. If c is the heat capacity, $\text{cal}/\text{g} \cdot {}^{\circ}\text{C}$, and ρ is the density, g/cm^3, we have, with t for time,

$$-kA \frac{\partial u}{\partial x} - \left(-kA \left(\frac{\partial u}{\partial x} + \frac{\partial}{\partial x}\left(\frac{\partial u}{\partial x}\right) dx \right) \right) = c\rho(A \, dx)\frac{\partial u}{\partial t}.$$

Simplifying, we have

$$k \frac{\partial^2 u}{\partial x^2} = c\rho \frac{\partial u}{\partial t}.$$

This is the basic mathematical model for unsteady-state flow. We have derived it for heat flow, but it applies equally to diffusion of material, flow of fluids (under conditions of laminar flow), flow of electricity in cables (the telegraph equations), etc.

The function that we call the solution to the problem must not only obey the differential equation given above, but must also satisfy an initial condition and a set of boundary conditions. For the one-dimensional heat flow problem we first consider, the initial condition will be the initial temperatures at all points along the rod,

$$u(x,t)\big|_{t=0} = u(x,0) = f(x).$$

The boundary conditions will describe the temperature at each end of the rod as functions of time. Our first examples will consider the case where these temperatures are held constant:

$$u(0,t) = c_1,$$
$$u(L,t) = c_2.$$

More general (and more practical) boundary conditions will involve not only the temperature but the temperature gradient,

$$A_1 u(0,t) + B_1 \frac{\partial u(0,t)}{\partial x} = F_1(t),$$

$$A_2 u(L,t) + B_2 \frac{\partial u(L,t)}{\partial x} = F_2(t).$$

1. THE EXPLICIT METHOD

Our approach to solving parabolic partial differential equations by a numerical method is to replace the partial derivatives by finite difference approximations. For the one-dimensional heat flow equation,

$$\frac{\partial^2 u}{\partial x^2} = \frac{c\rho}{k} \frac{\partial u}{\partial t}, \tag{1.1}$$

we use the relations

$$\frac{\partial^2 u}{\partial x^2} = \frac{u_{i+1}^j - 2u_i^j + u_{i-1}^j}{(\Delta x)^2} + O(\Delta x)^2 \tag{1.2}$$

$$\frac{\partial u}{\partial t} = \frac{u_i^{j+1} - u_i^j}{\Delta t} + O(\Delta t). \tag{1.3}$$

We use subscripts to denote position and superscripts for time. Note that the error terms are of different orders since a forward difference is used in Eq. (1.3). This introduces some special limitations, but when this is done, the procedure is simplified.

Substituting Eqs. (1.2) and (1.3) into (1.1) and solving for u_i^{j+1}, gives the equation for the forward difference method:

$$u_i^{j+1} = \frac{k\Delta t}{c\rho(\Delta x)^2} (u_{i+1}^j + u_{i-1}^j) + \left(1 - \frac{2k\Delta t}{c\rho(\Delta x)^2}\right) u_i^j. \tag{1.4}$$

We have solved for u_i^{j+1} in terms of the temperatures at time t_j in Eq. (1.4) in view of the normally known conditions for a parabolic partial differential equation. Equation (1.4) can give the values of u at each interior point at $t = t_1$ since the values at $t = t_0$ are given by the initial conditions. It then can be used to get values at t_2 using the values at t_1 as initial conditions, so we can step the solution forward in time. At the end points, the boundary conditions will determine u.

The relative size of the time and distance steps, Δt and Δx, affects Eq. (1.4). If the ratio of $\Delta t/(\Delta x)^2$ is chosen so $k\,\Delta t/c\rho(\Delta x)^2 = \frac{1}{2}$, the equation is simplified in that the last term vanishes and we have

$$u_i^{j+1} = \frac{1}{2}(u_{i+1}^j + u_{i-1}^j). \tag{1.5}$$

The phenomena of stability and convergence, which we discuss later, set a maximum value to the ratio of $\Delta t/(\Delta x)^2$ which is exactly the value that gives this particularly simple equation.   Because of its ease, Eq. (1.5) is often used to compute the approximate solution to unsteady-state flow problems.   Since each temperature is a function only of known values, and can be immediately determined, this method is called an *explicit method*.

Example. A large flat steel plate is 2 cm thick. If the initial temperatures (°C) within the plate are given as a function of the distance from one face by the equation

$$u(x,t)\big|_{t=0} = 100 \sin \frac{\pi x}{2},$$

find the temperatures as a function of x and t if both faces are maintained at 0°C.

Since the plate is large, we neglect lateral flow of heat relative to the flow perpendicular to the faces and hence use Eq. (1.1) for heat flow in one direction. For steel, $k = 0.13$ cal/sec·cm·°C, $c = 0.11$ cal/g·°C and $\rho = 7.8$ g/cm³. In order to use Eq. (1.5) as an approximation to the physical problem, we subdivide the total thickness into an integral number of spaces.   Let us use $x = 0.25$, giving eight subdivisions. Δt is then fixed by the relation

$$\frac{k\,\Delta t}{c\rho(\Delta x)^2} = \frac{1}{2}, \qquad \Delta t = \frac{(0.11)(7.8)(0.25)^2}{(2)(0.13)} = 0.206 \text{ second.}$$

The boundary conditions are

$$u(0,t) = 0, \qquad u(2,t) = 0.$$

The initial condition is

$$u(x,0) = 100 \sin \frac{\pi x}{2}.$$

Our calculations are conveniently recorded in a table, as in Table 1.1, where each row of figures is at a particular time.   We begin by filling in the initial con-

Table 1.1. Numerical Solution to Heat Flow Example

$t =$	$x = 0$	$x = 0.25$	$x = 0.50$	$x = 0.75$	$x = 1.00$	$x = 1.25$
0.0	0	38.3	70.7	92.4	100	92.4
0.206	0	35.35	65.35	85.35	92.4	85.35
0.412	0	32.68	60.35	78.88	85.35	78.88
0.619	0	30.18	55.78	72.86	78.88	72.86
0.825	0	27.89	51.52	67.33	72.86	67.33
1.031	0	25.76	47.61	62.19	67.33	62.19
1.238	0	23.80	43.98	57.47	62.19	57.47
1.444	0	21.99	40.64	53.08	57.47	53.08
1.650	0	20.32	37.54	49.06	53.08	49.06
1.856	0	18.77	34.69	45.31	49.06	45.31
2.062	0	17.34	32.04	41.88	45.31	41.88

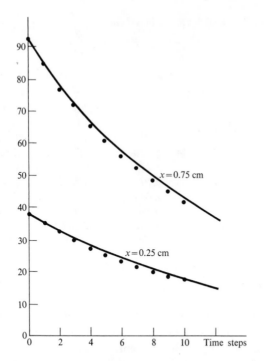

Fig. 1.1

ditions along the first row, at $t = 0.0$. The simple algorithm of Eq. (1.5) tells us that, at each interior point, the temperature at any point at the end of a time step is just the arithmetic average of the temperatures at the adjacent points at the beginning of that time step. The end temperatures are given by the boundary conditions. Because the temperatures are symmetrical on either side of the center line, we compute only for $x \le 1.0$. The temperature at $x = 1.25$ is the same as at $x = 0.75$.

In Figure 1.1, we compare some of the numerical results with the analytical solution, which is particularly simple for this problem:

$$u = 100e^{-\pi^2 kt/c\rho L^2} \sin \frac{\pi x}{L} = 100e^{-0.3738t} \sin \frac{\pi x}{2}.$$

Note that the numerical values are close to the curves which are drawn to represent the analytical solution. In general the errors of Table 1.1 are less than 2%. Unfortunately, this easy method is not always so accurate, as shown in the next example.

As a second example, we will solve a problem in diffusion, which is governed by the same mathematical equation as is heat conduction in a solid.

Example. A hollow tube 20 cm long is initially filled with air containing 2% of ethyl alcohol vapors. At the bottom of the tube is a pool of alcohol which evaporates into the stagnant gas above. (Heat transfers to the alcohol from the

Table 1.2. Diffusion of Alcohol Vapors in a Tube

$$u_i^{j+i} = \tfrac{1}{2}(u_{i-1}^j + u_{i+1}^j)$$

Time, seconds	$x = 0$	$x = 4$	$x = 8$	$x = 12$	$x = 16$	$x = 20$
0	2.0 0.0	2.0	2.0	2.0	2.0	2.0 10.0
67.2	0.0	1.00	2.00	2.00	6.00	10.0
134.4	0.0	1.00	1.50	4.00	6.00	10.0
201.6	0.0	0.75	2.50	3.75	7.00	10.0
268.8	0.0	1.25	2.25	4.75	6.875	10.0
336.0	0.0	1.125	3.00	4.562	7.375	10.0
403.2	0.0	1.500	2.844	5.188	7.281	10.0
470.4	0.0	1.422	3.344	5.062	7.594	10.0
537.6	0.0	1.672	3.242	5.469	7.531	10.0
604.8	0.0	1.621	3.570	5.386	7.734	10.0
Steady state	0.0	2.0	4.0	6.0	8.0	10.0

surroundings to maintain a constant temperature of 30°C, at which temperature the vapor pressure is 0.1 atm.)   At the upper end of the tube, the alcohol vapors dissipate to the outside air, so the concentration is essentially zero.   Considering only the effects of molecular diffusion, determine the concentration of alcohol as a function of time and the distance x measured from the top of the tube.

Molecular diffusion follows the law

$$\frac{\partial c}{\partial t} = D \frac{\partial^2 c}{\partial x^2},$$

where D is the diffusion coefficient, with units of cm^2/sec in the cgs system.   (This is the same as for the ratio $k/c\rho$, which is often termed *thermal diffusivity*.)   For ethyl alcohol, $D = 0.119$ cm^2/sec at 30°C, and the vapor pressure is such that 10 volume percent alcohol in air is present at the surface.

Subdivide the length of the tube into five intervals, so $\Delta x = 4$ cm.   Using the maximum value permitted for Δt yields

$$D \frac{\Delta t}{(\Delta x)^2} = \frac{1}{2} = 0.119 \frac{\Delta t}{(4)^2}, \qquad \Delta t = 67.2 \text{ sec.}$$

Our initial condition is $c(x,0) = 2.0$.   The boundary conditions are $c(0,t) = 0.0$, $c(20,t) = 10.0$.   The concentrations are measured by the percent of alcohol vapor in the air.

The computations are shown in Table 1.2.   Again, using Eq. (1.5), each interior value of c is given by the arithmetic average of concentrations on either side in the line above.   A little reflection is required to determine the proper concentrations to be used for $x = 0$ and $x = 20$ at $t = 0$.   While these are initially at 2%, we assume they are changed instantaneously to 0% and 10% because the

Analytical solution:

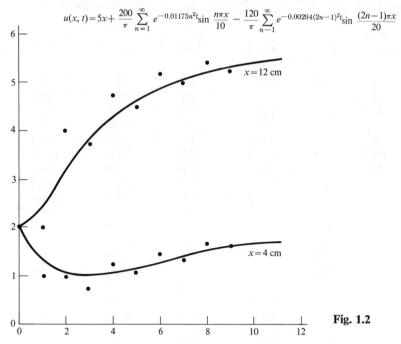

$$u(x, t) = 5x + \frac{200}{\pi} \sum_{n=1}^{\infty} e^{-0.01175n^2 t} \sin \frac{n\pi x}{10} - \frac{120}{\pi} \sum_{n-1}^{\infty} e^{-0.00294(2n-1)^2 t} \sin \frac{(2n-1)\pi x}{20}$$

x = 12 cm

x = 4 cm

Fig. 1.2

effective concentrations acting during the first time interval are not the initial values but the changed values. We have accordingly rewritten these values as shown in the first row of Table 1.2.

As time passes, the concentration will become a linear function of x, from 10% at $x = 20$ to 0% at $x = 0$. The calculated values approach to the steady-state concentrations as time passes. However, as Fig. 1.2 shows, the successive calculated values oscillate about the values calculated from the analytical solution.*

While the algorithm of Eq. (1.5) is simple to use, the results leave something to be desired in the way of accuracy. The oscillatory nature of the results, when a smooth trend of the concentration is expected, is also of concern. We can improve accuracy by smaller steps, but if we decrease Δx, the time steps must also decrease, because the ratio $D \Delta t/(\Delta x)^2$ cannot exceed $\frac{1}{2}$. Cutting Δx in half will require making Δt one-fourth of its previous value, giving a total of eight times as many calculations. With the mixed order of error terms, it is not obvious what reduction of error this would give, but they should be reduced about fourfold.

We can reduce Δt without change in Δx, however, for the method is stable for any value of the ratio less than $\frac{1}{2}$. Suppose we make $D \Delta t/(\Delta x)^2 = \frac{1}{4}$. The basic

---

* In this instance the analytical solution is not so simple; it is given by an infinite series shown in Fig. 1.2.

Table 1.3. Diffusion of Alcohol Vapors in a Tube

$$u_i^{j+1} = \tfrac{1}{4}(u_{i-1}^j + u_{i+1}^j) + \tfrac{1}{2}u_i^j$$

Time, seconds	$x = 0$	$x = 4$	$x = 8$	$x = 12$	$x = 16$	$x = 20$
0.0	2̶0̶ 0.0	2.0	2.0	2.0	2.0	2̶0̶ 10.0
33.6	0.0	1.50	2.00	2.00	4.00	10.0
67.2	0.0	1.25	1.875	2.50	5.00	10.0
100.8	0.0	1.094	1.875	2.969	5.625	10.0
134.4	0.0	1.015	1.953	3.360	6.055	10.0
168.0	0.0	0.996	2.070	3.682	6.368	10.0
201.6	0.0	1.015	2.204	3.950	6.604	10.0
235.2	0.0	1.058	2.343	4.177	6.790	10.0
268.8	0.0	1.115	2.480	4.372	6.939	10.0
302.4	0.0	1.177	2.612	4.541	7.062	10.0
336.0	0.0	1.241	2.736	4.689	7.166	10.0
369.6	0.0	1.304	2.850	4.820	7.255	10.0
403.2	0.0	1.364	2.956	4.936	7.332	10.0
436.8	0.0	1.421	3.053	5.040	7.400	10.0
470.4	0.0	1.474	3.142	5.133	7.460	10.0
504.0	0.0	1.522	3.223	5.217	7.513	10.0
537.6	0.0	1.567	3.296	5.292	7.561	10.0
Steady state	0.0	2.0	4.0	6.0	8.0	10.0

difference equation, (1.4), now becomes

$$u_i^{j+1} = \tfrac{1}{4}(u_{i+1}^j + u_{i-1}^j) + \tfrac{1}{2}u_i^j. \tag{1.6}$$

Table 1.3 summarizes the use of Eq. (1.6) on the same example as before. As shown more clearly by Fig. 1.3, this last calculation causes the calculated concentrations to follow smooth curves, and the error is reduced considerably. The poorest accuracy is near the beginning. The later values are close to the curve.

This initial poor accuracy is a result of the discontinuities in the boundary conditions. The abrupt change of concentrations at the ends of the tube makes the explicit method inaccurate. In the first example, the continuity of the initial temperatures plus continuity of its derivatives and the absence of discontinuities in the boundary conditions led to good accuracy throughout.

2. CRANK-NICOLSON METHOD

When the difference equation, Eq. (1.4), was derived, we noted that a mixed order of error was involved because a forward difference was used to replace the time derivative while a central difference was used for the distance derivative. The difference quotient, $(u_i^{j+1} - u_i^j)/\Delta t$ can be considered a central difference, however, if we take it as corresponding to the midpoint of the time interval. Suppose

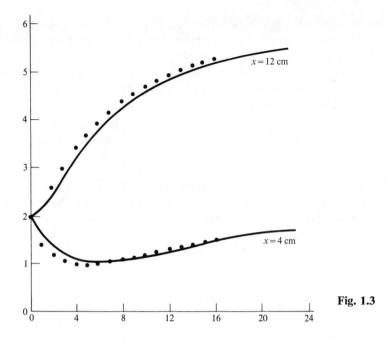

x = 12 cm

x = 4 cm

Fig. 1.3

we equate it to a central difference quotient for the second derivative with distance, also corresponding to the midpoint in time, by averaging difference quotients at the beginning and end of the time step:

$$\frac{\partial^2 u}{\partial x^2} = \frac{c\rho}{k}\frac{\partial u}{\partial t},$$

$$\frac{1}{2}\left(\frac{u_{i+1}^j - 2u_i^j + u_{i-1}^j}{(\Delta x)^2} + \frac{u_{i+1}^{j+1} - 2u_i^{j+1} + u_{i-1}^{j+1}}{(\Delta x)^2}\right) = \frac{c\rho}{k}\left(\frac{u_i^{j+1} - u_i^j}{\Delta t}\right).$$

central difference   central difference   central differ-

at t_j at t_{j+1} ence at $t_{j+\frac{1}{2}}$

When this is rearranged we get the Crank-Nicolson formula, where

$$r = k\,\Delta t/c\rho(\Delta x)^2,$$

$$-ru_{i-1}^{j+1} + (2 + 2r)u_i^{j+1} - ru_{i+1}^{j+1} = ru_{i-1}^j + (2 - 2r)u_i^j + ru_{i+1}^j.$$

Letting $r = 1$, we get some simplification:

$$-u_{i-1}^{j+1} + 4u_i^{j+1} - u_{i+1}^{j+1} = u_{i-1}^j + u_{i+1}^j. \qquad (2.1)$$

As we will discuss in the next section, one advantage of the Crank-Nicolson method is that it is stable for any value of r, although small values are more accurate. Values much larger than unity are not desirable.

Equation (2.1) is the usual formula for using the Crank-Nicolson method. Note that the new temperature u_i^{j+1} is not given directly in terms of known temperatures one time step earlier, but is a function of unknown temperatures at adjacent positions as well. It is therefore termed an *implicit method* in distinction from the explicit method of the previous section.

We illustrate the method by solving the problem of diffusion of alcohol vapors, the same as previously attacked by the explicit method. Again let us take $\Delta x = 4$ cm. Restating the problem, we have

$$\frac{\partial^2 u}{\partial x^2} = \frac{1}{D}\frac{\partial u}{\partial t}, \qquad u(x,0) = 2.0, \qquad \begin{cases} u(0,t) = 0, \\ u(20,t) = 10.0, \end{cases}$$

$$D = 0.119 \text{ cm}^2/\text{sec}.$$

If $D\,\Delta t/(\Delta x)^2 = 1$ and $\Delta x = 4$ cm, then $\Delta t = 134.4$ sec. For $t = 134.4$, at the end of the first time step, we write the equations at each point whose concentration is unknown. The effective concentrations at the two ends, 0% and 10%, respectively, are again used:

$$\begin{aligned}
-0.0 + 4u_1 - u_2 &= 0.0 + 2.0, \\
-u_1 + 4u_2 - u_3 &= 2.0 + 2.0, \\
-u_2 + 4u_3 - u_4 &= 2.0 + 2.0, \\
-u_3 + 4u_4 - 10.0 &= 2.0 + 10.0.
\end{aligned}$$

This method requires more work because, at each time step, we must solve a set of equations similar to the above. Fortunately, the system is tridiagonal. Since we will need to repeatedly solve the system with the same coefficient matrix (we transpose the constant values to the right side in the first and last equations), it is often desirable to invert the coefficient matrix. We then solve the equations by multiplying the vector whose components are the constant terms by the inverse matrix:

$$A = \begin{bmatrix} 4 & -1 & 0 & 0 \\ -1 & 4 & -1 & 0 \\ 0 & -1 & 4 & -1 \\ 0 & 0 & -1 & 4 \end{bmatrix},$$

$$A^{-1} = \begin{bmatrix} 0.2554 & 0.0718 & 0.01914 & 0.00478 \\ 0.0718 & 0.2871 & 0.07655 & 0.01914 \\ 0.01914 & 0.07655 & 0.2871 & 0.0718 \\ 0.00478 & 0.01914 & 0.0718 & 0.2554 \end{bmatrix}.$$

To solve $Ax = b$, we use

$$A^{-1}Ax = x = A^{-1}b.$$

In Table 2.1 are listed the results of calculations by the Crank-Nicolson method. For the first line of calculations the b-vector has components 2.0, 4.0, 4.0, and 22.0, and the calculated values are the components of $A^{-1}b$. In computing the second line, the components of b are 2.019, 4.052, 8.011, and 23.072. Succeeding lines use the proper sums of values from the line above to determine the b-vector.

When the calculated values are compared to the analytical results, it is observed that the errors of this method, in this example, are about the same as those made in the explicit method with $D\,\Delta t/(\Delta x)^2 = \frac{1}{4}$.

Table 2.1. Diffusion of Alcohol Vapors in a Tube—Crank-Nicolson Method

Time, seconds	$x = 0$	$x = 4$	$x = 8$	$x = 12$	$x = 16$	$x = 20$
0.0	2.0 *0.0*	2.0	2.0	2.0	2.0	2.0 *10.0*
134.4	0.0	0.980	2.019	3.072	5.992	10.0
268.8	0.0	1.070	2.363	4.305	6.555	10.0
403.2	0.0	1.276	2.861	4.762	6.962	10.0
537.6	0.0	1.471	3.165	5.115	7.159	10.0

Analytical values

Time, seconds	$x = 4$	$x = 12$
134.4	1.078	3.191
268.8	1.108	4.272
403.2	1.340	4.873
537.6	1.543	5.248

3. DERIVATIVE BOUNDARY CONDITIONS

In heat conduction problems, the most usual situation at the end points is not that they are held at a constant temperature, but that heat is lost by conduction or radiation at a rate proportional to some power of the temperature difference between the surface of the body and its surroundings. This leads to a relationship involving the space derivative of temperature at the surface. In the analytical solution of heat flow problems through Fourier series, this adds considerable complications in determining the coefficients, but our numerical technique requires only minor modifications.

The rate of heat loss from the surface of a solid is generally expressed as

$$\text{Rate of heat loss} = hA(u - u_0), \tag{3.1}$$

where A is the surface area, u is the surface temperature, u_0 is the temperature of the surrounding medium, and h is a coefficient of heat transfer. h is increased by motion of the surrounding medium. To facilitate heat flow, the surrounding medium is often caused to flow rapidly by mechanical means (as in some salt evaporators) or by proper design (as in vertical tube heat exchangers in distillation columns). This situation is called *forced convection*.

In other situations, h is also a function of the surface temperature, as in *natural convection*, wherein the motion of surrounding fluid is affected by thermal currents. Another important situation is heat loss by radiation, in which the rate of heat loss is proportional to the fourth power of the temperature difference between the surface and the surrounding surfaces to which heat is being radiated. In both these situations, the rate of heat loss is proportional to some power of the surface temperature. One way to approximate this is to use Eq. (3.1), absorbing the nonlinear aspects into the coefficient h, and to change h appropriately through the progress of the calculations.   Our examples will treat only the simpler situations where Eq. (3.1) applies directly.

As we shall see, heat loss from the surface by conduction leads to a derivative boundary condition.   Our procedure will be to replace the derivatives in both the differential equation and in the boundary conditions by difference quotients. We illustrate with a simple example.

Example.   An aluminum cube is 4 by 4 by 4 inches ($k = 0.52$, $c = 0.226$, $\rho = 2.70$ in cgs units).   All but one face is perfectly insulated, and the cube is initially at 1000°F.   Heat is lost from the uninsulated face to a fluid flowing past it according to the equation

$$\text{Rate of heat loss, Btu per sec} = hA(u - u_0),$$

where

$$h = 0.15 \text{ Btu /sec} \cdot \text{ft}^2 \cdot \text{°F},$$
$$A = \text{surface area in ft}^2,$$
$$u = \text{surface temperature, °F},$$
$$u_0 = \text{temperature of fluid, °F}.$$

If u_0, the temperature of fluid flowing past the aluminum cube, is constant at 70°F, find the temperatures inside the cube as a function of time.   While we could work the problem in cgs units, we elect to use English units, making suitable changes in k, c, and ρ.

Because of the insulation on the lateral faces, again the only direction in which heat will flow is perpendicular to the uninsulated face, and the equation is

$$\frac{\partial u}{\partial t} = \frac{k}{c\rho} \frac{\partial^2 u}{\partial x^2}.$$

The initial condition, with x representing distance from the uninsulated face, is

$$u(x,0) = 1000.$$

For boundary conditions, at the open surface we have

$$-kA \frac{\partial u}{\partial x}\bigg|_{x=0} = -hA(u - 70),$$

because the rate at which heat leaves the surface must be equal to the rate at which heat flows to the surface.   The negative sign on the left is because heat flows in a direction opposite to a positive gradient; on the right it occurs because heat is being lost in the direction of negative x.   At the other side of the cube, the rate of heat flow is zero because of insulation:

$$\frac{\partial u}{\partial x}\bigg|_{x=4} = 0.$$

We plan to use the explicit method with

$$k \,\Delta t /c\rho(\Delta x)^2 = \tfrac{1}{4}.^*$$

---

*   The ratio must be smaller than $\tfrac{1}{2}$ to give stability with the derivative end conditions.

Suppose we let $\Delta x = 1$ inch. To calculate Δt, we need k, c, and ρ expressed in the units of inches, pounds, seconds, and °F. We first convert units:

$$k = \left(0.52 \frac{\text{cal}}{\text{sec} \cdot \text{cm} \cdot °\text{C}}\right)\left(\frac{1 \text{ Btu}}{252 \text{ cal}}\right)\left(\frac{2.54 \text{ cm}}{1 \text{ in.}}\right)\left(\frac{1°\text{C}}{1.8°\text{F}}\right)$$

$$= 0.00291 \frac{\text{Btu}}{\text{sec} \cdot \text{in.} \cdot °\text{F}},$$

$$c = \left(0.226 \frac{\text{cal}}{\text{g} \cdot °\text{C}}\right)\left(\frac{1 \text{ Btu}}{252 \text{ cal}}\right)\left(\frac{454 \text{ g}}{1 \text{ lb}}\right)\left(\frac{1°\text{C}}{1.8°\text{F}}\right)$$

$$= 0.226 \frac{\text{Btu}}{\text{lb} \cdot °\text{F}},$$

$$\rho = \left(2.70 \frac{\text{g}}{\text{cm}^3}\right)\left(\frac{1 \text{ lb}}{454 \text{ g}}\right)\left(\frac{2.54 \text{ cm}}{1 \text{ in.}}\right)^3$$

$$= 0.0975 \frac{\text{lb}}{\text{in.}^3},$$

$$t = \frac{c\rho(\Delta x)^2}{4k} = \frac{(0.226)(0.0975)(1)^2}{(4)(0.00291)} = 1.89 \text{ sec.}$$

For the ratio of values of Δt and Δx that we have chosen, our differential equation becomes

$$u_i^{j+1} = \tfrac{1}{4}(u_{i+1}^j + u_{i-1}^j) + \tfrac{1}{2}u_i^j, \tag{3.2}$$

which is to be applied at every point where u is unknown. In this example, this includes the points at $x = 0$ and $x = 4$ as well as the interior points. To enable us to write Eq. (3.2), we extend the domain of u one step on either side of the boundary. Let x_R be a fictitious point to the right of $x = 4$, and let x_L be a fictitious point to the left of $x = 0$. If x_1 signifies $x = 0$, and if x_5 signifies $x = 4$, we have the relations from Eq. (3.2)

$$u_1^{j+1} = \tfrac{1}{4}(u_2^j + u_L^j) + \tfrac{1}{2}u_1^j,$$
$$u_5^{j+1} = \tfrac{1}{4}(u_R^j + u_4^j) + \tfrac{1}{2}u_5^j.$$

Now we use the boundary conditions to eliminate the fictitious points, writing them as central difference quotients:

$$-k\frac{\partial u}{\partial x}\bigg|_{x=0} \doteq -(0.00291)\left(\frac{u_2^j - u_L^j}{2(1)}\right)$$

$$= -\frac{0.15}{144}(u_1^j - 70),$$

$$-k\frac{\partial u}{\partial x}\bigg|_{x=4} \doteq -(0.00291)\left(\frac{u_R^j - u_4^j}{2(1)}\right)$$

$$= 0.$$

The 144 factor in the first equation changes h to a basis of in². Solving for u_L and u_R, we have

$$u_L = u_2 - 0.716u_1 + 50.1,$$
$$u_R = u_4,$$

and the set of equations which give the temperatures becomes

$$u_1^{j+1} = \tfrac{1}{2}u_2^j + 0.32u_1^j + 12.5,$$
$$u_2^{j+1} = \tfrac{1}{4}(u_1^j + u_3^j) + \tfrac{1}{2}u_2^j,$$
$$u_3^{j+1} = \tfrac{1}{4}(u_2^j + u_4^j) + \tfrac{1}{2}u_3^j, \qquad (3.3)$$
$$u_4^{j+1} = \tfrac{1}{4}(u_3^j + u_5^j) + \tfrac{1}{2}u_4^j,$$
$$u_5^{j+1} = \tfrac{1}{2}u_4^j + \tfrac{1}{2}u_5^j.$$

A similar treatment of the boundary conditions using the Crank-Nicolson method with $r = 1$ leads to the set of simultaneous equations

$$4.716u_1^{j+1} - 2u_2^{j+1} = 2u_2^j - 0.716u_1^j,$$
$$-u_1^{j+1} + 4u_2^{j+1} - u_3^{j+1} = u_1^j + u_3^j,$$
$$-u_2^{j+1} + 4u_3^{j+1} - u_4^{j+1} = u_2^j + u_4^j, \qquad (3.4)$$
$$-u_3^{j+1} + 4u_4^{j+1} - u_6^{j+1} = u_3^j + u_5^j,$$
$$-2u_4^{j+1} + 4u_5^{j+1} = 2u_4^j.$$

For the set of equations in (3.4), Δt will be 7.56 sec. We advance the solution one time step at a time by repeatedly solving either (3.3) or (3.4), using the proper values of u_i^j.

4. STABILITY AND CONVERGENCE CRITERIA

We have previously stated that in order to ensure stability and convergence in the explicit method, the ratio $k \, \Delta t / c\rho(\Delta x)^2$ must be $\tfrac{1}{2}$ or less. The implicit Crank-Nicolson method has no such limitation. In this section, these phenomena and criteria will be studied in more detail.

By *convergence*, we mean that the results of the method approach the analytical values as Δt and Δx both approach zero. By *stability*, we mean that errors made at one stage of the calculations do not cause increasingly large errors as the computations are continued, but rather will eventually damp out.

We shall first discuss convergence, limiting ourselves to the simple case of the unsteady-state heat flow equation in one dimension:*

$$\frac{\partial U}{\partial t} = \frac{k}{c\rho} \frac{\partial^2 U}{\partial x^2}. \qquad (4.1)$$

Let us use the symbol U to represent the exact solution to Eq. (4.1), and u to represent the numerical solution. At the moment we assume that u is free of round-

---

* We could have treated the simpler equation $\partial U/\partial T = \partial^2 U/\partial X^2$ without loss of generality, for with the change of variables $X = \sqrt{c\rho}\, x$, $T = kt$, the two equations are seen to be identical.

off errors, so the only difference between U and u is the error made by replacing Eq. (4.1) by the difference equation.   Let $e_i^j = U_i^j - u_i^j$, at the point $x = x_i$, $t = t_j$. By the explicit method, Eq. (4.1) becomes

$$u_i^{j+1} = r(u_{i+1}^j + u_{i-1}^j) + (1 - 2r)u_i^j, \tag{4.2}$$

where $r = k \, \Delta t / c\rho(\Delta x)^2$.   Substituting $u = U - e$ into Eq. (4.2), we get

$$e_i^{j+1} = r(e_{i+1}^j + e_{i-1}^j) + (1 - 2r)e_i^j - r(U_{i+1}^j + U_{i-1}^j) - (1 - 2r)U_i^j + U_i^{j+1}. \tag{4.3}$$

By using Taylor series expansions, we have

$$U_{i+1}^j = U_i^j + \left(\frac{\partial U}{\partial x}\right)_{i,j} \Delta x + \frac{(\Delta x)^2}{2} \frac{\partial^2 U(\xi_1, t_j)}{\partial x^2}, \qquad x_i < \xi_1 < x_{i+1},$$

$$U_{i-1}^j = U_i^j - \left(\frac{\partial U}{\partial x}\right)_{i,j} \Delta x + \frac{(\Delta x)^2}{2} \frac{\partial^2 U(\xi_2, t_j)}{\partial x^2}, \qquad x_{i-1} < \xi_2 < x_i,$$

$$U_i^{j+1} = U_i^j + \Delta t \, \frac{\partial U(x_i, \eta)}{\partial t}, \qquad t_j < \eta < t_{j+1}.$$

Substituting these into Eq. (4.3) and simplifying, remembering that $r(\Delta x)^2 = k\Delta t / c\rho$, we get

$$e_i^{j+1} = r(e_{i+1}^j + e_{i-1}^j) + (1 - 2r)e_i^j + \Delta t \left[\frac{\partial U(x_i, \eta)}{\partial t} - \frac{k}{c\rho} \frac{\partial^2 U(\xi, t_j)}{\partial x^2}\right], \quad x_{i-1} < \xi < x_{i+1}. \tag{4.4}$$

Let E^j be the magnitude of the maximum error in the row of calculations for $t = t_j$, and let $M > 0$ be an upper bound for the magnitude of the expression in brackets in Eq. (4.4).   If $r \leq \frac{1}{2}$, all the coefficients in Eq. (4.4) are positive (or zero) and we may write the inequality

$$\left|e_i^{j+1}\right| \leq 2rE^j + (1 - 2r) E^j + M \, \Delta t = E^j + M \, \Delta t.$$

This is true for all the e_i^{j+1} at $t = t_{j+1}$, so

$$E^{j+1} \leq E^j + M \, \Delta t.$$

Since this is true at each time step,

$$E^{j+1} \leq E^j + M \, \Delta t \leq E^{j-1} + 2M \, \Delta t \leq \cdots \leq E^0 + (j+1)M \, \Delta t = E^0 + Mt_{j+1}$$

$$= Mt_{j+1},$$

because E^0, the errors at $t = 0$ are zero, since U is given by the initial conditions.
   Now, as $\Delta x \to 0$, $\Delta t \to 0$ if $k \, \Delta t / c\rho(\Delta x)^2 \leq \frac{1}{2}$, and $M \to 0$, because, as both Δx and Δt get smaller,

$$\left[\frac{\partial U(x_i, \eta)}{\partial t} - \frac{k}{c\rho} \frac{\partial^2 U(\xi, t_j)}{\partial x^2}\right] \to \left(\frac{\partial U}{\partial t} - \frac{k}{c\rho} \frac{\partial^2 U}{\partial x^2}\right)_{i,j} = 0.$$

This last is by virtue of Eq. (4.1), of course.   Consequently, we have shown that

Table 4.1. Propagation of Errors—Explicit Method

t	End point x_1	x_2	x_3	x_4	End point x_5
t_0	0	0	0	0	0
t_1	0	e	0	0	0
t_2	0	0	0.50e	0	0
t_3	0	0.25e	0	0.25e	0
t_4	0	0	0.25e	0	0
t_5	0	0.125e	0	0.125e	0
t_6	0	0	0.125e	0	0
t_7	0	0.062e	0	0.062e	0
t_8	0	0	0.062e	0	0

the explicit method is convergent for $r \leq \frac{1}{2}$, because the errors approach zero as Δt and Δx are made smaller.

For the solution to the heat flow equation by the Crank-Nicolson method, the analysis of convergence may be made by similar methods. The treatment is more complicated, but it can be shown that each E^{j+1} is no greater than a finite multiple of E^j plus a term that vanishes as both Δx and Δt become small, and this is independent of r. Hence, since the initial errors are zero, the finite difference solution approaches the analytical solution as $\Delta t \to 0$ and $\Delta x \to 0$, requiring only that r stay finite.

Let us begin our discussion of stability with a numerical example. Since the heat flow equation is linear, if two solutions are known, their sum is also a solution. We are interested in what happens to errors made in one line of the computations as the calculations are continued, and because of the additivity feature, the effect of a succession of errors is just the sum of the effects of the individual errors. We follow, then, a single error, which most likely occurred due to round-off. If this single error does not grow in magnitude, we will call the method *stable*, for then the cumulative effect of all errors affects the later calculations no more than a linear combination of the previous errors.

Table 4.1 illustrates the principle. We have calculated for the simple case where the boundary conditions are fixed, so that the errors at the end points are zero. We assume that a single error of size e occurs at $t = t_1$ and $x = x_2$. The explicit method, $k \, \Delta t / c\rho(\Delta x)^2 = \frac{1}{2}$, was used. The original error quite obviously dies out. As an exercise, it is left to the student to show that with $r > 0.5$, errors have an increasingly large effect on later computations. Table 4.2 shows that errors damp out for the Crank-Nicolson method with $r = 1$ even more rapidly than in the explicit method with $r = 0.5$.

In order to discuss stability in a more analytical sense, we need some material from linear algebra. In Chapter 8 we discussed eigenvalues and eigenvectors of

Table 4.2. Propagation of Errors—Crank-Nicolson Method

t	x_1	x_2	x_3	x_4	x_5
t_0	0	0	0	0	0
t_1	0	e	0	0	0
t_2	0	$0.071e$	$0.286e$	$0.071e$	0
t_3	0	$0.107e$	$0.143e$	$0.107e$	0
t_4	0	$0.049e$	$0.092e$	$0.049e$	0
t_5	0	$0.036e$	$0.053e$	$0.036e$	0
t_6	0	$0.022e$	$0.033e$	$0.022e$	0
t_7	0	$0.013e$	$0.020e$	$0.013e$	0
t_8	0	$0.008e$	$0.013e$	$0.008e$	0

a matrix.   We recall that for the matrix A and vector x, if

$$Ax = \lambda x,$$

then the scalar λ is an eigenvalue of A and x is the corresponding eigenvector.   If the N eigenvalues of the N by N matrix A are all different, then the corresponding N eigenvectors are linearly independent, and any N-component vector can be written uniquely in terms of them.

Consider an unsteady-state heat flow problem with fixed boundary conditions. Suppose we subdivide into $N + 1$ subintervals so there are N unknown values of the temperature being calculated at each time step.   Think of these N values as the components of a vector.   Our algorithm for the explicit method (Eq. 1.4) can be written as the matrix equation

$$\begin{bmatrix} u_1^{j+1} \\ u_2^{j+1} \\ \cdot \\ \cdot \\ \cdot \\ u_N^{j+1} \end{bmatrix} = \begin{bmatrix} (1-2r) & r & & & \\ r & (1-2r) & r & & \\ & & \cdot & & \\ & & & \cdot & \\ & & & r & (1-2r) \end{bmatrix} \begin{bmatrix} u_1^{j} \\ u_2^{j} \\ \cdot \\ \cdot \\ \cdot \\ u_N^{j} \end{bmatrix} \qquad (4.5)$$

or

$$u^{j+1} = Au^j,$$

where A represents the coefficient matrix and u^j and u^{j+1} are the vectors whose N components are the successive calculated values of temperature.   The components of u^0 are the initial values from which we begin our solution.   The successive rows of our calculations are

$$u^1 = Au^0,$$
$$u^2 = Au^1 = A^2u^0,$$
$$\vdots$$
$$u^j = Au^{j-1} = A^2u^{j-2} = \cdots = A^ju^0.$$

(The superscripts on the A are here exponents; on the vectors they indicate time.)

Suppose that errors are introduced into u^0, so that it becomes $\bar{u}^0$. We shall follow the effects of this error through the calculations. The successive lines of calculation are now

$$\bar{u}^j = A\bar{u}^{j-1} = \cdots = A^j\bar{u}^0.$$

Let us define the vector e^j as $u^j - \bar{u}^j$ so that e^j represents the errors in u^j caused by the errors in u^0. We have

$$e^j = u^j - \bar{u}^j = A^j u^0 - A^j\bar{u}^0 = A^j e^0. \tag{4.6}$$

This shows that errors are propagated by using the same algorithm as that by which the temperatures are calculated, as was implicitly assumed earlier in this section.

Now the N eigenvalues of A are distinct (see below) so that its N eigenvectors $x_1, x_2, \ldots, x_N$ are independent, and

$$Ax_1 = \lambda_1 x_1,$$
$$Ax_2 = \lambda_2 x_2,$$
$$\vdots$$
$$Ax_N = \lambda_N x_N.$$

We now write the error vector e^0 as a linear combination of the x_i:

$$e^0 = c_1 x_1 + c_2 x_2 + \cdots + c_N x_N,$$

where the c's are constants. Then e^1 is, in terms of the x_i,

$$e^1 = Ae^0 = \sum_{i=1}^{N} Ac_i x_i = \sum_{i=1}^{N} c_i Ax_i = \sum_{i=1}^{N} c_i \lambda_i x_i,$$

and for e^2,

$$e^2 = Ae^1 = \sum_{i=1}^{N} Ac_i \lambda_i x_i = \sum_{i=1}^{N} c_i \lambda_i^2 x_i.$$

(Again, the superscripts on vectors indicate time; on λ they are exponents.) After i steps, Eq. (4.6) can be written

$$e^j = \sum_{i=1}^{N} c_i \lambda^j x_i.$$

If the magnitudes of all of the eigenvalues are less than or equal to unity, errors will not grow as the computations proceed, i.e., the computational scheme is stable. This then is the analytical condition for stability: that the largest eigenvalue of the coefficient matrix for the algorithm be one or less in magnitude.

The eigenvalues of matrix A (Eq. 4.5) can be shown to be (note they are all distinct):

$$1 - 4r \sin^2 \frac{n\pi}{2N}, \qquad n = 1, 2, \ldots, N.$$

We will have stability for the explicit scheme if

$$-1 \leq 1 - 4r \sin^2 \frac{n\pi}{2N} \leq 1.$$

The limiting value of r is given by

$$-1 \leq 1 - 4r \sin^2 \frac{n\pi}{2N},$$

$$r \leq \tfrac{1}{2} / \sin^2 \frac{n\pi}{2N}$$

Hence, if $r \leq \tfrac{1}{2}$, the explicit scheme is stable.

The Crank-Nicolson scheme, in matrix form, is

$$
\begin{bmatrix}
(2+2r) & -r & & & \\
-r & (2+2r) & -r & & \\
 & & \cdot & & \\
 & & & \cdot & \\
 & & & -r & (2+2r)
\end{bmatrix}
\begin{bmatrix}
u_1^{j+1} \\
u_2^{j+1} \\
\cdot \\
\cdot \\
u_N^{j+1}
\end{bmatrix}
=
\begin{bmatrix}
(2-2r) & r & & & \\
r & (2-2r) & r & & \\
 & & \cdot & & \\
 & & & \cdot & \\
 & & & r & (2-2r)
\end{bmatrix}
\begin{bmatrix}
u_1^{j} \\
u_2^{j} \\
\cdot \\
\cdot \\
u_N^{j}
\end{bmatrix},
$$

or $Au^{j+1} = Bu^j$.

We can write

$$u^{j+1} = (A^{-1}B)u^j,$$

so that stability is given by the magnitudes of the eigenvalues of $A^{-1}B$. These are

$$\frac{2 - 4r \sin^2 (n\pi/2N)}{2 + 4r \sin^2 (n\pi/2N)}, \qquad n = 1, 2, \ldots, N.$$

Clearly all the eigenvalues are no greater than one in magnitude for any positive value of r.

With derivative boundary conditions, a similar analysis shows that the Crank-Nicolson method is stable for any positive value of r. For the explicit scheme, $r = \tfrac{1}{2}$ leads to instability with a finite surface coefficient. Smith (1965) shows that the limitation on r for stability is

$$r \leq \frac{1}{2 + P \Delta x},$$

where P is the ratio of surface coefficient to conductivity, h/k.

5. PARABOLIC EQUATIONS IN TWO OR MORE DIMENSIONS

In principle, we can readily extend the above methods to higher space dimensions, especially when the region is rectangular. The heat flow equation in two directions is

$$\frac{\partial u}{\partial t} = \frac{k}{c\rho} \left(\frac{\partial^2 u}{\partial x^2} + \frac{\partial^2 u}{\partial y^2} \right).$$

Taking $\Delta x = \Delta y$, and letting $r = k\,\Delta t / c\rho(\Delta x)^2$, we find that the explicit scheme becomes

$$u_{i,j}^{k+1} - u_{i,j}^{k} = r(u_{i+1,j}^{k} - 2u_{i,j}^{k} + u_{i-1,j}^{k} + u_{i,j+1}^{k} - 2u_{i,j}^{k} + u_{i,j-1}^{k})$$

or

$$u_{i,j}^{k+1} = r(u_{i+1,j}^{k} + u_{i-1,j}^{k} + u_{i,j+1}^{k} + u_{i,j-1}^{k}) + (1 - 4r)u_{i,j}^{k}.$$

In this scheme, the maximum value permissible for r in the simple case of constant end conditions is $\frac{1}{4}$. (Note that this corresponds again to the numerical value that gives a particularly simple formula.) In the more general case with $\Delta x \neq \Delta y$, the criterion is

$$\frac{k\,\Delta t}{c\rho\big((\Delta x)^2 + (\Delta y)^2\big)} \le \frac{1}{8}.$$

The analogous equation in three dimensions, with equal grid spacing each way, has the coefficient $(1 - 6r)$, and $r \le \frac{1}{6}$ is required for convergence and stability.

The difficulty with the use of the explicit scheme is that the restrictions on Δt require inordinately many rows of calculations. One then looks for a method in which Δt can be made larger without loss of stability. In one dimension, the Crank-Nicolson method was such a method. In the two-dimensional case, using averages of central difference approximations to give $\partial^2 u/\partial x^2$ and $\partial^2 u/\partial y^2$ at the mid-value of time, we get

$$u_{i,j}^{k+1} - u_{i,j}^{k} = \frac{r}{2}\Big[u_{i+1,j}^{k+1} - 2u_{i,j}^{k+1} + u_{i-1,j}^{k+1} + u_{i+1,j}^{k} - 2u_{i,j}^{k} + u_{i-1,j}^{k}$$

$$+ u_{i,j+1}^{k+1} - 2u_{i,j}^{k+1} + u_{i,j-1}^{k+1} + u_{i,j+1}^{k} - 2u_{i,j}^{k} + u_{i,j-1}^{k} \Big].$$

The problem now is that a set of $(M)(N)$ simultaneous equations must be solved at each time step, where M is the number of unknown values in the x-direction and N in the y-direction. Furthermore, the coefficient matrix is no longer tridiagonal, so the solution to each set of equations is slower.

The advantage of a tridiagonal matrix is retained in the alternating-direction-implicit scheme (ADI) proposed by Peaceman and Rachford (1955). It is widely used in modern computer programs for the solution of parabolic partial differential equations. In this method, we average a central difference approximation to $\partial^2 u/\partial x^2$ written at the beginning of the interval with a similar expression for $\partial^2 u/\partial y^2$ written at the end:

$$u_{i,j}^{k+1} - u_{i,j}^{k} = \frac{r}{2}\Big[\underbrace{u_{i+1,j}^{k} - 2u_{i,j}^{k} + u_{i-1,j}^{k}}_{\text{from }\partial^2 u/\partial x^2\text{ at start}} + \underbrace{u_{i,j+1}^{k+1} - 2u_{i,j}^{k+1} + u_{i,j-1}^{k+1}}_{\text{from }\partial^2 u/\partial y^2\text{ at end}} \Big].$$

The obvious bias in this formula is balanced by reversing the order for the second derivative approximations in the next time span:

$$u_{i,j}^{k+2} - u_{i,j}^{k+1} = \frac{r}{2}\Big[\underbrace{u_{i+1,j}^{k+2} - 2u_{i,j}^{k+2} + u_{i-1,j}^{k+2}}_{\text{from }\partial^2 u/\partial x^2\text{ at end}} + \underbrace{u_{i,j+1}^{k+1} - 2u_{i,j}^{k+1} + u_{i,j-1}^{k+1}}_{\text{from }\partial^2 u/\partial y^2\text{ at start}} \Big].$$

The compensation of errors produced by this alternation of direction gives a scheme that is convergent and stable for all values of r, although accuracy requires r to not be too large.   The three-dimensional analog alternates three ways, returning to each of the three formulas after every third step.   (Unfortunately the three-dimensional case is not stable for all fixed values of $r > 0$.)   When the formulas are rearranged, in each case tridiagonal coefficient matrices result.

When the region in which the heat flow equation is satisfied is not rectangular, one may perturbate the boundary to make it agree with a square mesh or interpolate from boundary points to estimate u at adjacent mesh points as discussed in Chapter 9 for elliptic equations.   The frequency with which circular or spherical regions occur makes it worthwhile to mention the heat equation in polar and spherical coordinates.   The basic equation

$$\frac{\partial u}{\partial t} = \frac{k}{c\rho} \nabla^2 u$$

becomes in polar coordinates (r,θ)

$$\frac{\partial u}{\partial t} = \frac{k}{c\rho} \left(\frac{\partial^2 u}{\partial r^2} + \frac{1}{r} \frac{\partial u}{\partial r} + \frac{1}{r^2} \frac{\partial^2 u}{\partial \theta^2} \right),$$

and in spherical coordinates (r,ϕ,θ)

$$\frac{\partial u}{\partial t} = \frac{k}{c\rho} \left(\frac{\partial^2 u}{\partial r^2} + \frac{2}{r} \frac{\partial u}{\partial r} + \frac{1}{r^2} \frac{\partial^2 u}{\partial \theta^2} + \frac{\cot \phi}{r} \frac{\partial u}{\partial \theta} + \frac{1}{r^2 \sin^2\theta} \frac{\partial^2 u}{\partial \phi^2} \right).$$

Using finite difference approximations to convert these to difference equations is straightforward except at the origin where $r = 0$.   For this point, consider $\nabla^2 u$ in rectangular coordinates, so that, in two dimensions,

$$\nabla^2 u = \frac{u_{i+1,j} + u_{i-1,j} + u_{i,j+1} + u_{i,j-1} - 4u_{i,j}}{(\Delta r)^2}, \qquad u_{ij} \text{ at } r = 0.$$

This is exactly the same as the expression for the Laplacian in Chapter 9, Eq. (2.3). This expression for $\nabla^2 u$ is obviously independent of the orientation of the axes. We get the best value by using the average value of all points which are a distance Δr from the origin, so that for $r = 0$,

$$\nabla^2 u = \frac{4(u_{av} - u_0)}{(\Delta r)^2} \qquad \text{at } r = 0.$$

The corresponding relation for spherical coordinates is

$$\nabla^2 u = \frac{6(u_{av} - u_0)}{(\Delta r)^2} \qquad \text{at } r = 0.$$

6. PROGRAMS TO SOLVE PARABOLIC EQUATIONS

Two computer programs are presented as examples of how parabolic partial differential equations in one space dimension can be solved on a computer.   Program 1 uses the explicit method of Section 1; Program 2 employs the implicit Crank-

Nicolson method of Section 2.   In the first example, both ends are held at a constant temperature.   In the second, however, derivative end conditions are permitted.

For Program 1, the fundamental difference equation which approximates the differential equation is Eq. (1.4),

$$u_i^{j+1} = r(u_{i+1}^j + u_{i-1}^j) + (1 - r)u_i^j,$$

where

$$r = \frac{k \, \Delta t}{c\rho(\Delta x)^2}.$$

This relation is applied at each interior point.   The constant end conditions, $u(0,t)$ and $u(L,t)$, are read in.   At the end of each time step, the interior temperatures are computed and a line of temperatures is printed.

The program is tested by computing the temperatures in a 10-cm long bar with $k = 0.53$, $c = 0.226$, $\rho = 2.70$, $\Delta x = 2.0$, $r = 0.25$ (all in cgs units).   The bar is initially at 20°C at all points, and temperatures are caused to change by suddenly cooling one end to 0°C and heating the other end to 100°C.

Program 2, using the Crank-Nicolson method, is more elaborate in that derivative end conditions are permitted, in the form

$$\frac{\partial u}{\partial x} = \alpha(u - v),$$

where u is the temperature at the end of the bar and v is the temperature of the surroundings.   This equation simulates the loss of heat by convection.   The program generates the coefficients of the set of simultaneous equations that are to be solved, which are

At left end: $-(4 + 2\alpha \, \Delta x)u_1^{j+1} + 2u_2^{j+1} = -2u_2^j + 2\alpha \, \Delta x u_1^j - 4\alpha \, \Delta x v;$

Interior points: $u_{i-1}^{j+1} - 4u_i^{j+1} + u_{i+1}^{j+1} = -u_{i-1}^j - u_{i+1}^j;$

At right end: $2u_{n-1}^{j+1} - (4 + 2\alpha \, \Delta x)u_n^{j+1} = -2u_{n-1}^j + 2\alpha \, \Delta x u_n^j - 4\alpha \, \Delta x v.$

The coefficients are compressed into an n by 4 array and a subroutine similar to the program to solve a tridiagonal system in Chapter 7 is employed.   Temperatures are not printed after each time step, but only after about 100-second intervals.

Program 1

```
ZZJOB 5                        CSC   001      GERALD. C. F.
ZZFORX5
C    PROGRAM TO COMPUTE THE TEMPERATURE DISTRIBUTION AS A FUNCTION OF
C    TIME IN A ONE DIMENSIONAL HEAT FLOW SYSTEM WITH CONSTANT END
C    CONDITIONS. USING THE EXPLICIT METHOD.
C    INITIAL TEMPERATURES ARE READ IN. ALSO FIXED TEMPERATURES AT EACH END
     DIMENSION U(50,2)
C    READ IN SYSTEM PARAMETERS.  L = TOTAL LENGTH. CON = THERMAL CONDUCT-
C    IVITY. CAP = HEAT CAPACITY. DENS = DENSITY. DELX = DISTANCE INTERVAL.
C    ALL MUST BE IN CONSISTENT UNITS.
C    ALSO READ IN RATIO = (CON * DELT) / (CAP * DENS * DELX SQUARE). AND
C    FINAL TIME TO WHICH COMPUTATIONS WILL CONTINUE.
     READ 100.EL. CON. CAP. DENS. DELX. RATIO. TFINAL
C    READ IN INTIAL TEMPERATURES AFTER FIRST COMPUTING NUMBER OF
```

(Continued)

Program 1 Cont.

```
C  U VALUES IN THE ARRAY.
       NUMB =EL/DELX + 1.
       READ 100, (U(I,1) , I = 1, NUMB)
C  PRINT HEADING AND INITIAL TEMPERATURES.
       T = 0.
       PRINT 200, DELX
       PRINT 201, T, (U(I,1), I = 1, NUMB)
C  READ IN BOUNDARY TEMPERATURES.
       READ 100, U(1,1), U(NUMB,1)
C  COMPUTE SET OF TEMPERATURES AFTER ONE DELT INTERVAL.
       DELT = RATIO*CAP*DENS*(DELX**2)/CON
       NM1 = NUMB - 1
     5 T = T + DELT
       DO 10 I = 2,NM1
    10 U(I,2) = (1. - 2.*RATIO)*U(I,1) + RATIO*(U(I+1,1) + U(I-1,1))
C  SET NEW TEMPERATURES INTO OLD ARRAY AND PRINT.
       DO 20 I = 2,NM1
    20 U(I,1) = U(I,2)
       PRINT 201, T, (U(I,1), I = 1,NUMB)
       IF (T - TFINAL) 5, 30, 30
    30 CALL EXIT
   100 FORMAT (10F8.0)
   200 FORMAT (1H1, 30X, 60H UNSTEADY STATE TEMPERATURE DISTRIBUTION BY E
      1XPLICIT METHOD  /30X, 40H   TEMPERATURES ARE GIVEN AT SPACING OF
      2 F6.1, 10H CM. APART   )
   201 FORMAT (1H0, 8H AT T =  F10.3, 8H SECONDS /(1H , 10F10.3))
       END
10.0      0.52      0.226     2.70      2.0       0.25      100.0
20.       20.       20.       20.       20.       20.
0.0       100.0
ZZZZ
```

              UNSTEADY STATE TEMPERATURE DISTRIBUTION BY EXPLICIT METHOD
              TEMPERATURES ARE GIVEN AT SPACING OF   2.0 CM. APART

AT T =   0.000 SECONDS
    20.000   20.000   20.000   20.000   20.000   20.000

AT T =   1.173 SECONDS
     0.000   15.000   20.000   20.000   40.000   100.000

AT T =   2.346 SECONDS
     0.000   12.500   18.750   25.000   50.000   100.000

AT T =   3.520 SECONDS
     0.000   10.937   18.750   29.687   56.250   100.000

AT T =   4.693 SECONDS
     0.000   10.156   19.531   33.593   60.546   100.000

AT T =   5.867 SECONDS
     0.000   9.960   20.703   36.816   63.671   100.000

AT T =   7.040 SECONDS
     0.000   10.156   22.045   39.501   66.040   100.000

AT T =   8.214 SECONDS
     0.000   10.589   23.437   41.772   67.895   100.000

AT T =   9.387 SECONDS
     0.000   11.154   24.809   43.719   69.390   100.000

AT T =   10.561 SECONDS
     0.000   11.779   26.123   45.409   70.625   100.000

AT T =   11.734 SECONDS
     0.000   12.420   27.358   46.891   71.665   100.000

AT T =   12.908 SECONDS
     0.000   13.049   28.507   48.201   72.555   100.000

(Continued)

Program 1 Cont.

```
AT T =     14.081 SECONDS
    0.000    13.651    29.566       49.366      73.328     100.000

AT T =     15.254 SECONDS
    0.000    14.217    30.538       50.407      74.005     100.000

AT T =     16.428 SECONDS
    0.000    14.743    31.425       51.339      74.604     100.000

AT T =     17.601 SECONDS
    0.000    15.227    32.233       52.177      75.137     100.000

AT T =     18.775 SECONDS
    0.000    15.672    32.967       52.931      75.612     100.000

AT T =     19.948 SECONDS
    0.000    16.078    33.634       53.610      76.039     100.000

AT T =     80.968 SECONDS
    0.000    19.978    39.965       59.965      79.978     100.000

AT T =     82.142 SECONDS
    0.000    19.980    39.968       59.968      79.980     100.000

AT T =     83.315 SECONDS
    0.000    19.982    39.971       59.971      79.982     100.000

AT T =     84.489 SECONDS
    0.000    19.984    39.974       59.974      79.984     100.000

AT T =     85.662 SECONDS
    0.000    19.985    39.976       59.976      79.985     100.000

AT T =     86.836 SECONDS
    0.000    19.987    39.979       59.979      79.987     100.000

AT T =     88.009 SECONDS
    0.000    19.988    39.981       59.981      79.988     100.000

AT T =     89.183 SECONDS
    0.000    19.989    39.982       59.982      79.989     100.000

AT T =     90.356 SECONDS
    0.000    19.990    39.984       59.984      79.990     100.000

AT T =     91.529 SECONDS
    0.000    19.991    39.986       59.985      79.991     100.000

AT T =     92.703 SECONDS
    0.000    19.992    39.987       59.987      79.992     100.000

AT T =     93.876 SECONDS
    0.000    19.992    39.988       59.988      79.992     100.000

AT T =     95.050 SECONDS
    0.000    19.993    39.989       59.989      79.993     100.000

AT T =     96.223 SECONDS
    0.000    19.994    39.990       59.990      79.994     100.000

AT T =     97.397 SECONDS
    0.000    19.994    39.991       59.991      79.994     100.000

AT T =     98.570 SECONDS
    0.000    19.995    39.992       59.992      79.995     100.000

AT T =     99.744 SECONDS
    0.000    19.995    39.993       59.993      79.995     100.000

AT T =    100.917 SECONDS
    0.000    19.996    39.993       59.993      79.996     100.000
```

Program 2

```
ZZJOB 5                              CSC   001     GERALD, C. F.
ZZFOR 5
*LIST PRINTER
*LDISK
        SUBROUTINE TRDG77 (X,N)
        DIMENSION X(50,4)
        DO 10 I=2,N.
        X(I,2)=X(I,2)-X(I,1)/X(I-1,2)*X(I-1,3)
     10 X(I,4)=X(I,4)-X(I,1)/X(I-1,2)*X(I-1,4)
        NMI= N - 1
        DO 20 I = 1,NMI
        M = N - I
     20 X(M,4) = X(M,4) - X(M,3)/X(M+1,2)*X(M+1,4)
        DO 30 I = 1,N
     30 X(I,4) = X(I,4)/X(I,2)
        RETURN
        END
ZZFORX5
*LIST PRINTER
C    THIS PROGRAM DETERMINES TEMPERATURES AS A FUNCTION OF TIME IN A
C    ONE-DIMENSIONAL SYSTEM LOSING HEAT FROM BOTH ENDS ACCORDING TO
C         DU/DX = ALPHA * (U - V)
C    WHERE V IS THE CONSTANT AMBIENT TEMPERATURE.
C    THE INITIAL TEMPERATURES ARE COMPUTED BY AN ARITHMETIC STATEMENT
C    FUNCTION.
C    A SUBROUTINE IS EMPLOYED TO SOLVE THE SET OF EQUATIONS WHICH
C    RESULT FROM APPLYING THE CRANK-NICOLSON METHOD.
C    THE TEMPERATURES ARE OUTPUT ABOUT EVERY 100 SECONDS UNTIL THE
C    MIDPOINT TEMPERATURE DROPS TO APPROXIMATELY 60 DEGREES.
        F(X) = 50. + .5*X
        DIMENSION U(50,4),  USTART(50)
        READ 100, C, RHO, COND, NOINT, ALPHA, V, XL
C    COMPUTE INITIAL VALUES OF TEMPS, DELTA X, AND DELTA T
        T = 0.
        TOUT = 100.
        XINT = NOINT
        N = NOINT +1
        DX = XL/XINT
        DT = C*RHO*DX**2/COND
        USTART(1)=F(0.)
        DO 10 I=1,NOINT
        XI = I
     10 USTART(I+1) = F(XI*DX)
        PRINT 199
        PRINT 200, T, (USTART(I), I = 1,N)
C    GET COEFFICIENTS OF SYSTEM
     21 U(1,1) = 0.
        U(1,2) = -4.-2.*ALPHA*DX
        U(1,3) = 2.
        U(N,1) = 2.
        U(N,2) = U(1,2)
        U(N,3) = 0.
        DO 20 I = 2,NOINT
        U(I,1) = 1.
        U(I,2) = -4.
     20 U(I,3) = 1.
        U(1,4) = -2.*USTART(2)+2.*ALPHA*DX*USTART(1)-4.*ALPHA*DX*V
        U(N,4)=-2.*USTART(N-1)+2.*ALPHA*DX*USTART(N)-4.*ALPHA*DX*V
        DO 30 I = 2,NOINT
     30 U(I,4) = -USTART(I-1) - USTART(I+1)

C    COMPUTE VALUES.  PRINT OUT ABOUT EVERY 100 SECONDS.
        T = T + DT
        CALL TRDG77 (U,N)
C    PUT NEW VALUES INTO USTART AND SEE IF TIME TO PRINT OUT.
        DO 40 I = 1,N
     40 USTART(I) = U(I,4)
        IF (T - TOUT) 21, 22, 22
     22 PRINT 200, T, (U(I,4), I = 1,N)
        TOUT = TOUT + 100.
        IF (USTART(11) - 60.) 23,21,21
```

(Continued)

Program 2 Cont.

```
   23 CALL EXIT
  100 FORMAT (3F10.0, I10, 3F10.0)
  199 FORMAT(/57H1SOLUTION TO HEAT FLOW PROBLEM BY CRANK-NICOLSON METHOD
     1  /)
  200 FORMAT (/5H T = F6.1, 8H SECONDS / (1H 10F8.2))
      END
.0928      8.89        .918            20   .04       20.          100.
ZZJOB 5
ZZDUP
*DELETTRDG77
ZZZZ
```

SOLUTION TO HEAT FLOW PROBLEM BY CRANK-NICOLSON METHOD

```
T =     0.0 SECONDS
  50.00    52.50    55.00    57.50    60.00    62.50    65.00    67.50    70.00    72.50
  75.00    77.50    80.00    82.50    85.00    87.50    90.00    92.50    95.00    97.50
 100.00

T =   112.3 SECONDS
  43.77    48.31    52.36    55.95    59.15    62.07    64.79    67.40    69.95    72.47
  74.96    77.41    79.79    82.01    83.92    85.23    85.55    84.34    81.08    75.38
  67.08

T =   202.2 SECONDS
  42.43    46.79    50.85    54.59    58.04    61.23    64.19    66.98    69.63    72.15
  74.52    76.69    78.58    80.04    80.87    80.84    79.68    77.16    73.07    67.35
  60.03

T =   314.5 SECONDS
  41.37    45.56    49.54    53.30    56.82    60.12    63.19    66.04    68.65    71.01
  73.06    74.74    75.96    76.58    76.48    75.50    73.53    70.46    66.24    60.87
  54.44

T =   404.4 SECONDS
  40.73    44.81    48.71    52.42    55.92    59.19    62.22    64.99    67.47    69.63
  71.40    72.72    73.51    73.67    73.11    71.74    69.48    66.29    62.15    57.10
  51.21

T =   516.7 SECONDS
  40.05    43.99    47.77    51.38    54.77    57.93    60.82    63.42    65.69    67.57
  69.02    69.98    70.38    70.15    69.25    67.62    65.22    62.03    58.08    53.40
  48.07

T =   606.6 SECONDS
  39.52    43.36    47.04    50.53    53.81    56.84    59.60    62.03    64.12    65.80
  67.03    67.75    67.93    67.51    66.44    64.71    62.28    59.17    55.39    50.98
  46.01

T =   718.9 SECONDS
  38.85    42.55    46.09    49.43    52.55    55.41    57.98    60.21    62.08    63.53
  64.53    65.04    65.02    64.43    63.25    61.47    59.07    56.08    52.51    48.41
  43.84

T =   808.8 SECONDS
  38.29    41.88    45.30    48.52    51.50    54.22    56.64    58.72    60.43    61.73
  62.58    62.96    62.82    62.15    60.92    59.13    56.79    53.91    50.51    46.63
  42.34

T =   921.1 SECONDS
  37.57    41.01    44.28    47.34    50.16    52.71    54.95    56.86    58.40    59.53
  60.23    60.47    60.23    59.50    58.25    56.49    54.23    51.49    48.29    44.67
  40.69

T =  1011.0 SECONDS
  36.98    40.30    43.44    46.38    49.07    51.49    53.61    55.39    56.80    57.83
  58.43    58.59    58.29    57.53    56.28    54.56    52.37    49.74    46.70    43.27
  39.51
```

PROBLEMS

Section 1

1. For molecular diffusion of matter due to concentration differences, the fundamental equation is

$$\text{Rate of flow of material} = -D\frac{\partial C}{\partial x},$$

where D is the so-called diffusion constant, and C is the volumetric concentration. Show that the equation for unsteady-state diffusion can be written

$$\frac{\partial^2 C}{\partial x^2} = \alpha\frac{\partial C}{\partial t}.$$

What is the relation of α to D?

2. The parameters of the basic equation for unsteady-state heat transfer are dimensional. If it is desired to measure u in °F and x in inches, how must the units of k, c, and ρ be chosen in

$$\frac{\partial^2 u}{\partial x^2} = \frac{c\rho}{k}\frac{\partial u}{\partial t}?$$

3. If C is measured in percent by volume, t is in minutes, and x is in inches, what must be the units of diffusivity D in Problem 1 to be consistent with these units?

4. Solve for the temperatures at $t = 2.062$ seconds in the 2-cm-thick steel slab of Section 1 if the initial temperatures are given by the relation

$$u(x,0) = 100 - 100|x - 1|.$$

Use the explicit method with $\Delta x = 0.25$ cm.

5. Solve for the temperatures in an aluminum rod 10 inches long, with the outer curved surface insulated so that heat flows in only one direction. The initial temperature is linear from 0°C at one end to 100°C at the other, when suddenly the hot end is brought to 0°C, and the cold end is held at 0°C. Use $\Delta x = 1$ inch and an appropriate value of Δt so that $k\,\Delta t/c\rho(\Delta x)^2 = \frac{1}{2}$. Look up values of k, c, and ρ in a handbook. Carry out the solution for 10 time steps.

6. Repeat Problem 5 with $\Delta x = 0.5$ inch, and compare the temperatures at points 1 inch, 3 inches, and 6 inches from the cold end in the two calculations.

7. Repeat Problem 5 with $\Delta x = 1$ inch, and Δt such that $k\,\Delta t/c\rho(\Delta x)^2 = \frac{1}{4}$. Compare results with Problems 5 and 6.

8. Repeat computations with Δx as given below for the diffusion example of Section 1, and compare to the analytical solution at $x = 4$ and $x = 12$ cm. Carry them out until $t = 250$ sec.

   (a) With $\Delta x = 2$ cm, $k\,\Delta t/c\rho(\Delta x)^2 = \frac{1}{2}$,
   (b) With $\Delta x = 4$ cm, $k\,\Delta t/c\rho(\Delta x)^2 = 0.1$.

Section 2

9. Solve Problem 4 by the Crank-Nicolson method, $r = 1$.

10. Solve Problem 5 by the Crank-Nicolson method, $r = 1$. Compare results with Problem 7.

11. The methods of Sections 1 and 2 can be applied readily to more complicated situations. For example, if heat is being generated at various points along a bar at a rate which is a function of x, the unsteady state heat equation becomes

$$k \frac{\partial^2 u}{\partial x^2} - cp \frac{\partial u}{\partial t} = f(x).$$

Solve this equation where $f(x) = x$ cal/cm$^3 \cdot$ sec., subject to conditions $u(0,t) = 0$, $u(1,t) = 0$, $u(x,0) = 0$. Take $\Delta x = 0.2$, $k = 0.37$, $cp = 0.433$ in cgs units. (These are properties of magnesium.) Use the Crank-Nicolson method, $r = 1$, and solve for five time steps.

Section 3

12. Use the set of equations in Eq. (3.2) to find the solution to the example in Section 3 through eight time steps.

13. Solve the set of equations in Eq. (3.4) to find the solution to the example in Section 3 by the Crank-Nicolson method. Find temperatures at $t = 15.12$ sec.

14. Solve the example of Section 3 except with two opposite faces of the cube losing heat at a rate equal to $0.15A(u - 70)$, where u is the surface temperature in °F, and A is the area. Use $\Delta x = 1$ inch, and employ the explicit method with $k \, \Delta t / cp(\Delta x)^2 = \frac{1}{4}$.

15. Solve Problem 14 using the Crank-Nicolson method with $r = 1$. Compare results by the two methods of solution.

16. Heat is added to one end of a 2-foot-long bar of copper at a rate given by

$$- hA(u - u_0),$$

and lost at the other end at a similar rate. $u_0 = 500$°F at the hot end, and $u_0 = 60$°F at the cold end, with resistance to heat flow being such that $h = 0.3$ Btu/sec. $\cdot$ ft^2 $\cdot$ °F at both ends. Find the time required for the midpoint of the bar to reach 100°F. The bar is initially at 60°F at all points.

Section 4

17. Demonstrate by performing calculations similar to those given in Table 4.1 that the explicit method is unstable with $k \, \Delta t / cp(\Delta x)^2 = 0.6$.

18. Demonstrate by performing calculations similar to those given in Table 4.1 that the explicit method with $k \, \Delta t / cp(\Delta x)^2 = \frac{1}{4}$ has errors that damp out more rapidly than those in Table 4.1.

19. Demonstrate by performing calculations similar to those in Table 4.1, except that at both $x = x_1$ and $x = x_5$ the gradient is zero ($\partial u / \partial x = 0$ instead of $u = $ constant), that the explicit method is still stable with $k \, \Delta t / cp(\Delta x)^2 = \frac{1}{2}$. Note, however, how

much more slowly an error damps out, and that the error at a later step becomes a linear combination of earlier errors.

20. Demonstrate by performing calculations similar to those in Table 4.2 that the Crank-Nicolson method is still stable even though the value of $k \, \Delta t / cp(\Delta x)^2$ is taken as 10. You will probably wish to invert the matrix of coefficients.

Section 5

21. A rectangular plate 2 by 3 inches is initially at 50°. At $t = 0$, one 2-inch edge is suddenly raised to 100°, and one 3-inch edge is suddenly cooled to 0°. The temperature on these two edges is then held constant at these temperatures. The other two faces are perfectly insulated. Use a 1-inch grid to subdivide the plate and write the ADI equations for each of the six points where unknown temperatures are involved. Use $r = 2$, and solve the equations for three time intervals.

22. Suppose the cube used for the example problem in Section 3 (4 by 4 by 4 inches) had heat flowing in all three directions, such as by having three adjacent faces lose heat by conduction to the flowing liquid. Set up the equations to solve for the temperature at any point. Do this in terms of the surrounding temperatures one Δt previously, using the explicit method with $\Delta x = \Delta y = \Delta z = 1$. How many time steps are needed to reach $t = 15.12$ sec? How many equations are involved in each stage of the calculation?

23. Repeat Problem 22 for the Crank-Nicolson method, $r = 1$.

24. Repeat Problem 22 using the ADI method.

Section 6

25. Use the first computer program to solve Problem 4, except extend the solution until the center temperature cools to 25°. Run with different values of Δx and observe the effect on the time for the center temperature to reach 25°. Utilize $r = 0.5$.

26. Use the first computer program to solve Problem 8.

27. Use the second computer program to solve Problem 15, except carry out the calculations for 10 time steps. You will need to modify the program to print out temperatures after each time step instead of at 100-sec intervals.

28. Modify Program 1 to allow derivative end conditions similar to those incorporated in Program 2. Test with a suitable problem.

29. Write a program to solve the heat flow equation in two space dimensions, utilizing the ADI method of Section 5.

11
hyperbolic
partial
differential
equations

The third classification of partial differential equations, hyperbolic differential equations, includes the "wave equation" which is fundamental to the study of vibrating systems. It is instructive to outline the derivation of the simple wave equation in one dimension. Imagine an elastic string stretched between two fixed end points and set to vibrating as sketched below. We make a number of simplifying assumptions: the string is perfectly elastic and we neglect gravitational forces, so that the only force is the tension force in the direction of the string; the string is uniform in density and thickness and has a weight per unit length of w lb/ft; the lateral displacement of the string is so small that the tension can be considered to be a constant value of T lb; and the slope of the string is hence small enough that $\sin \alpha \doteq \tan \alpha$, where α is the angle of inclination.

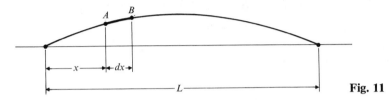

Fig. 11

We take $x = 0$ at the left end. L is the total length of the string. Focus attention on the element of length dx between points A and B. The uniform tension T acts at each end. We are interested in how y, the lateral displacement, varies with time t and with distance x along the string.

Element from x to $x+dx$

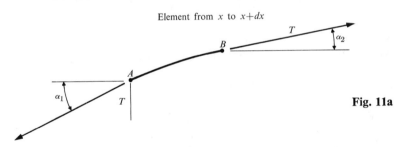

Fig. 11a

The forces acting in the y-direction are the vertical components of the two tensions. We take the upward direction as positive:

$$\text{Upward force at left end} = -T \sin \alpha_1 \doteq -T \tan \alpha_1 = -T\left(\frac{\partial y}{\partial x}\right)_A,$$

$$\text{Upward force at right end} = +T \sin \alpha_2 \doteq +T \tan \alpha_2 = T\left(\frac{\partial y}{\partial x}\right)_B$$

$$= T\left(\left(\frac{\partial y}{\partial x}\right)_A + \frac{\partial}{\partial x}\left(\frac{\partial y}{\partial x}\right)dx\right),$$

$$\text{Net force} = T\frac{\partial^2 y}{\partial x^2}dx.$$

Partials are used to express the slope because y is a function of both t and x. We use Newton's law, and equate the force to mass times acceleration in the vertical direction. Our simplifying assumptions permit us to use $w \, dx$ as the weight:

$$\left(\frac{w \, dx}{g}\right) \frac{\partial^2 y}{\partial t^2} = T \frac{\partial^2 y}{\partial x^2} \, dx, \qquad \frac{\partial^2 y}{\partial t^2} = \frac{Tg}{w} \frac{\partial^2 y}{\partial x^2}.$$

This is the wave equation in one dimension. The conditions imposed on the solution are the end conditions $[ay + b \, (\partial y / \partial x) = f(t)]$ at $x = 0$ and $x = L$, and the initial conditions at $t = 0$. Initial conditions specifying both $y = f(x)$ and the velocity $\partial y / \partial t = g(x)$ are usual for this problem.

1. SOLVING THE WAVE EQUATION BY FINITE DIFFERENCES

We attack the problem of solving the one-dimensional wave equation in the usual way by replacing derivatives by difference quotients. We use superscripts to denote time, and subscripts for position:

$$\frac{\partial^2 y}{\partial t^2} = \frac{Tg}{w} \frac{\partial^2 y}{\partial x^2},$$

$$\frac{y_i^{j+1} - 2y_i^j + y_i^{j-1}}{(\Delta t)^2} = \frac{Tg}{w} \left(\frac{y_{i+1}^j - 2y_i^j + y_{i-1}^j}{(\Delta x)^2} \right).$$

Solving for the displacement at the end of the current interval, at $t = t_{j+1}$, we get

$$y_i^{j+1} = \frac{Tg(\Delta t)^2}{w(\Delta x)^2} (y_{i+1}^j + y_{i-1}^j) - y_i^{j-1} + 2 \left(1 - \frac{Tg(\Delta t)^2}{w(\Delta x)^2} \right) y_i^j.$$

Note that selecting the ratio $Tg(\Delta t)^2 / w(\Delta x)^2 = 1$ gives some simplification, though y_i^{j+1} is still a function of conditions at both t_j and at t_{j-1}.

$$y_i^{j+1} = y_{i+1}^j + y_{i-1}^j - y_i^{j-1}, \qquad \Delta t = \frac{\Delta x}{\sqrt{Tg/w}}. \tag{1.1}$$

Equation (1.1) is the usual way that the wave equation is solved numerically. We are of course interested in the validity of choosing the ratio at unity. Our finite difference replacements to the derivatives do not here have mixed error terms as they did in the explicit method for the heat equation, but as we shall later show after discussing the method of characteristics, if the ratio is greater than one, we cannot be sure of convergence. Stability also sets a limit of unity to the ratio. It is surprising to find that if one uses a value of $w(\Delta t)^2 / Tg(\Delta x)^2$ of less than one, the results are less accurate, while with the ratio equal to one we can get exact analytical answers!

There is still a problem in applying Eq. (1.1). We know y at $t = t_0 = 0$ from the initial condition, but to compute y at $t = \Delta t = t_1$, we need values at $t = t_{-1}$. We may at first be bothered by a need to know displacements *before* the start of the problem, but if we imagine the function $y = y(x,t)$ to be extended backward in time, the term t_{-1} makes good sense. Since we will ordinarily get periodic functions for y versus t at a given point, we can consider zero time as an arbitrary point at which we know the displacement and velocity, within the duration of an ongoing process.

One commonly used way to get values for the fictitious point at $t = t_{-1}$ is by using the specification for the initial velocity $\partial y / \partial t = g(x)$ at $t = 0$. Using a central difference approximation, we have

$$\frac{\partial y}{\partial t}(x_i, 0) \doteq \frac{y_i^1 - y_i^{-1}}{2\Delta t} = g(x_i),$$

$$y_i^{-1} = y_i^1 - 2g(x_i)\,\Delta t \qquad \text{at } t = 0 \text{ only.} \tag{1.2}$$

Equation (1.2) is valid only at $t = 0$; substituting into Eq. (1.1) gives us, for $t = t_1$,

$$y_i^1 = \tfrac{1}{2}(y_{i+1}^0 + y_{i-1}^0) + g(x_i)\Delta t. \tag{1.3}$$

After computing the first line with Eq. (1.3), we use Eq. (1.1) thereafter.* If the boundary conditions involve derivatives, we rewrite them in terms of difference quotients to incorporate them in our difference equation.

Example. A banjo string is 80 cm long and weighs 1.0 gram. It is stretched with a tension of 40,000 grams. At a point 20 cm from one end it is pulled 0.6 cm from the equilibrium position and then released. Find the displacement of points along the string as a function of time. How long does it take for one complete cycle of motion? From this compute the frequency with which it vibrates.

Let $\Delta x = 10$ cm. In Table 1.1 we show the results of using Eqs. (1.3) and (1.1). Because the string is just released from its initially displaced position, the initial velocity at all points is zero, and Eq. (1.3) becomes simply

$$y_i^1 = \tfrac{1}{2}(y_{i+1}^0 + y_{i-1}^0).$$

The initial conditions imply that y is linear in x from $y = 0$ at $x = 0$ to $y = 0.6$ at $x = 20$ and also linear to $x = 80$. The size of time steps is given by

$$\frac{Tg(\Delta t)^2}{w(\Delta x)^2} = 1 = \frac{(40{,}000)(980)(\Delta t)^2}{(1.0)(10)^2}, \qquad \Delta t = 0.00179 \text{ sec.}$$

After we have completed calculations for eight time steps, we observe that the y-values are reproducing the original steps but with negative signs and the end of the string reversed, i.e., half a cycle has been completed in eight steps. For a complete cycle, 16 Δt's are needed. The frequency of vibration is then

$$f = \frac{1}{(16)(0.00179)} = 350 \text{ cycles/sec.}$$

(This value is exactly the same as given by the standard formula, $f = \dfrac{1}{2L}\sqrt{Tg/w}$.)

---

* We shall later discuss a more accurate way to begin the solution. Equation (1.3) is satisfactory when the initial velocity is zero.

Table 1.1. Solution to Banjo String Example

t	0	10	20	30	40	50	60	70	80
0	0.0	0.3	0.6	0.5	0.4	0.3	0.2	0.1	0.0
Δt	0.0	0.3	0.4	0.5	0.4	0.3	0.2	0.1	0.0
$2\Delta t$	0.0	0.1	0.2	0.3	0.4	0.3	0.2	0.1	0.0
$3\Delta t$	0.0	−0.1	0.0	0.1	0.2	0.3	0.2	0.1	0.0
$4\Delta t$	0.0	−0.1	−0.2	−0.1	0.0	0.1	0.2	0.1	0.0
$5\Delta t$	0.0	−0.1	−0.2	−0.3	−0.2	−0.1	0.0	0.1	0.0
$6\Delta t$	0.0	−0.1	−0.2	−0.3	−0.4	−0.3	−0.2	−0.1	0.0
$7\Delta t$	0.0	−0.1	−0.2	−0.3	−0.4	−0.5	−0.4	−0.3	0.0
$8\Delta t$	0.0	−0.1	−0.2	−0.3	−0.4	−0.5	−0.6	−0.3	0.0
$9\Delta t$	0.0	−0.1	−0.2	−0.3	−0.4	−0.5	−0.4	−0.3	0.0
$10\Delta t$	0.0	−0.1	−0.2	−0.3	−0.4	−0.3	−0.2	−0.1	0.0
$11\Delta t$	0.0	−0.1	−0.2	−0.3	−0.2	−0.1	0.0	0.1	0.0
$12\Delta t$	0.0	−0.1	−0.2	−0.1	0.0	0.1	0.2	0.1	0.0
$13\Delta t$	0.0	−0.1	0.0	0.1	0.2	0.3	0.2	0.1	0.0
$14\Delta t$	0.0	0.1	0.2	0.3	0.4	0.3	0.2	0.1	0.0
$15\Delta t$	0.0	0.3	0.4	0.5	0.4	0.3	0.2	0.1	0.0
$16\Delta t$	0.0	0.3	0.6	0.5	0.4	0.3	0.2	0.1	0.0

The solution to our example as given in Table 1.1 is exactly the analytical solution, as we shall now demonstrate. We shall compare our finite difference equation (Eq. 1.1), to the D'Alembert solution to the wave equation.

2. COMPARISON TO THE D'ALEMBERT SOLUTION

The one-dimensional wave equation, with $\sqrt{Tg/w} = c$, is

$$\frac{\partial^2 y}{\partial t^2} = c^2 \frac{\partial^2 y}{\partial x^2}. \tag{2.1}$$

By direct substitution, it is readily shown that for any arbitrary functions F and G, Eq. (2.1) is solved by

$$y(x,t) = F(x + ct) + G(x - ct). \tag{2.2}$$

The demonstration is easy since

$$\frac{\partial y}{\partial t} = F' \frac{\partial(x + ct)}{\partial t} + G' \frac{\partial(x - ct)}{\partial t} = cF' - cG',$$

$$\frac{\partial^2 y}{\partial t^2} = c^2 F'' - c(-c)G'' = c^2 F'' + c^2 G'', \tag{2.3}$$

$$\frac{\partial y}{\partial x} = F' \frac{\partial(x + ct)}{\partial x} + G' \frac{\partial(x - ct)}{\partial x} = F' + G',$$

$$\frac{\partial^2 y}{\partial x^2} = F'' + G''. \tag{2.4}$$

Equation (2.1) results immediately from Eqs. (2.3) and (2.4).

The solution to the problem is found then, if we can find a pair of functions of the form of (2.2) whose sum matches the initial conditions and the boundary conditions for the problem. If the initial conditions are

$$y(x,0) = f(x),$$

$$\frac{\partial y}{\partial t}(x,0) = g(x),$$

it is again readily shown that $y(x,t)$ is given by

$$y(x,t) = \tfrac{1}{2}[f(x + ct) + f(x - ct)] + \frac{1}{2c} \int_{x-ct}^{x+ct} g(v) \, dv. \tag{2.5}$$

Note that we can rewrite Eq. (2.5) in the form of Eq. (2.2). In Eq. (2.5) we have changed to the dummy variable v under the integral sign. The demonstration parallels that above when we recall how to differentiate an integral:

$$\frac{\partial}{\partial t} \left[\frac{1}{2c} \int_{x-ct}^{x+ct} g(v) \, dv \right] = \frac{1}{2c} [cg(x + ct) - (-c)g(x - ct)],$$

$$\frac{\partial^2}{\partial t^2} \left[\frac{1}{2c} \int_{x-ct}^{x+ct} g(v) \, dv \right] = \frac{\partial}{\partial t} \left[\frac{1}{2} g(x + ct) + \frac{1}{2} g(x - ct) \right] = \frac{1}{2} cg' - \frac{1}{2} cg' = 0,$$

$$\frac{\partial}{\partial x} \left[\frac{1}{2c} \int_{x-ct}^{x+ct} g(v) \, dv \right] = \frac{1}{2c} [g(x + ct) - g(x - ct)],$$

$$\frac{\partial^2}{\partial x^2} \left[\frac{1}{2c} \int_{x-ct}^{x+ct} g(v) \, dv \right] = \frac{\partial}{\partial x} \left[\frac{1}{2c} g(x + ct) - \frac{1}{2c} g(x - ct) \right] = \frac{1}{2c} g' - \frac{1}{2c} g' = 0.$$

Equation (2.5) gives the value of y at any time t provided that y is known at points ct to the right and left of the point at $t = 0$, and in terms of the integral of the initial velocity between the lateral points. Hence it is useful to find the displacement of points in the interior portions of a vibrating string.

We are now in a position to verify that our numerical procedure, given by Eq. (1.1), gives the exact solution (except for round-off) provided that two lines of correct y-values are known. The algorithm we use for the numerical procedure is

$$y_i^{j+1} = y_{i-1}^j + y_{i+1}^j - y_i^{j-1}. \tag{2.6}$$

We now show that the solution to this difference equation is a solution to the differential equation.   Consider the function

$$y_i^j = F(x_i + ct_j) + G(x_i - ct_j). \tag{2.7}$$

Because we use

$$\frac{Tg(\Delta t)^2}{w(\Delta x)^2} = 1 = \frac{c^2(\Delta t)^2}{(\Delta x)^2},$$

$$\Delta x = c \, \Delta t.$$

Write x_i and t_j in terms of starting values, x_0, and $t_0 = 0$:

$$x_i = x_0 + i \, \Delta x,$$

$$ct_j = c(t_0 + j \, \Delta t) = cj \, \Delta t = j \, \Delta x.$$

Substituting into Eq. (2.7), we get,

$$y_i^j = F(x_0 + i \, \Delta x + j \, \Delta x) + G(x_0 + i \, \Delta x - j \, \Delta x)$$
$$= F\big(x_0 + (i + j)\Delta x\big) + G\big(x_0 + (i - j)\Delta x\big).$$

Use this relation to rewrite the right side of the difference equation, Eq. (2.6):

$$y_{i-1}^j + y_{i+1}^j - y_i^{j-1} = F\big(x_0 + (i-1+j)\Delta x\big) + G\big(x_0 + (i-1-j)\Delta x\big)$$
$$+ F\big(x_0 + (i+1+j)\Delta x\big) + G\big(x_0 + (i+1-j)\Delta x\big)$$
$$- F\big(x_0 + (i+j-1)\Delta x\big) - G\big(x_0 + (i-j+1)\Delta x\big)$$
$$= F\big(x_0 + (i+j+1)\Delta x\big) + G\big(x_0 + (i-j-1)\Delta x\big) = y_i^{j+1}. \tag{2.8}$$

Equation (2.8) shows that if the previous two lines of the numerical solution are of the form of Eq. (2.7) (and hence must be exact solutions of the wave equation in view of Eq. (2.2), it then follows that the values on the next line are exact solutions also, because they also are of the form of Eq. (2.7).

In order that our simple algorithm give the analytical solution, it is necessary only that two lines of the computation be correct.   The first line is correct because it is given by the initial conditions.   While Eq. (1.3) is frequently recommended for giving the second line in terms of the initial velocity, it is sometimes inaccurate because it assumes that the initial velocity and the average velocity from t_{-1} to t_1 are the same, and this may not be true.

Equation (2.5) offers a more exact way to get the second line of y-values.   If we employ that relationship at $t = \Delta t$, we have

$$y(x, \Delta t) = y_i^1 = \frac{1}{2} [f(x_i + \Delta x) + f(x_i - \Delta x)] + \frac{1}{2c} \int_{x_i - \Delta x}^{x_i + \Delta x} g(v) \, dv$$

$$= \frac{1}{2} [y_{i+1}^0 + y_{i-1}^0] + \frac{1}{2c} \int_{x_{i-1}}^{x_{i+1}} g(v) \, dv. \tag{2.9}$$

Equation (2.9) differs from Eq. (1.3) only in the last term.   We now see why the banjo string example of Section 1 gave correct values in spite of using Eq. (1.3): in the example $g(x)$ was everywhere zero, so the form of the last term was unimportant.   In the next example we illustrate the difference between the two different ways to begin the solution.

Table 2.1. Comparison of Ways to Begin the Solution

All y values should be multiplied by Lk/c

Eq. (1.3): $y_i^1 = \frac{1}{2}(y_{i-1}^0 + y_{i+1}^0) + g(x_i)\Delta t$, $\Delta x = L/4$

t	$x = 0$	$x = 0.25L$	$x = 0.5L$	$x = 0.75L$	$x = L$
0	0	0	0	0	0
$L/4c$	0	0.1768	0.25	0.1768	0

		Analytical Solution			
$L/4c$	0	0.1592	0.2251	0.1592	0

Eq. (1.3) with $\Delta x = L/8$

t	$x = 0$	$x = 0.125L$	$x = 0.25L$	$x = 0.375L$	$x = 0.5L$
0	0	0	0	0	0
$L/8c$	0	0.0478	0.0884	0.1155	0.125
$L/4c$	0	0.0884	0.1633	0.2134	0.2310

Eq. (2.9): $y_i^1 = \dfrac{1}{2c} \displaystyle\int_{x_i-1}^{x_i+1} g(z)\, dz$, Simpson's rule, $\Delta x = 0.25L$

t	$x = 0$	$x = 0.25L$	$x = 0.5L$
$L/4c$	0	0.1595	0.2256

Eq. (2.9): Simpson's rule, $\Delta x = 0.125L$

$L/4c$	0	0.1592	0.2251

Example. A string whose length is L units is initially in its equilibrium position. It is set into motion by striking it so that the initial velocity is given by $\partial y/\partial t = k \sin(\pi x/L)$. If the ends are fixed, find the displacements at $t = L/4c$ sec later $(c = \sqrt{Tg/w})$.

Subdivide the length L into four intervals, so $\Delta x = L/4$, $\Delta t = \Delta x/c = L/4c$. The displacements we require are after one time step.   Table 2.1 summarizes the computations.   The values at $t = L/4c$ by Eq. (1.3) disagree by 10 percent from the analytical values, which are given by

$$y\left(x, \frac{L}{4c}\right) = \frac{1}{2c} \int_{x-c(L/4c)}^{x+c(L/4c)} k \sin \frac{\pi v}{L}\, dv = \frac{\sqrt{2}\, kL}{2c\pi} \sin \frac{\pi x}{L}.$$

Equation (1.3) gives improved values when Δx is cut in half to $L/8$; it now requires two steps to reach $t = L/4c$, of course.   This is also shown in Table 2.1.

Table 2.2. Propagation of Single Error in Numerical Solution to Wave Equation

Initially error-free values	0	0	0	0	0	0	0
	0	0	0	0	0	0	0
Error made here	0	0	1	0	0	0	0
	0	1	0	1	0	0	0
	0	0	0	0	1	0	0
	0	1	0	0	0	1	0
	0	0	1	0	0	0	0
	0	0	0	1	0	−1	0
	0	0	0	0	−2	0	0
	0	0	0	−1	0	1	0
	0	0	−1	0	0	0	0
	0	−1	0	0	0	1	0

Using Eq. (2.9), and evaluating the integral with Simpson's rule, $\Delta x = L/4$, gives nearly correct answers (0.2 percent errors), while with $\Delta x = L/8$ in the integration, the results are accurate to four decimal places.

3. STABILITY OF THE NUMERICAL METHOD

Since we will ordinarily solve the wave equation numerically only with $Tg(\Delta t)^2 / w(\Delta x)^2 = 1$, it is sufficient to demonstrate stability for that scheme. We assume a set of error-less computations have been made when a single error of size e occurs. We trace the effects of the single error. If the error does not have increasingly great effect on subsequent calculations, we call the method *stable*.

This simple procedure is adequate because, since the problem is linear, the principle of superposition lets us add the effects of all errors together and lets us add these errors to the true solution to obtain the actual results. Table 2.2 demonstrates the principle, assuming that the displacement of the end points is specified so that these are always free of error.

As the arrows indicate, the wave equation propagates disturbances in opposite directions, with reflections occurring at fixed ends, with a reversal of sign on reflection.

4. METHOD OF CHARACTERISTICS

The properties of the solution to the wave equation are further elucidated by considering the "characteristic curves" of the equation. This also will permit us to extend our numerical method to more general hyperbolic equations.

Consider the second-order partial differential equation in two variables x and t:

$$au_{xx} + bu_{xt} + cu_{tt} + e = 0. \tag{4.1}$$

Here we have used the subscript notation to represent partial derivatives. The coefficients a, b, c, and e may be functions of x, t, u_x, u_t, and u, so the equation is very general.* We take $u_{xt} = u_{tx}$. To facilitate manipulations, let

$$p = \frac{\partial u}{\partial x} = u_x, \qquad q = \frac{\partial u}{\partial t} = u_t.$$

Write out the differentials of p and q:

$$dp = \frac{\partial p}{\partial x} dx + \frac{\partial p}{\partial t} dt = u_{xx} dx + u_{xt} dt,$$

$$dq = \frac{\partial q}{\partial x} dx + \frac{\partial q}{\partial t} dt = u_{xt} dx + u_{tt} dt.$$

Solving these last equations for u_{xx} and u_{tt}, respectively, we have

$$u_{xx} = \frac{dp - u_{xt} dt}{dx} = \frac{dp}{dx} - u_{xt} \frac{dt}{dx},$$

$$u_{tt} = \frac{dq - u_{xt} dx}{dt} = \frac{dq}{dt} - u_{xt} \frac{dx}{dt}.$$

Substituting in Eq. (4.1) and rearranging, we obtain

$$-au_{xt} \frac{dt}{dx} + bu_{xt} - cu_{xt} + a\frac{dp}{dx} + c\frac{dq}{dt} + e = 0.$$

Now multiplying by $-dt/dx$, we get

$$u_{xt}\left[a\left(\frac{dt}{dx}\right)^2 - b\left(\frac{dt}{dx}\right) + c\right] - \left[a\frac{dp}{dx}\frac{dt}{dx} + c\frac{dq}{dx} + e\frac{dt}{dx}\right] = 0.$$

Suppose, in the xt-plane, we define curves such that the first bracketed expression is zero. On such curves, the original differential equation is equivalent to setting the second bracketed expression equal to zero. That is, if

$$am^2 - bm + c = 0, \tag{4.2}$$

where $m = dt/dx$, then the solution to the original equation (Eq. 4.1) can be found by solving

$$am\, dp + c\, dq + e\, dt = 0. \tag{4.3}$$

We have elected to write Eq. (4.3) in the form of differentials.

The curves whose slope m is given by Eq. (4.2) are called the characteristics of the differential equation. Since the equation is a quadratic, it may have one, two, or no real solutions, depending on whether $b^2 - 4ac = 0$. The value of this discriminant is the usual basis for classifying partial differential equations. If $b^2 - 4ac < 0$, the equation is called *elliptic*, and there are no (real) characteristics. If

---

* When the coefficients are independent of the function u or its derivatives, it is linear. If they are functions of u, u_x, or u_t (but not u_{xx} or u_{tt}), it is called quasi-linear.

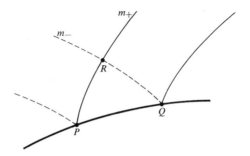

Fig. 4.1

$b^2 - 4ac = 0$, there is a single characteristic at any point, and the equation is termed *parabolic*. When $b^2 - 4ac > 0$, at every point there will be a pair of characteristic curves whose slopes are given by the two distinct, real roots of Eq. (4.2). Such equations are called *hyperbolic*, and our present discussion considers only this type.

Along the characteristics, the solution has special and desirable properties. For example, discontinuities in the initial conditions are propagated along them. On the characteristic curves, the numerical solution can be developed in the general case also.

We shall now outline a method of solving equations of the form of Eq. (4.1) by numerical integration along the characteristics. We visualize the initial conditions as specifying the function u on some curve in the tx-plane,* as well as its normal derivative. Consider two points, P and Q on this initial curve (Fig. 4.1). When Eq. (4.1) is hyperbolic, there are two characteristic curves through each point. The rightmost curve through P intersects the leftmost curve through Q, and these curves are such that their slopes are given by the appropriate roots of Eq. (4.2). Call m_+ the values of the slope on curve PR, and m_- the values on curve QR.

Since these curves are characteristics, the solution to the problem can be found by solving Eq. (4.3) along them.

Our procedure will be to first find point R (perhaps only as a first approximation if a, b, or c involve the unknown function u). This is done by solving the equation

$$\Delta t = \left(\frac{dt}{dx}\right)_{av} \Delta x = m_{av}\Delta x$$

applied over the arcs PR and QR simultaneously. We will use a procedure resembling the Euler predictor-corrector method, when m is a function of u, by predicting with m_{av} taken as equal to m_+ at P or m_- at Q to start the solution. We correct by using the arithmetic average of m at the end points of each arc as soon as the value of m at R can be evaluated.

---

* This curve must not itself be one of the characteristics, or advancing the solution is impossible.

We then integrate Eq. (4.3) in the form

$$am\,\Delta p + c\,\Delta q + e\,\Delta t = 0,$$

starting first from P and then from Q, using the appropriate values of m for each. This will estimate p and q at point R.   Finally we evaluate the function u at R from

$$du = \frac{\partial u}{\partial x}\,dx + \frac{\partial u}{\partial t}\,dt,$$

used in the form

$$\Delta u = p\,\Delta x + q\,\Delta t.$$

It may be necessary to iterate this procedure to get improved values at R. Average values are used for all varying quantities.   The calculations are repeated for a second point S which is the intersection of characteristics through Q and another initial point W, and then continued in a like manner to evaluate u throughout the region in the xt-plane as desired.   We illustrate with an example.

Example. Solve the quasi-linear equation, with conditions as shown, by numerical integration along the characteristics. (This might be a vibrating string with tension related to the displacement u and subject to an external lateral force.)

$$\frac{\partial^2 u}{\partial x^2} - u\frac{\partial^2 u}{\partial t^2} + (1 - x^2) = 0, \qquad u(x,0) = x(1 - x), \qquad u_t(x,0) = 0, \tag{4.4}$$

$$u(0,t) = 0, \qquad u(1,t) = 0.$$

We shall advance the solution beyond the start from P, at $x = 0.2$, and Q, at $x = 0.4$, to one new point R.   Comparing Eq. (4.4) to the standard form,

$$au_{xx} + bu_{xt} + cu_{tt} + e = 0,$$

we have $a = 1$, $b = 0$, $c = -u$, $e = 1 - x^2$.   We first compute u, p, and q at points P and Q:

$$u = x(1 - x),$$
$$u_P = 0.2(1 - 0.2) = 0.16,$$
$$u_Q = 0.4(1 - 0.4) = 0.24,$$

$$p = \frac{\partial u}{\partial x} = 1 - 2x,$$
$$p_P = 1 - 2(0.2) = 0.6,$$
$$p_Q = 1 - 2(0.4) = 0.2,$$

$$q = \frac{\partial u}{\partial t} = 0,$$
$$q_P = 0,$$
$$q = 0.$$

To locate point R, we need the slope m of the characteristic:

$$am^2 - bm + c = 0,$$

$$m = \frac{b \pm \sqrt{b^2 - 4ac}}{2a},$$

$$m = \frac{\pm \sqrt{4u}}{2} = \pm \sqrt{u}.$$

Since m depends on the solution u, we shall need to find point R through the predictor-corrector approach. In the first trial, use the initial values over the whole arc, i.e., take $m_+ = m_P$ and $m_- = m_Q$.

$$m_+ = \sqrt{u_P} = \sqrt{0.16} = 0.4,$$
$$m_- = \sqrt{u_Q} = -\sqrt{0.24} = -0.490.$$

We now estimate the coordinates of R by solving simultaneously

$$t_R = m_+(x_R - x_P) = 0.4(x_R - 0.2),$$
$$t_R = m_-(x_R - x_Q) = -0.490(x_R - 0.4).$$

These give

$$x_R = 0.310, \qquad t_R = 0.044.$$

We write Eq. 4.3 along each characteristic, again using the initial values of m since m at R is still unknown:

$$am \, \Delta p + c \, \Delta q + e \, \Delta t = 0,$$

$$(1)(0.4)(p_R - 0.6) + (-0.16)(q_R - 0) + \left(1 - \frac{0.04 + 0.096}{2}\right)(0.044) = 0,$$

$$(1)(-0.490)(p_R - 0.2) + (-0.24)(q_R - 0) + \left(1 - \frac{0.16 + 0.096}{2}\right)(0.044) = 0.$$

In these equations we used the arithmetic average of x^2 in the last terms. Solving simultaneously we get

$$p_R = 0.399, \qquad q_R = -0.246.$$

As a first approximation for u at R, then,

$$\Delta u = p \, \Delta x + q \, \Delta t,$$

$$u_R - 0.16 = \frac{0.6 + 0.399}{2}(0.310 - 0.2) + \frac{0 - 0.246}{2}(0.044 - 0),$$

$$u_R = 0.2095.$$

The last computation was along PR. We could have alternatively proceeded along QR. If this is done,

$$u_R - 0.24 = \frac{0.2 + 0.399}{2}(0.310 - 0.4) + \frac{0 - 0.246}{0}(0.044 - 0),$$

$$u_R = 0.2076.$$

The two values should be close to each other. Let us use the average value, 0.2086, as our initial estimate of u_R. We now repeat the work. In getting the coordinates of R, we now use average values of the slopes,

$$t_R = \frac{0.4 + \sqrt{0.2086}}{2}(x_R - 0.2),$$

$$t_R = \frac{-0.490 - \sqrt{0.2086}}{2}(x_R - 0.4),$$

$$x_R = 0.305, \qquad t_R = 0.045.$$

$$(1)\left(\frac{0.4 + \sqrt{0.2086}}{2}\right)(p_R - 0.6) - \left(\frac{0.16 + 0.2086}{2}\right)(q_R - 0)$$

$$+ \left(1 - \frac{0.04 + 0.0930}{2}\right)(0.045) = 0,$$

$$(1)\left(\frac{-0.490 - \sqrt{0.2086}}{2}\right)(p_R - 0.2) - \left(\frac{0.24 + 0.2086}{2}\right)(q_R - 0)$$

$$+ \left(1 - \frac{0.16 + 0.0930}{2}\right)(0.045) = 0,$$

$$p_R = 0.398, \qquad q_R = -0.242.$$

$$u_R = 0.16 + \frac{0.6 + 0.398}{2}(0.305 - 0.2) + \frac{0 - 0.242}{2}(0.045 - 0),$$

$$u_R = 0.2071 \quad \text{(along } PR\text{)},$$

$$u_R = 0.24 + \frac{0.2 + 0.398}{2}(0.305 - 0.4) + \frac{0 - 0.242}{2}(0.045 - 0),$$

$$u_R = 0.2063 \quad \text{(along } QR\text{)}.$$

The average value is 0.2067.

Another round of calculations gives $u_R = 0.2066$, which checks the previous value sufficiently. This method is of course very tedious by hand.

For the simple wave equation,

$$u_{tt} = c^2 u_{xx},$$

the slopes of the characteristics are

$$m = \pm \frac{1}{c},$$

and the characteristics are the lines

$$t = \pm \frac{1}{c}(x - x_i).$$

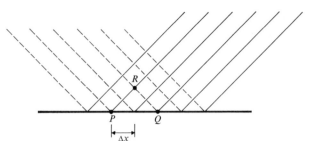

Fig. 4.2

Consider the curves from points P and Q, taken on the line for $t = 0$ (Fig. 4.2). The network of points used in the explicit finite-difference method of Section 1 are seen to be the intersections of characteristics through pairs of points spaced $2\,\Delta x$ apart. The finite-difference method, with $Tg(\Delta x)^2/w(\Delta t)^2 = 1$, will be found to be the equivalent of integration along the characteristics, lending further support to its giving exact answers.

5. A PROGRAM FOR THE SIMPLE WAVE EQUATION

Program 1 uses the easy algorithm of Section 1, but starts the solution by use of Eq. (2.9). The integral of the initial velocity is computed by Simpson's $\frac{1}{3}$ rule with the interval from x_{i-1} to x_{i+1} subdivided into eight panels. The initial displacement $f(x)$ and the initial velocity $g(x)$ are each defined by arithmetic statement functions. The program is tested by calculating the example problem of Section 2, in which a string initially at rest is set vibrating by giving it an initial velocity of $g(x) = \sin(\pi x/L)$. Input values are $L = 20$, $\Delta x = 1$, $c = \sqrt{Tg/w} = 1$, so that five time steps are here required to reach the time $t = L/4c$.

Program 1

```
ZZJOB  5                        CSC   001    GERALD. C. F.
ZZFORX5
*LIST PRINTER
C    PROGRAM TO SOLVE THE WAVE EQUATION IN ONE DIMENSION.
C    F(X) DEFINES INITIAL DISPLACEMENT. G(X) DEFINES THE INITIAL
C    VELOCITY. DISPLACEMENTS AFTER FIRST TIME STEP  ARE COMPUTED USING A
C    SIMPSONS RULE INTEGRATION.
C    SET UP ARRAY TO STORE DISPLACEMENT AT THREE TIME INTERVALS.
         DIMENSION Y(3,21)
C    DEFINE FUNCTIONS
         F(X) = 0.
         G(X) = SINF (3.1415926*X/EL)
C    READ IN PARAMTERS
         READ 100,EL, DELX, C
C    PRINT HEADING AND INITIAL DISPLACEMENT
         PRINT 200
         T = 0.
         X = 0.
         DO 10 I = 1,21
         Y(1,I) = F(X)
   10 X = X + DELX
         PRINT 201, T, (Y(1,I), I = 1,21)
```

(Continued)

Program 1 Cont.

```
C   COMPUTE DISPLACEMENTS AFTER FIRST TIME STEP AND PRINT
       Y(2,1) = 0.
       Y(2,21) = 0.
       T = T + DELX/C
       X = 0.
       DO 20 I = 2,20
       SUM = G(X) + G(X + 2.*DELX) + 4.*G(X + 0.25*DELX)
       DO 25 J = 1,3
       SUM = SUM + 2.*G(X + 0.5*DELX) + 4.*G(X + 0.75*DELX)
    25 X = X + 0.5*DELX
       X = X - 0.5*DELX
    20 Y(2,I) = 0.5*(Y(1,I-1) + Y(1,I+1)) + SUM/(2.*C)
       PRINT 201, T, (Y(2,I), I = 1,21)
C   COMPUTE DISPLACEMENTS AFTER ADDITIONAL TIME STEPS AND PRINT.
C   THE ARRAY MUST BE REARRANGED EACH TIME.
    15 DO 30 I = 2,20
    30 Y(3,I) = Y(2,I-1) + Y(2,I+1) - Y(1,I)
       DO 40 I = 2,20
       Y(1,I) = Y(2,I)
    40 Y(2,I) = Y(3,I)
       T = T + DELX/C
       PRINT 201, T, (Y(2,I), I = 1,21)
       IF (T - 40.) 15, 50, 50
    50 CALL EXIT
   100 FORMAT (3F10.0)
   200 FORMAT (45H1 FINITE DIFFERENCE SOLUTION TO WAVE EQUATION )
   201 FORMAT (9H0 AT T = ,F5.1, 19H DISPLACEMENTS ARE //(1H ,10F10.2))
       END
20.        1.         1.
ZZZZ
```

```
    FINITE DIFFERENCE SOLUTION TO WAVE EQUATION

AT T =     0.0 DISPLACEMENTS ARE

    0.00     0.00     0.00     0.00     0.00     0.00     0.00     0.00     0.00     0.00
    0.00     0.00     0.00     0.00     0.00     0.00     0.00     0.00     0.00     0.00
    0.00

AT T =     1.0 DISPLACEMENTS ARE

    0.00     1.86     3.69     5.42     7.02     8.45     9.66    10.64    11.36    11.80
   11.95    11.80    11.36    10.64     9.66     8.45     7.02     5.42     3.69     1.86
    0.00

AT T =     2.0 DISPLACEMENTS ARE

    0.00     3.69     7.29    10.71    13.87    16.69    19.09    21.03    22.45    23.31
   23.60    23.31    22.45    21.03    19.09    16.69    13.87    10.71     7.29     3.69
    0.00

AT T =     3.0 DISPLACEMENTS ARE

    0.00     5.42    10.71    15.74    20.38    24.52    28.05    30.90    32.98    34.25
   34.68    34.25    32.98    30.90    28.05    24.52    20.38    15.74    10.71     5.42
    0.00

AT T =     4.0 DISPLACEMENTS ARE

    0.00     7.02    13.87    20.38    26.39    31.75    36.32    40.00    42.70    44.35
   44.90    44.35    42.70    40.00    36.32    31.75    26.39    20.38    13.87     7.02
    0.00

AT T =     5.0 DISPLACEMENTS ARE

    0.00     8.45    16.69    24.52    31.75    38.19    43.70    48.13    51.37    53.35
   54.01    53.35    51.37    48.13    43.70    38.19    31.75    24.52    16.69     8.45
    0.00
```

(Continued)

Program 1 Cont.

```
AT T =    6.0 DISPLACEMENTS ARE

   0.00    9.66   19.09   28.05   36.32   43.70   50.00   55.06   58.77   61.04
  61.80   61.04   58.77   55.06   50.00   43.70   36.32   28.05   19.09    9.66
   0.00

AT T =    7.0 DISPLACEMENTS ARE

   0.00   10.64   21.03   30.90   40.00   48.13   55.06   60.64   64.73   67.22
  68.06   67.22   64.73   60.64   55.06   48.13   40.00   30.90   21.03   10.64
   0.00

AT T =    8.0 DISPLACEMENTS ARE

   0.00   11.36   22.45   32.98   42.70   51.37   58.77   64.73   69.09   71.76
  72.65   71.76   69.09   64.73   58.77   51.37   42.70   32.98   22.45   11.36
   0.00

AT T =    9.0 DISPLACEMENTS ARE

   0.00   11.80   23.31   34.25   44.35   53.35   61.04   67.22   71.76   74.52

AT T =   12.0 DISPLACEMENTS ARE

   0.00   11.36   22.45   32.98   42.70   51.37   58.77   64.73   69.09   71.76
  72.65   71.76   69.09   64.73   58.77   51.37   42.70   32.98   22.45   11.36
   0.00

AT T =   13.0 DISPLACEMENTS ARE

   0.00   10.64   21.03   30.90   40.00   48.13   55.06   60.64   64.73   67.22
  68.06   67.22   64.73   60.64   55.06   48.13   40.00   30.90   21.03   10.64
   0.00

AT T =   14.0 DISPLACEMENTS ARE

   0.00    9.66   19.09   28.05   36.32   43.70   50.00   55.06   58.77   61.04
  61.80   61.04   58.77   55.06   50.00   43.70   36.32   28.05   19.09    9.66
   0.00

AT T =   15.0 DISPLACEMENTS ARE

   0.00    8.45   16.69   24.52   31.75   38.19   43.70   48.13   51.37   53.35
  54.01   53.35   51.37   48.13   43.70   38.19   31.75   24.52   16.69    8.45
   0.00

AT T =   16.0 DISPLACEMENTS ARE

   0.00    7.02   13.87   20.38   26.39   31.75   36.32   40.00   42.70   44.35
  44.90   44.35   42.70   40.00   36.32   31.75   26.39   20.38   13.87    7.02
   0.00

AT T =   17.0 DISPLACEMENTS ARE

   0.00    5.42   10.71   15.74   20.38   24.52   28.05   30.90   32.98   34.25
  34.68   34.25   32.98   30.90   28.05   24.52   20.38   15.74   10.71    5.42
   0.00

AT T =   18.0 DISPLACEMENTS ARE

   0.00    3.69    7.29   10.71   13.87   16.69   19.09   21.03   22.45   23.31
  23.60   23.31   22.45   21.03   19.09   16.69   13.87   10.71    7.29    3.69
   0.00

AT T =   19.0 DISPLACEMENTS ARE

   0.00    1.86    3.69    5.42    7.02    8.45    9.66   10.64   11.36   11.80
  11.95   11.80   11.36   10.64    9.66    8.45    7.02    5.42    3.69    1.86
   0.00
```

(Continued)

Program 1 Cont.

```
AT T =    20.0 DISPLACEMENTS ARE

0.00       0.00       0.00       0.00       0.00       0.00       0.00       0.00       0.00       0.00
0.00       0.00       0.00       0.00       0.00       0.00       0.00       0.00       0.00       0.00
0.00

AT T =    21.0 DISPLACEMENTS ARE

0.00      -1.86      -3.69      -5.42      -7.02      -8.45      -9.66     -10.64     -11.36     -11.80
-11.95    -11.80     -11.36     -10.64      -9.66      -8.45      -7.02      -5.42      -3.69      -1.86
0.00
```

PROBLEMS

Section 1

1. If the banjo string in the example of Section 1 is tightened, its frequency of vibration is increased.   Likewise, if the length of the vibrating portion of the string is shortened, as by holding it against one of the frets with the finger, the pitch is raised.   What would the frequency be if the tension is made 42,000 grams, and the effective length is 70 cm?   Determine the answer by finding the number of Δt steps for the original displacement to be duplicated and compare to

$$f = \frac{1}{2L}\sqrt{\frac{Tg}{w}} \, .$$

2. A vibrating string system has $Tg/w = 4$ cm^2/sec^2.   Divide the length L into intervals so that $\Delta x = L/8$ cm.   Find the displacement for $t = 0$ up to $t = L$ if both ends are fixed, and the initial conditions are:

(a) $y = \dfrac{x(L-x)}{L^2}$, $\dfrac{\partial y}{\partial t} = 0.$

(b) The string is displaced $+1$ units at $L/4$ and -1 units at $5L/8$, $\partial y/\partial t = 0$.

(c) $y = 0$, $\dfrac{\partial y}{\partial t} = \dfrac{-x(L-x)}{L^2}$.   (Use Eq. 1.2 to begin the solution.)

(d) The string is displaced 1 unit at $L/2$, $\partial y/\partial t = -y$.

In part (a), compare to the analytical solution:

$$y = \frac{4}{\pi^3}\sum_{n=1}^{\infty}\frac{1}{n^3}(1 - \cos n\pi)\cos n\pi t \sin n\pi x.$$

3. A function u satisfies the equation

$$\frac{\partial^2 u}{\partial x^2} = \frac{\partial^2 u}{\partial t^2},$$

with boundary conditions $u = 0$ at $x = 0$, $u = 0$ at $x = 1$ for all values of t, and with initial conditions $u = \sin \pi x$, $\partial u/\partial t = 0$ at $t = 0$ for $0 \le x \le 1$.   Solve by the finite difference method and show the results are the same as the analytical solution:

$$U(x,t) = \sin \pi x \cos \pi t.$$

4. Show that appropriate changes of variables transforms

$$\frac{\partial^2 y}{\partial t^2} = \frac{Tg}{w} \frac{\partial^2 y}{\partial x^2}$$

to the normalized equation

$$\frac{\partial^2 Y}{\partial T^2} = \frac{\partial^2 Y}{\partial X^2}.$$

5. The ends of the vibrating string need not be fixed. Solve $\partial^2 y/\partial t^2 = \partial^2 y/\partial x^2$ with initial conditions $y = 0$, $\partial y/\partial t = 0$, for $0 \le x \le 1$, and end conditions $y = 0$ at $x = 0$, $y = \sin \pi t/4$, $\partial y/\partial x = 0$ at $x = 1$.

Section 2

6. Verify that the analytical solution given to Problem 3 satisfies (a) the differential equation, (b) the initial conditions, and the boundary conditions, and (c) Eq. (2.5). What are the functions F and G of Eq. (2.2) for this solution?

7. Since Eq. (1.3) is sometimes inaccurate when the initial velocity is not zero, the solutions to parts (c) and (d) of Problem 2 are not exact. Repeat these problems getting more accurate y-values by using Eq. (2.9) and evaluating the integral by Simpson's $\frac{1}{3}$ rule.

8. Find the solution for 10 time steps:

$$\frac{\partial^2 y}{\partial t^2} = \frac{1}{4} \frac{\partial^2 y}{\partial x^2}.$$

Initial conditions are $y = 1 - |1 - 2x|$, $\partial y/\partial t = -\frac{1}{2}y$. Boundary conditions are $y = 0$ at both $x = 0$ and $x = 1$. Take $\Delta x = 0.2$.

Section 3

9. Demonstrate that if $Tg(\Delta t)^2/w(\Delta x)^2 > 1$, the finite difference method for the wave equation will propagate an error with increasing effect by tracing the growth of the error similarly to Table 4.2. Take the simple case with ends fixed.

10. Trace the magnitude of an isolated error in solving the wave equation with $Tg(\Delta t)^2/w(\Delta x)^2 = 1$, similar to Table 4.2, but for the case when the left-hand end is fixed and the right-hand end moves according to

$$y(1,t) = \sin\frac{\pi t}{4}, \qquad \frac{\partial y}{\partial x} = 0 \qquad \text{at } x = 1.$$

Section 4

11. For the partial differential equation

$$au_{xx} + bu_{xt} + cu_{tt} + e = 0,$$

sketch the characteristic curves through the point $x = 0.5$, $t = 0$ when

(a) $a = 1$, $b = 2$, $c = 1$ (b) $a = 1$, $b = 2$, $c = -3$

(c) $a = 1, b = 2, c = 3$ (d) $a = 1, b = 2, c = -1$

(e) $a = t^2, b = xt, c = -2x^2$

12. Get part of the solution by integrating along characteristics for the equation

$$\partial^2 u / \partial x^2 + \partial^2 u / \partial x \partial t - \partial^2 u / \partial t^2 = 1,$$

subject to initial conditions

$$u = 0, \qquad \partial u / \partial t = x(1 - x).$$

Find the solution at three points in the xt-plane, where the characteristics through (0.4,0), (0.5,0) and (0.6,0) intersect.

13. Solve the equation

$$u_{xx} + xtu_{xt} - u_{tt} = 0,$$

with initial conditions

$$u = 2x, \qquad \partial u / \partial t = 0.$$

Find the solution at several points in the region of the xt-plane bounded by the lines $x = 0.4, x = 0.6$.

14. Continue the solution of the example problem of Section 4, $u_{xx} - uu_{tt} + 1 - x^2 = 0$, $u(x,0) = x(1 - x)$, $u_t(x,0) = 0$, $u(0,t) = 0$, $u(1,t) = 0$, by finding the solution at the intersection of characteristics through points at $x = 0(0.2)1$, $t = 0$. Then find the solution at the intersection of characteristics through (0.2,0) and (0.6,0) as an example of how the solution can be advanced in time.

Section 5

15. Utilize the computer program to solve Problem 8, except make $\Delta x = 0.1$ and compute for 50 time steps.

16. Write a computer program to solve problems similar to Problem 5, in which the displacements and slopes at each end are functions of time. Further, allow the initial velocity to be nonzero. Test by solving Problem 5.

17. Write a computer program to calculate the function at a point which is the intersection of two characteristics according to the procedure of Section 4.

12
numerical double interpolation and multiple integration

After we have solved a partial differential equation by the methods of the previous chapters we end up with a tabulation of the function at the nodes of our spatial network and at successive increments of time. Frequently we wish the value of the function at points other than at the nodes because of some critical significance at that point, and/or perhaps at some intermediate value of the time. We cannot always avoid this problem by subdividing so that our critical point is at a node because this usually gives a nonintegral number of spaces. However, we should be able to interpolate from the nearby known values, although the problem of interpolation in a tabulation of two or more variables takes more effort than in a one-way table as considered in Chapter 2.

In this chapter we will also consider the somewhat related problem of multiple integration. The need for this may arise in a manner analogous to the torsion problem. The twisting moment T causing a twist of θ radians per unit length of a bar of constant cross section is related to the torsion function ψ by the equation

$$T = 2E_s\theta \iint \psi \, dx \, dy,$$

where E_s is the shear modulus. We determine ψ by solving the Poisson equation

$$\frac{\partial^2\psi}{\partial x^2} + \frac{\partial^2\psi}{\partial y^2} + 2 = 0$$

over the cross section of the bar subject to $\psi = 0$ on the boundary. Hence to solve the torsion problem for a nonstandard section, we solve the Poisson equation and then need to perform a double integration over the region using the values of ψ known at the intersection of the lines which subdivided our region.

Alternatively we may wish to derive an algorithm for multiple integration of a function of several variables to be used in a computer program. Numerical methods are the usual starting point for this. Here we will have a choice of methods including Gaussian quadrature if we are able to evaluate the integrand at any point.

1. POLYNOMIAL INTERPOLATION IN TWO DIMENSIONS

When the function u is a polynomial function of two variables x and y, say of degree three in x and of degree two in y, we would have

$$\begin{aligned}u = f(x,y) = a_0 &+ a_1x + a_2y + a_3x^2 + a_4xy + a_5y^2 + a_6x^3 \\ &+ a_7x^2y + a_8xy^2 + a_9x^3y + a_{10}x^2y^2 + a_{11}x^3y^2.\end{aligned} \quad (1.1)$$

The functional relation is seen to involve many terms. If we were concerned with four independent variables (three space dimensions plus time), even low degree polynomials would be quite intractable. Except for special purposes, such as when we need an explicit representation, perhaps to permit ready differentiation at an arbitrary point, we can avoid such complications by handling each variable separately. We shall treat only this case.

Note the immediate simplification of Eq. (1.1) if we let y take on a constant value, say $y = c$. Combining the y factors with the coefficients, we get

$$u|_{y=c} = b_0 + b_1 x + b_2 x^2 + b_3 x^3.$$

This will be our attack in interpolating at the point (a,b) in a table of two variables—hold one variable constant, say $y = y_1$, and the table becomes a single-variable problem. The methods of Chapter 2 then apply to give $f(a,y_1)$. If we repeat this at various values of y, $y = y_2, y_3, \ldots, y_n$, we will get a table with x constant at the value $x = a$ and with y varying. We then interpolate at $y = b$.

Example. Estimate $f(1.6, 0.33)$ from the values in Table 1.1. Use quadratic interpolation in the x-direction and cubic interpolation for y. We select one of the variables to hold constant, say x. This choice is arbitrary since we will get the same result except for differences due to round-off if we had chosen to hold y constant. We decide to interpolate for y within the three rows of the table at $x = 1.0$, 1.5, and 2.0 since the desired value at $x = 1.6$ is most nearly centered within this set.

We may use either Lagrangian interpolation or derive the interpolated values using the lozenge diagram, Fig. 5.1, Chapter 2, after making difference tables. Let us use the later method:

	y	u	Δu	$\Delta^2 u$	$\Delta^3 u$
	0.2	0.640			
			0.363		
	0.3	1.003		−0.007	
$x = 1.0$			0.356		−0.005
	0.4	1.359		−0.012	
			0.344		
	0.5	1.703			
	0.2	0.990			
			0.534		
	0.3	1.524		−0.013	
$x = 1.5$			0.521		−0.004
	0.4	2.045		−0.017	
			0.504		
	0.5	2.549			
	0.2	1.568			
			0.816		
	0.3	2.384		−0.023	
$x = 2.0$			0.793		−0.004
	0.4	3.177		−0.027	
			0.766		
	0.5	3.943			

Table 1.1. Tabulation of a Function of Two Variables

$$u = f(x,y)$$

x \ y	0.1	0.2	0.3	0.4	0.5	0.6
0.5	0.165	0.428	0.687	0.942	1.190	1.431
1.0	0.271	0.640	1.003	1.359	1.703	2.035
1.5	0.447	0.990	1.524	2.045	2.549	3.031
2.0	0.738	1.568	2.384	3.177	3.943	4.672
2.5	1.216	2.520	3.800	5.044	6.241	7.379
3.0	2.005	4.090	6.136	8.122	10.030	11.841
3.5	3.306	6.679	9.986	13.196	16.277	19.198

We need the subtables from $y = 0.2$ to $y = 0.5$ since, for a cubic interpolation, four points are required and $y = 0.33$ is most nearly centered within this group. Using any convenient path through the tables with coefficients given by the lozenge, we arrive at the results

	x	u	Δu	$\Delta^2 u$
	1.0	1.1108		
			0.5710	
$y = 0.33$	1.5	1.6818		0.3717
			0.9427	
	2.0	2.6245		

In the last tabulation we carry one extra decimal to guard against round-off errors. Interpolating again, we get $x = 1.8406$, which we report as $x = 1.841$.

The function tabulated in Table 1.1 is $f(x,y) = e^x \sin y + y - 0.1$, so the true value is $f(1.6, 0.33) = 1.8350$. Our error of -0.006 occurs because quadratic interpolation for x is inadequate in view of the large second difference. In retrospect, it would have been better to use quadratic interpolation for y, since the third differences of the y-subtables are small, and let x take on a third-degree relationship.

It is instructive to observe which of the values in Table 1.1 entered into our computation. The shaded rectangle covers these values. This is the "region of fit" for the interpolating polynomial that we have used. The principle of choosing values so that the point at which the interpolating polynomial is used is centered in the region of fit obviously applies here in exact analogy to the one-way table situation. It also applies to tables of three and four variables in the same way. Of course, the labor of interpolating in such multidimensional cases soon becomes astronomical.

A rectangular region of fit is not the only possibility. We may change the degree of interpolating as we subtabulate the different rows or columns. Intuitively, it would seem best to use higher-degree polynomials for the rows near the interpolating point, decreasing the degree as we get farther away. The coefficient of the error term, when this is done, will be found to be minimized thereby, though for multidimensional interpolating polynomials the error term is quite complex. The region of fit will be diamond shaped when such tapered degree functions are used.

We may adapt the Lagrangian form of interpolating polynomial to the multi-dimensional case also. It is perhaps easiest to imagine a similar process to the above example. Holding one variable constant, we write a series of Lagrangian polynomials for interpolation at the given value of the other variable, and then combine these values in a final Lagrange form. The net result is a Lagrangian polynomial in which the function factors are replaced by Lagrangian polynomials. The resulting expression for the above example would be

$$
\frac{(y - 0.3)(y - 0.4)(y - 0.5)}{(0.2 - 0.3)(0.2 - 0.4)(0.2 - 0.5)}
$$
$$
\times \left[\frac{(x - 1.5)(x - 2.0)}{(1.0 - 1.5)(1.0 - 2.0)}(0.640) + \frac{(x - 1.0)(s - 2.0)}{(1.5 - 1.0)(1.5 - 2.0)}(0.990) + \frac{(x - 1.0)(x - 1.5)}{(2.0 - 1.0)(2.0 - 1.5)}(1.568) \right]
$$
$$
+ \frac{(y - 0.2)(y - 0.4)(y - 0.5)}{(0.3 - 0.2)(0.3 - 0.4)(0.3 - 0.5)}
$$
$$
\times \left[\frac{(x - 1.5)(x - 2.0)}{(1.0 - 1.5)(1.0 - 2.0)}(1.003) + \frac{(x - 1.0)(x - 2.0)}{(1.5 - 1.0)(1.5 - 2.0)}(1.534) + \frac{(x - 1.0)(x - 1.5)}{(2.0 - 1.0)(2.0 - 1.5)}(2.384) \right]
$$
$$
+ \frac{(y - 0.2)(y - 0.3)(y - 0.5)}{(0.4 - 0.2)(0.4 - 0.3)(0.4 - 0.5)}
$$
$$
\times \left[\frac{(x - 1.5)(x - 2.0)}{(1.0 - 1.5)(1.0 - 2.0)}(1.359) + \frac{(x - 1.0)(x - 2.0)}{(1.5 - 1.0)(1.5 - 2.0)}(2.045) + \frac{(x - 1.0)(x - 1.5)}{(2.0 - 1.0)(2.0 - 1.5)}(3.177) \right]
$$
$$
+ \frac{(y - 0.2)(y - 0.3)(y - 0.4)}{(0.5 - 0.2)(0.5 - 0.3)(0.5 - 0.4)}
$$
$$
\times \left[\frac{(x - 1.5)(x - 2.0)}{(1.0 - 1.5)(1.0 - 2.0)}(1.703) + \frac{(x - 1.0)(x - 2.0)}{(1.5 - 1.0)(1.5 - 2.0)}(2.549) + \frac{(x - 1.0)(x - 1.5)}{(2.0 - 1.0)(2.0 - 1.5)}(3.943) \right].
$$

The equation is easy to write, but its evaluation by hand is laborious. If one is writing a computer program for interpolation in such multivariate situations, the Lagrangian form is recommended. There is a special advantage in that equal spacing in the table is not required. The Lagrangian form is also perhaps the most straightforward way to write out the polynomial as an explicit function.

The cubic spline (described in the next chapter) could also be used to inter-polate in multivariate tables. Here again holding all but one variable fixed so that a series of one-way table problems is solved is perhaps the best approach. The computational effort to employ spline interpolation would be terribly severe.

2. NUMERICAL DOUBLE INTEGRATION

We consider first the case when the limits of integration are constants. In the calculus we learned that a double integral may be evaluated as an iterated integral;

in other words, we may write

$$\iint\limits_{A} f(x,y)\, dA = \int_a^b \left(\int_c^d f(x,y)\, dy \right) dx = \int_c^d \left(\int_a^b f(x,y)\, dx \right) dy. \qquad (2.1)$$

In Eq. (2.1) the rectangular region A is bounded by the lines $x = a$, $x = b$, $y = c$, $y = d$. In computing the iterated integrals, we hold x constant while integrating with respect to y (or vice-versa in the second case).

Adapting the numerical integration formulas of Chapter 3 is quite straightforward. Recall that any of the integration formulas is just a linear combination of functional values with varying values of the independent variable. In other words, a quadrature formula is just a weighted sum of certain functional values. The inner integral is written then as a weighted sum of function values with one variable held constant. We then add together a weighted sum of these sums. If the function is known only at the nodes of a rectangular grid through the region, we are constrained to use these values. The Newton-Cotes formulas are a convenient set to employ. There is no reason why the same formula must be used in each direction, although it is often particularly convenient to do so.

We illustrate this technique by evaluating the integral of the function of Table 1.1 over the rectangular region bounded by $x = 1.5$, $x = 3.0$, $y = 0.2$, $y = 0.6$. Let us use the trapezoidal rule in the x-direction and Simpson's $\frac{1}{3}$ rule in the y-direction. (Since the number of panels in the x-direction is not even, Simpson's rule does not apply readily.) It is immaterial as to which integral we evaluate first. Suppose we start with x constant:

$$y = 0.2, \qquad \int_{1.5}^{3.0} f(x,y)\, dx = \int_{1.5}^{3.0} f(x,0.2)\, dx = \frac{h}{2}\left(f_1 + 2f_2 + 2f_3 + f_4 \right)$$

$$= \frac{0.5}{2}\left(0.990 + 2(1.568) + 2(2.520) + 4.090 \right) = 3.1800,$$

$$y = 0.3, \qquad \int_{1.5}^{3.0} f(x,0.3)\, dx = \frac{0.5}{2}\left(1.524 + 2(2.384) + 2(3.800) + 6.136 \right) = 5.0070.$$

Similarly, at

$$y = 0.4, \qquad I = 6.6522,$$
$$y = 0.5, \qquad I = 8.2368,$$
$$y = 0.6, \qquad I = 9.7435.$$

We now sum these in y-direction according to Simpson's rule:

$$f(x,y)\, dx = \frac{0.1}{3}\left(3.1800 + 4(5.0070) + 2(6.6522) + 4(8.2368) + 9.7435 \right) = 2.640.$$

(In this example our answer does not check well with the analytical value of 2.5344 because the x-intervals are large. We could improve our estimate somewhat by fitting a higher degree polynomial than the first to provide the integration formula. We can even use values outside the range of integration for this. The lozenges of Figs. 3.1 and 3.2 of Chapter 4 are one source of formulas. The methods of undetermined coefficients is an alternative source.)

The previous example shows that double integration by numerical means reduces to a double summation of weighted function values. The calculations we have just made could be written in the form

$$\int f(x,y) \, dx \, dy = \sum_{j=1}^{m} v_j \sum_{i=1}^{n} w_i f_{ij} =$$

$$= \frac{\Delta y}{3} \frac{\Delta x}{2} [(f_{1,1} + 2f_{2,1}$$

$$+ 2f_{3,1} + f_{4,1}) + 4(f_{1,2} + 2f_{2,2} + 2f_{3,2} + f_{4,2})$$

$$+ \cdots + (f_{1,5} + 2f_{2,5} + 2f_{3,5} + f_{4,5})].$$

It is tempting to write this in the pictorial operator form that we found convenient in solving partial differential equations:

$$\int f(x,y) \, dx \, dy = \frac{\Delta y}{3} \frac{\Delta x}{2} \begin{Bmatrix} 1 & 4 & 2 & 4 & 1 \\ 2 & 8 & 4 & 8 & 2 \\ 2 & 8 & 4 & 8 & 2 \\ 1 & 4 & 2 & 4 & 1 \end{Bmatrix} f_{i,j}.$$

Other combinations of Newton-Cotes formulas give similar results. It is probably easiest for hand calculation to use these pictorial integration operators. It is readily adapted to any desired combination of integration formulas. Except for the difficulty of representation beyond two dimensions, this operator technique also applies to triple and quadruple integrals.

There is an alternative representation to such pictorial operators that is easier to translate into a computer program. We also derive it somewhat differently. Consider the numerical integration formula for one variable,

$$\int_{-1}^{1} f(x) \, dx \doteq \sum_{i=1}^{n} a_i x_i. \tag{2.2}$$

We have seen in Chapters 4 and 5 that such formulas can be made exact if $f(x)$ is any polynomial of a certain degree. Assume that Eq. (2.2) holds for polynomials up to degree s.*

We now consider the multiple integral formula

$$\int_{-1}^{1} \int_{-1}^{1} \int_{-1}^{1} f(x,y,z) \, dx \, dy \, dz \stackrel{?}{=} \sum_{i=1}^{n} \sum_{j=1}^{n} \sum_{k=1}^{n} a_i a_j a_k f(x_i, y_j, z_k). \tag{2.3}$$

We wish to show that Eq. (2.3) is exact for all polynomials in x, y, and z up to degree s. Such a polynomial is a linear combination of terms of the form $x^\alpha y^\beta z^\gamma$, where α, β, and γ are nonnegative integers whose sum is equal to s or less. If we can prove that Eq. (2.3) holds for the general term of this form, it will then hold for the polynomial.

To do this we assume that

$$f(x,y,z) = x^\alpha y^\beta z^\gamma.$$

---

* For Newton-Cotes formulas $s = n - 1$ for n even and $s = n$ for n odd. For Gaussian quadrature formulas, $s = 2n - 1$, and the x_i may be unevenly spaced.

Then, since the limits are constants and the integrand is factorable,

$$I = \int_{-1}^{1} \int_{-1}^{1} \int_{-1}^{1} x^{\alpha} y^{\beta} z^{\gamma} \, dx \, dy \, dz$$

$$= \left(\int_{-1}^{1} x^{\alpha} \, dx \right) \left(\int_{-1}^{1} y^{\beta} \, dy \right) \left(\int_{-1}^{1} z^{\gamma} \, dz \right).$$

Replacing each term according to Eq. (2.2), we get,

$$I = \left(\sum_{i=1}^{n} a_i x_i \right) \left(\sum_{j=1}^{n} a_j y_j \right) \left(\sum_{k=1}^{n} a_k z_k \right)$$

$$= \sum_{i=1}^{n} a_i x_i \sum_{j=1}^{n} a_j y_j \sum_{k=1}^{n} a_k z_k. \tag{2.4}$$

We need now an elementary rule about the product of summations. We illustrate it for a simple case. We assert that

$$\left(\sum_{i=1}^{3} u_i \right) \left(\sum_{j=1}^{2} v_j \right) = \sum_{i=1}^{3} \left(\sum_{j=1}^{2} u_i v_j \right)$$

$$= \sum_{i=1}^{3} \sum_{j=1}^{2} u_i v_j.$$

The last equality is purely notational. We prove the first by expanding both sides:

$$\left(\sum_{i=1}^{3} u_i \right) \left(\sum_{j=1}^{2} v_j \right) = \sum_{i=1}^{3} u_i \sum_{j=1}^{2} v_i$$

$$\doteq (u_1 + u_2 + u_3)(v_1 + v_2)$$

$$= u_1 v_1 + u_1 v_2 + u_2 v_1 + u_2 v_2 + u_3 v_1 + u_3 v_2,$$

$$\sum_{i=1}^{n} \sum_{j=1}^{n} u_i v_j = (u_1 v_1 + u_1 v_2) + (u_2 v_1 + u_2 v_2) + (u_3 v_1 + u_3 v_2).$$

On removing parentheses we see they are the same. Using this principle, we can write Eq. (2.4) in the form

$$I = \sum_{i=1}^{n} \sum_{j=1}^{n} \sum_{k=1}^{n} a_i a_j a_k x_i y_j z_k, \tag{2.5}$$

which shows that the questioned equality of Eq. (2.3) is valid, and we can write a program for a triple integral by three nested DO loops. The coefficients a_i are chosen from any numerical integration formula. If the three one-variable formulas corresponding to Eq. (2.3) are not identical, an obvious modification of Eq. (2.5) applies. In some cases a change of variable is needed to correspond to Eq. (2.2).

If we are evaluating a multiple integral numerically where the integrand is a known function, our choice of the form of Eq. (2.2) is wider. Of higher efficiency than the Newton-Cotes formulas is Gaussian quadrature. Since it also fits the pattern of Eq. (2.2), the formula of Eq. (2.5) applies. We illustrate with a simple example.

Example. Evaluate

$$I = \int_0^1 \int_{-1}^0 \int_{-1}^1 yze^x \, dx \, dy \, dz$$

by Gaussian quadrature using a three-term formula for x and two-term formulas for y and z. We first make the changes of variables to adjust the limits for y and z to $(-1,1)$:

$$y = \tfrac{1}{2}(u + 1), \qquad dy = \tfrac{1}{2}du,$$
$$z = \tfrac{1}{2}(v - 1), \qquad dz = \tfrac{1}{2}dv.$$

Our integral becomes

$$I = \frac{1}{16} \int_{-1}^1 \int_{-1}^1 \int_{-1}^1 (u + 1)(v - 1)e^x \, dx \, du \, dv.$$

The two- and three-point Gaussian formulas are, from Chapter 3,

$$\int_{-1}^1 f(x) \, dx = (1)f(-0.5774) + (1)f(0.5774),$$

$$\int_{-1}^1 f(x) \, dx = (\tfrac{5}{9})f(-0.7746) + (\tfrac{8}{9})f(0) + (\tfrac{5}{9})f(0.7746).$$

The integral is then

$$I = \frac{1}{16} \sum_{i=1}^2 \sum_{j=1}^2 \sum_{k=1}^3 a_i a_j b_k (u_i + 1)(v_j - 1)e^{x_k},$$

$$a_1 = 1 \qquad a_2 = 1,$$
$$b_1 = \tfrac{5}{9}, \qquad b_2 = \tfrac{8}{9}, \qquad b_3 = \tfrac{5}{9},$$

and values of u, v, and x as above.

A few representative terms of the sum are

$$I = \tfrac{1}{16}[(1)(1)(\tfrac{5}{9})(-0.5774 + 1)(-0.5774 - 1)e^{-0.7446}$$
$$+ (1)(1)(\tfrac{8}{9})(-0.5774 + 1)(-0.5774 - 1)e^0$$
$$+ (1)(1)(\tfrac{5}{9})(-0.5774 + 1)(-0.5774 - 1)e^{0.7746}$$
$$+ (1)(1)(\tfrac{5}{9})(0.5774 + 1)(-0.5774 - 1)e^{-0.7746}$$
$$+ \cdots].$$

On evaluating, we get $I = -0.58758$. The analytical value is $-\tfrac{1}{4}(e - e^{-1}) = -0.58760$.

3. ERRORS IN MULTIPLE INTEGRATION AND EXTRAPOLATIONS

The error term of a one-variable quadrature formula is an additive one just like the other terms in the linear combination (although of special form). It would seem reasonable that it would go through the multiple summations in a similar fashion, so we should expect error terms for multiple integration which are analogous to the one-dimensional case. We illustrate that this is true for double integration using the trapezoidal rule in both directions, with uniform spacings, choosing n intervals in the x-direction and m in the y-direction.

From Chapter 4 we have

$$\text{Error of } \int_a^b f(x) \, dx = -\frac{b-a}{12} h^2 f''(\xi) = O(h^2),$$

$$h = \Delta x = \frac{b-a}{n}.$$

In developing Romberg integration, we found that the error term could be written as

$$\text{Error} = O(h^2) = Ah^2 + O(h^4) \doteq Ah^2 + Bh^4,$$

where A is a constant and the value of B depends on a fourth derivative of the function. Appending this error term to the trapezoidal rule, we get (equivalent to Eq. 2.2)

$$\int_a^b f(x) \, dx \bigg|_{y=y_j} = \frac{h}{2}[f_{0,j} + 2f_{1,j} + 2f_{2,j} + \cdots + f_{n,j}] + A_j h^2 + B_j h^4.$$

Summing these in the y-direction and retaining only the error terms, we have

$$\int_c^d \int_a^b f(x,y) \, dx \, dy = \frac{k}{2} \frac{h}{2} \sum_{i=0}^n \sum_{j=0}^m a_i a_j f_{i,j}$$

$$+ \frac{k}{2}(A_0 + 2A_1 + 2A_2 + \cdots + A_m)h^2$$

$$+ \frac{k}{2}(B_0 + 2B_1 + 2B_2 + \cdots + B_m)h^4 \qquad (3.1)$$

$$+ \bar{A}k^2 + \bar{B}k^4,$$

$$k = \Delta y = \frac{d-c}{m}.$$

In Eq. (3.1), $\bar{A}$ and $\bar{B}$ are the coefficients of the error term for y. The coefficients A and B for the error terms in the x-direction may be different for each of the $(m+1)$ y-values, but each of the sums in parentheses in Eq. (3.1) is $2n$ times some average value of A or B, so the error terms become

$$\text{Error} = \frac{k}{2}(nA_{av})h^2 + \frac{k}{2}(nB_{av})h^4 + \bar{A}k^2 + \bar{B}k^4. \qquad (3.2)$$

Since both Δx and Δy are constant, we may take $\Delta y = k = \alpha \Delta x = \alpha h$, where $\alpha = \Delta y / \Delta x$, and Eq. (3.2) can be written, with $nh = (b-a)$,

$$\text{Error} = \left(\frac{b-a}{2} A_{av}\alpha\right)h^2 + \left(\frac{b-a}{2} B_{av}\alpha\right)h^4 + A\alpha^2 h^2 + B\alpha^4 h^4$$

$$= K_1 h^2 + K_2 h^4.$$

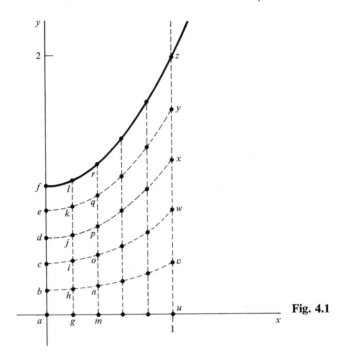

Fig. 4.1

K_2 will depend on fourth-order partial derivatives.   Our expectation that the error term of double integration by numerical means is of the same form as for single integration is confirmed.

Since this is true, a Romberg integration may be applied to multiple integration whereby we extrapolate to an $O(h^4)$ estimate from two trapezoidal computations at a 2 to 1 interval ratio.   From two such $O(h^4)$ computations we may extrapolate to one of $O(h^6)$ error.   The procedure is a direct application of the formulas of Chapter 3.

4. MULTIPLE INTEGRATION WITH VARIABLE LIMITS

If the limits of integration are not constant, so that the region in the xy-plane upon which the integrand $f(x,y)$ is to be summed is not rectangular, we must modify the above procedure.   Let us consider a simple example to illustrate the method. Evaluate

$$\int \int f(x,y) \, dy \, dx$$

over the region bounded by the lines $x = 0$, $x = 1$, $y = 0$, and the curve $y = x^2 + 1$. The region is sketched in Fig. 4.1.   If we draw vertical lines spaced at $\Delta x = 0.2$ apart, shown as dashed lines in Fig. 4.1, it is obvious that we can approximate the inner integral at constant x-values along any one of the vertical lines (including $x = 0$ and $x = 1$).   If we use the trapezoidal rule with five panels for each of these, we get the series of sums

$$S_1 = \frac{h_1}{2}\left(f_a + 2f_b + 2f_c + 2f_d + 2f_e + f_f\right),$$

$$S_2 = \frac{h_2}{2}\left(f_g + 2f_h + 2f_i + 2f_j + 2f_k + f_l\right),$$

$$S_3 = \frac{h_3}{2}\left(f_m + 2f_n + \cdots\right),$$

$$\vdots$$

$$S_6 = \frac{h_6}{2}\left(f_u + 2f_v + 2f_w + 2f_x + 2f_y + f_z\right).$$

The subscripts here indicate the values of the function at the points so labeled in Fig. 4.1.   The values of the h_i are not the same in the above equations, but in each they are the vertical distances divided by 5.   The combination of these sums to give an estimate of the double integral will then be

$$\text{Integral} = \frac{0.2}{2}\left(S_1 + 2S_2 + 2S_3 + 2S_4 + 2S_5 + S_6\right).$$

To be even more specific, suppose $f(x,y) = xy$.   Then

$$S_1 = \frac{1.0/5}{2}\left(0 + 0 + 0 + 0 + 0 + 0\right) = 0,$$

$$S_2 = \frac{1.04/5}{2}\left(0 + 0.0832 + 0.1664 + 0.2496 + 0.3328 + 0.208\right) = 0.1082,$$

$$S_3 = \frac{1.16/5}{2}\left(0 + 0.1856 + 0.3712 + 0.5568 + 0.7428 + 0.464\right) = 0.2692,$$

$$S_4 = \frac{1.36/5}{2}\left(0 + 0.3264 + 0.6528 + 0.9792 + 1.3056 + 0.816\right) = 0.5549,$$

$$S_5 = \frac{1.64/5}{2}\left(0 + 0.5248 + 1.0496 + 1.5744 + 2.0992 + 1.312\right) = 1.0758,$$

$$S_6 = \frac{2.0/5}{2}\left(0 + 0.8 + 1.6 + 2.4 + 3.2 + 2.0\right) = 2.0$$

$$\text{Integral} = \frac{0.2}{2}\left(0 + 0.2164 + 0.5384 + 1.1098 + 2.1516 + 2.0\right)$$

$$= 0.6016 \qquad \text{versus analytical value of } 0.6666.$$

The extension of this to more complicated regions and the adaptation to use Simpson's rule should be obvious.   If the functions that define the region are not single-valued, one must divide the region into subregions to avoid the problem, but this must also be done to integrate analytically.

Program 1

```
ZZJOB 5                          CSC   001      GERALD, C. F.
ZZFOR 5
*LDISK
*LIST PRINTER
      FUNCTION SUM(XA, XB, Y, DELX, N)
C   THIS SUBPROGRAM SUMS ACROSS ONE ROW
      SUM = DBLFCN(XA, Y) + DBLFCN(XB, Y)
      X = XA
      DO 10 I = 2,N
      X = X + DELX
   10 SUM = SUM + 2.*DBLFCN(X, Y)
      RETURN
      END
ZZFOR 5
*LDISK
*LIST PRINTER
*LDISK
      FUNCTION DBLFCN(U,V)
C   DEFINE THE INTEGRAND BY AN ARITHMETIC STATEMENT FUNCTION
      DBLFCN = 1./EXPF(U**2*V**2)
      RETURN
      END
ZZFORX5
*LIST PRINTER
      DIMENSION ACCUM(6,6)
C   DOUBLE INTEGRATION PROGRAM USING ROMBERG METHOD.
C   A SUBPROGRAM CALLED SUM IS USED WITH THIS PROGRAM
C   PRINT HEADING
      PRINT 200
C   INPUT PARAMETERS FOR THE INTEGRATION
      READ 100, XA, XB, YA, YB, TOL
C   FIRST COMPUTE THE INITIAL ESTIMATE OF THE INTEGRAL
C   USING FOUR SUBDIVISION EACH WAY
      DELX = (XB - XA)/4.
      DELY = (YB - YA)/4.
      N = 4
      Y = YA
      ACCUM(1,1) = SUM(XA, XB, Y, DELX, N)
      DO 10 I = 2,N
      Y = Y + DELY
   10 ACCUM(1,1) = ACCUM(1,1) + 2.*SUM(XA, XB, Y, DELX, N)
      ACCUM(1,1) = (ACCUM(1,1) + SUM(XA, XB, YB,DELX, N))*DELX*DELY/4.
C   HALVE THE VALUES OF DELX, DELY.  RECOMPUTE INTEGRAL, EXTRAPOLATE AND
C   TEST IF TOLERANCE IS MET
C   REPEAT UP TO FIVE TIMES
      DO 20 J = 1,5
      DELX = DELX/2.
      DELY = DELY/2.
      N = 2*N
      Y = YA
      ACCUM(J+1,1) = SUM(XA, XB, Y, DELX, N)
      DO 30 I = 2,N
      Y = Y + DELY
   30 ACCUM(J+1,1) = ACCUM(J+1,1) + 2.*SUM(XA, XB, Y, DELX, N)
      ACCUM(J+1,1) = (ACCUM(J+1,1)+ SUM(XA, XB, YB, DELX, N))*DELX*
     1 DELY/4.
C   EXTRAPOLATION
      DO 40 K = 1,J
   40 ACCUM(J+1,K+1) = ACCUM(J+1,K) + 1./(4.**K - 1.)*
     1 (ACCUM(J+1,K) - ACCUM(J,K))

      IF (ABSF(ACCUM(J+1,J+1) - ACCUM(J+1,J)) - TOL) 50, 50, 20
   20 CONTINUE
C   IF TOLERANCE NOT MET AFTER FIVE EXTRAPOLATIONS, PRINT
C   CALCULATED VALUE AND NOTE
      J = 5
      PRINT 201, TOL
      GO TO 60
   50 PRINT 202, TOL, J
      J = J + 1
   60 DO 70 L = 1,J
```

(Continued)

Program 1 Cont.

```
 70 PRINT 203, (ACCUM(L,I), I = 1,L)
    PRINT 204
    CALL EXIT
100 FORMAT(4F10.2, E14.7)
200 FORMAT(1H1, 37H DOUBLE INTEGRATION BY ROMBERG METHOD )
201 FORMAT(1H0, 14H TOLERANCE OF E14.7, 31H NOT MET AFTER 5 EXTRAPOLAT
   1IONS /)
202 FORMAT(1H0, 14H TOLERANCE OF E14.7, 15H WAS MET AFTER I3,
   1 15H EXTRAPOLATIONS /)
203 FORMAT(1H , 6F12.6)
204 FORMAT(1H0, 37H FINAL VALUE PRINTED IS BEST ESTIMATE )
    END
-1.0       1.0        -1.0      1.0                    2.E-5
ZZZZ
```

DOUBLE INTEGRATION BY ROMBERG METHOD

TOLERANCE OF   2.0000000E-05 WAS MET AFTER   2 EXTRAPOLATIONS

```
    3.560183
    3.607943     3.623863
    3.619811     3.623767     3.623761
```

FINAL VALUE PRINTED IS BEST ESTIMATE

5. A DOUBLE INTEGRATION PROGRAM

Program 1 integrates a function of two variables over a rectangular region.   The Romberg method is used, extrapolating from an initial estimate, which evaluates the function at 25 points using the trapezoidal rule, together with a second estimate with the Δx and Δy intervals halved.   The interval halving is continued until agreement in the estimates within a certain tolerance is attained.   In computing the various trapezoidal rule estimates of the integral, the weighted sum of the functional values along a row of points at constant y-value is obtained in a function subprogram.   These partial sums are summed in the main program.   Because of the peculiarities of the particular Fortran compiler used, the integrand function is supplied by a second function subprogram.   To change the integrand, only the one card that defines this function (DBLFCN) in this second subprogram need be changed.   The program is tested by evaluating

$$\int_{-1}^{1} \int_{-1}^{1} e^{-x^2 y^2} \, dx \, dy.$$

If the program were to be modified to obtain triple or higher integrals, it would be convenient to accumulate the lower-order partial sums in a cascade of function subprograms, totaling the final sum within the main program.

PROBLEMS

Section 1

1. In Section 1, the assertion is made that the order of interpolation makes no difference. Demonstrate that this is true by interpolating within the data of Table 1.1 to find the values at $y = 0.33$ (within rows with x constant at 1.0, 1.5, and 2.0), using cubic

interpolation formulas that fit the table at $y = 0.2, 0.3, 0.4,$ and 0.5. Then interpolate within these three values to determine $f(1.6, 0.33)$, and compare to the value 1.841 obtained in the text.

2. After the example of Section 1 was finished, it was observed that a cubic in x and a quadratic in y would have been preferable. Do this to obtain $f(1.6, 0.33)$ and compare to the true value, 1.8350. Use the best "region of fit."

3. Interpolate in the function $f(x,y)$ tabulated in Section 1 to find $f(2.9, 0.21)$ employing a quartic formula in x and a quadratic in y.

4. Table 1.1 in Chapter 10 gives the results of solving the unsteady-state heat flow equation in one dimension. Interpolate to find the temperature at $x = 0.6, t = 1.00$. By forming the difference tables, one is usually permitted to choose the degrees of the interpolation polynomials so that the last differences included are not "large," but the limited number of x-values is here quite restrictive. Use a cubic in x and a quartic in t.

5. Interpolation from a table of values for a function of three or more variables is done similarly to interpolation in a two-way table. Construct a table for $f(x,y,z) = xe^{y+z}$ for $x = 1.0(0.2)1.4$, $y = 0.2(0.2)0.6$, $z = 0.2(0.2)0.6$. Use these values to interpolate at $x = 1.3, y = 0.35, z = 0.46$ and compare to the exact value.

6. The example of Section 1 used a rectangular region of fit when a more nearly circular region would appear to have advantages. Interpolate from the data in Table 1.1 to evaluate $f(2.3, 0.31)$ by a set of polynomials that fits the table at $x = 1.5$ and 2.0 when $y = 0.2$ and when $y = 0.4$, and at $x = 0.5(0.5)2.5$ when $y = 0.3$. Do this by forming a series of difference tables. In this instance it is very awkward to interpolate first with x held constant, but there is no problem if we begin with y constant.

7. Repeat Problem 6, but now use the Lagrangian form of interpolating polynomials.

Section 2

8. In connection with the first example of Section 2 it was stated that it is immaterial which integral of a double integral (with constant limits) we integrate first. Confirm this by evaluating $\int_{1.5}^{3.0} \int_{0.2}^{0.6} f(x,y)dy\, dx$, with $f(x,y)$ given by Table 1.1, performing the integration first with respect to y.

9. Write the pictorial operators that result by using:

   (a) Simpson's $\frac{1}{3}$ rule in the y-direction and the trapezoidal rule in the x-direction.

   (b) Simpson's $\frac{1}{3}$ rule in both directions.

   (c) Simpson's $\frac{3}{8}$ rule in both directions.

   (d) What conditions are placed on the number of panels in each direction by the methods employed in parts (a), (b), and (c)?

10. Since Simpson's $\frac{1}{3}$ rule is accurate for a cubic, evaluation of the triple integral below using this rule should be exact. Confirm this by evaluating both numerically

and analytically:

$$\int_0^1 \int_0^1 \int_0^1 x^3 y \; z^2 \; dx \; dy \; dz.$$

Use Eq. (2.5) adapted to this case.

11. Draw a pictorial operator (three dimensions) to represent the formula used in Problem 10. It is perhaps easiest to do this with three widely separated planes on which one indicates the coefficients (see illustration).

12. Solve Problem 10 by use of two-term Gaussian quadrature formulas for each variable. This too will give the exact answer since two-term Gaussian formulas are exact for polynomials up to degree three.

13. Evaluate $\displaystyle\int_{0.1}^{0.7} \int_{-.2}^{0.6} e^x \sin y \; dy \; dx$

   (a) Using the trapezoidal rule in both directions, $\Delta x = \Delta y = 0.1$.
   (b) Using Simpson's $\frac{1}{3}$ rule in both directions, $\Delta x = \Delta y = 0.1$.
   (c) Using Gaussian quadrature, three-term formulas in both directions.

Section 3

14. Solve Problem 13 by extrapolating from the trapezoidal rule evaluations using $\Delta x = \Delta y = 0.2$, and $\Delta x = \Delta y = 0.1$. This should give the same result as part (b) of Problem 13. Does it?

15. The analytical value for the integral in Problem 13 is 0.140585. Compare the estimated errors with the actual errors for parts (a) and (b) of Problem 13.

Section 4

16. Integrate $\displaystyle\iint \sin x \sin y \; dx \; dy$ over the region defined by that portion of the unit circle that lies in the first quadrant. Integrate first with respect to y holding x constant with $\Delta x = 0.25$. Subdivide the vertical lines into four panels.

   (a) Use the trapezoidal rule.
   (b) Use Simpson's $\frac{1}{3}$ rule.

17. The order of integration in multiple integration may usually be changed. Evaluate the integral of Problem 16 by integrating first with respect to x, holding y constant. The integral then becomes

$$\int_0^1 \int_0^{\sqrt{1-x^2}} \sin x \sin y \; dy \; dx.$$

18. Reverse the order of integration to solve the example problem of Section 4.

19. Integrate the function $e^{-x^2y^2}$ over the region bounded by the two parabolas $y = x^2$ and $y = 2x^2 - 1$. Note that the integrand is an even function and that the region is symmetrical about the y-axis, so that the integral over half the region may be evaluated and then doubled. Choose reasonable values for Δx and Δy.

Section 5

20. Write a program that performs double interpolation in a table, using polynomials that fit at points within a rectangular region within the table. Test it by solving Problem 4.

21. Write a program that performs triple integration, utilizing the Trapezoidal Rule in each direction. Test it by solving Problem 10. Choose a value of Δx small enough to give six decimal place accuracy. How small does this have to be?

22. Use the program given in this chapter to solve Problem 13. Use a tolerance of 1.0×10^{-6}.

23. Modify the computer program given in this chapter to permit triple integration by the Romberg method.

13
curve fitting
and approximation
of functions

In an early chapter we studied how to fit a polynomial to a set of data, with the implicit assumption that the data were free of error (except round-off) so that it was appropriate to match the interpolating polynomial exactly at the data points. In the case of experimental data, this assumption as to accuracy is often not true. Each data point is subject to experimental errors which, in the case of complicated measurements, may be relatively large in magnitude. Again, in the previous chapter, the true function which relates the data was generally unknown, while, in the case of experimental results, the form of the function is frequently known from the physical laws which apply.

We wish to consider the problem of finding the "best" curve which represents data which are subject to error. "Best" is in quotation marks because the criterion of goodness of fit is to some degree arbitrary, although the least squares criterion is commonly applied. We shall also consider in this chapter the most efficient way in which functions can be represented by a polynomial or a ratio of polynomials where the efficiency is determined by making the maximum error smallest for a given number of parameters required in the rational function. This topic is of special importance in connection with the library function subroutines for digital computers.

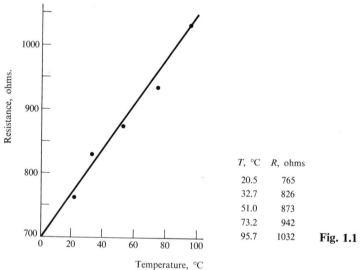

T, °C	R, ohms
20.5	765
32.7	826
51.0	873
73.2	942
95.7	1032

Fig. 1.1

Temperature, °C

1. LEAST SQUARE APPROXIMATIONS

Suppose we wish to fit a curve to an approximate set of data, such as from the calibration of a resistance thermometer by a student in his physics laboratory. He has recorded the temperature and resistance measurements as shown in Fig. 1.1, where the graph suggests a linear relationship. We want to suitably determine the constants a and b in the equation relating resistance R and temperature T,

$$R = aT + b, \tag{1.1}$$

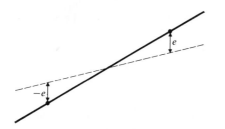

Fig. 1.2

so that in subsequent use an unknown temperature can be measured by determin-ing the resistance of the device.   The line as sketched by eye represents the data fairly well, but if we replotted the data and asked someone else to draw a line, he would rarely get exactly the same line.   One of our requirements for fitting a curve to data is that the process be unambiguous.   We would also like, in some sense, for the errors between the points and the line to be minimized.   The errors are measured by the distances from the points to the line—how these distances are measured depends on whether or not both variables are subject to error.   We shall assume that the error of reading the temperatures in Fig. 1.1 is negligible so that all the errors are in the resistance measurements, and use vertical distances. (If both were subject to error, we might use perpendicular distances, and would modify the following.   Thereby the problem also becomes considerably more complicated.   We shall treat only the simpler case.)

We might first suppose we could minimize the errors by making their sum a minimum, but this is not an adequate criterion.   Consider the case of only two points (Fig. 1.2).   Obviously, the best line passes through each, but any line that passes through the midpoint of the segment connecting them has a sum of errors equal to zero.

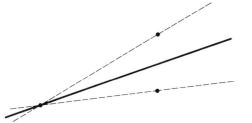

Fig. 1.3

Then what about making the sum of the magnitudes of the errors a minimum? This also is inadequate as the case of three points shows (Fig. 1.3).   Assume two of the points are at the same x-value (which is not an abnormal situation since frequently experiments are duplicated).   The best line obviously will pass through the average of the duplicated tests.   However, any line that falls between the dotted lines shown will have the same sum of the magnitudes of the vertical distances.   Since we wish an unambiguous result, we cannot use this as a basis for our work.

We might accept the criterion that we make the magnitude of the maximum error a minimum (the so-called *minimax* criterion), but for the problem at hand this is rarely done.*   This criterion is awkward because the absolute value function has no derivative at the origin, and it also is felt to give undue importance to a single large error.   The usual criterion is to minimize the sum of the squares of the errors, the "least squares" principle.

In addition to giving a unique result for a given set of data, the least squares method is also in accord with the maximum likelihood principle of statistics.   If the measurement errors have a so-called normal distribution and if the standard deviation is constant for all the data, the line determined by minimizing the sum of squares can be shown to have values of slope and intercept which have maximum likelihood of occurrence.

Let Y_i represent an experimental value, and let y_i be a value from the equation

$$y_i = ax_i + b,$$

where x_i is a particular value of the variable assumed free of error.   Let $e_i = Y_i - y_i$.   The least squares criterion requires that

$$S = e_1^2 + e_2^2 + \cdots + e_N^2 = \sum_{i=1}^{N} e_i^2$$

$$= \sum_{i=1}^{N} (Y_i - ax_i - b)^2$$

be a minimum. N is the number of x, Y pairs.   We reach the minimum by proper choice of the parameters a and b, so they are the "variables" of the problem.   At a minimum for S, the two partial derivatives $\partial S/\partial a$ and $\partial S/\partial b$ will both be zero. Hence, remembering that the x_i and Y_i are data points unaffected by our choice of values for a and b, we have

$$\frac{\partial S}{\partial a} = 0 = \sum_{i=1}^{N} 2(Y_i - ax_i - b_i)(-x_i),$$

$$\frac{\partial S}{\partial b} = 0 = \sum_{i=1}^{N} 2(Y_i - ax_i - b_i)(-1).$$

Dividing each of these equations by -2 and expanding the summation, we get the so-called normal equations

$$a \sum x_i^2 + b \sum x_i = \sum x_i Y_i,$$
$$a \sum x_i + bN = \sum Y_i. \tag{1.2}$$

All the summations in Eq. (1.2) are from $i = 1$ to $i = N$.   Solving these equations simultaneously gives the values for slope and intercept a and b.

For the data in Fig. 1.1 we find that $N = 5$, $\sum T_i = 273.1$, $\sum T_i^2 = 18,607.27$, $\sum R_i = 4438$, $\sum T_i R_i = 254,932.5$.   Our normal equations are then

$$18,607.27a + 273.1b = 254,932.5,$$
$$273.1a + 5b = 4438.$$

---

*   We will use this criterion later in this chapter, however.

From these we find $a = 3.395$, $b = 702.2$, and hence write Eq. (1.1) as

$$R = 702 + 3.39T.$$

2. FITTING NONLINEAR CURVES BY LEAST SQUARES

In many cases, of course, data from experimental tests are not linear, so we need to fit some other function than a first-degree polynomial to them. Popular forms that are tried are the exponential forms

$$y = ax^b$$

or

$$y = ae^{bx}.$$

We can develop normal equations for these analogously to the above by setting the partial derivatives equal to zero. Such nonlinear simultaneous equations are much more difficult to solve* than linear equations. Because of this, the exponential forms are usually linearized by taking logarithms before determining the parameters:

$$\ln y = \ln a + b \ln x$$

or

$$\ln y = \ln a + bx.$$

We now fit the new variable $z = \ln y$ as a linear function of $\ln x$ or x as described above. Here we do not minimize the sum of squares of the deviations of Y from the curve, but rather the deviations of $\ln Y$. In effect, this amounts to minimizing the squares of the percentage errors, which itself may be a desirable feature. An added advantage of the linearized forms is that plots of the data on either log-log or semi-log graph paper show at a glance whether these forms are suitable by whether a straight line represents the data when so plotted.

Because polynomials can be readily manipulated, fitting such functions to data that do not plot linearly is common. We now consider this case. It will turn out that the normal equations are linear for this situation, which is an added advantage. In the development, we use n as the degree of the polynomial and N as the number of data pairs. Obviously if $N = n + 1$, the methods of Chapter 2 apply, so we will always have $N > n + 1$ in the following.

We assume the functional relationship

$$y = a_0 + a_1x + a_2x^2 + \cdots + a_nx^n, \tag{2.1}$$

with errors defined by

$$e_i = Y_i - y_i = Y_i - a_0 - a_1x_i - a_2x_i^2 - \cdots - a_nx_i^n.$$

We again use Y_i to represent the experimental value corresponding to x_i, with x_i free of error. We minimize the sum of squares,

$$S = \sum_{i=1}^{N} e_i^2 = \sum_{i=1}^{N} (Y_i - a_0 - a_1x_i - a_2x_i^2 - \cdots - a_nx_i^n)^2.$$

---

* They are treated briefly in Chapter 7.

Table 2.1. Data to Illustrate Curve Fitting by a Quadratic

x_i	0.0	0.2	0.4	0.7	0.9	1.0
Y_i	1.016	0.768	0.648	0.401	0.272	0.193

$$N = 6$$
$$x_i = 3.2 \qquad\qquad Y_i = 3.298$$
$$x_i^2 = 2.50 \qquad\qquad x_i Y_i = 1.1313$$
$$x_i^3 = 2.144 \qquad\qquad x_i^2 Y_i = 0.74421$$
$$x_i^4 = 1.9234$$

At the minimum, all the partial derivatives $\partial S/\partial a_0$, $\partial S/\partial a_1$, $\cdots$, $\partial S/\partial a_n$ vanish. Writing the equations for these gives $n + 1$ equations:

$$\frac{\partial S}{\partial a_0} = 0 = \sum_{i=1}^{N} 2(Y_i - a_0 - a_1 x_i - \cdots - a_n x_i^n)(-1),$$

$$\frac{\partial S}{\partial a_1} = 0 = \sum_{i=1}^{N} 2(Y_i - a_0 - a_1 x_i - \cdots - a_n x_i^n)(-x_i),$$

$$\vdots$$

$$\frac{\partial S}{\partial a_n} = 0 = \sum_{i=1}^{N} 2(Y_i - a_0 - a_1 x_i - \cdots - a_n x_i^n)(-x_i^n).$$

Dividing each by -2 and rearranging gives the $n + 1$ normal equations to be solved simultaneously:

$$a_0 N + a_1 \sum x_i + a_2 \sum x_i^2 + \cdots + a_n \sum x_i^n = \sum Y_i,$$
$$a_0 \sum x_i + a_1 \sum x_i^2 + a_2 \sum x_i^3 + \cdots + a_n \sum x_i^{n+1} = \sum x_i Y_i,$$
$$a_0 \sum x_i^2 + a_1 \sum x_i^3 + a_2 \sum x_i^4 + \cdots + a_n \sum x_i^{n+2} = \sum x_i^2 Y_i, \quad (2.1)$$
$$\vdots$$
$$a_0 \sum x_i^n + a_1 \sum x_i^{n+1} + a_2 \sum x_i^{n+2} + \cdots + a_n \sum x_i^{2n} = \sum x_i^n Y_i.$$

All the summations in Eq. (2.2) run from 1 to N.

Solving large sets of linear equations is not a simple task. Methods for this are the subject of Chapter 7. These particular equations have an added difficulty in that they have the undesirable property known as *ill-conditioning*. The result of this is that round-off errors in solving them cause unusually large errors in the solutions, which of course are the desired values of the coefficients a_i in Eq. (2.1). Up to $n = 4$ or 5 the problem is not too great (i.e., double precision arithmetic in computer solutions is only desirable and not essential), but beyond this point special methods are needed. Such special methods use orthogonal polynomials in an equivalent form of Eq. (1.1). We shall not pursue this matter further* although we shall treat one form of orthogonal polynomials later in this chapter in connection with representation of functions. From the point of view of the ex-

---

* Ralston (1965) is a good source of further information.

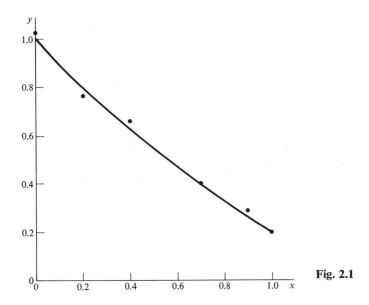

Fig. 2.1

perimentalist, functions more complex than fourth-degree polynomials are rarely needed, and when they are, the problem can often be handled by fitting a series of polynomials to subsets of the data.

We illustrate the use of Eqs. (2.2) to fit a quadratic to the data of Table 2.1. Figure 2.1 shows a plot of the data. (The data are actually a perturbation of the relation $y = 1 - x + 0.2x^2$. It will be of interest to see how well we approximate this function.) To set up the normal equations, we need the sums tabulated in Table 2.1. An electronic desk calculator can give these as accumulated totals directly. We need to solve the set of equations

$$6a_0 + 3.2a_1 + 2.50a_2 = 3.298,$$
$$3.2a_0 + 2.50a_1 + 2.144a_2 = 1.1313,$$
$$2.50a_0 + 2.144a_1 + 1.9234a_2 = 0.74421.$$

The result is $a_0 = 0.9986$, $a_1 = -1.0060$, $a_2 = 0.2103$, so the least squares method gives

$$y = 0.999 - 1.006x + 0.210x^2.$$

Compare this to $y = 1 - x + 0.2x^2$. We do not expect to reproduce the coefficients exactly because of the errors in the data.

In the general case, we may wonder what degree of polynomial should be used. As we use higher-degree polynomials, we of course will reduce the deviations of the points from the curve until, when the degree of the polynomial, n, equals $N - 1$, there is an exact match (assuming no duplicate data at the same x-value) and we have the interpolating polynomials of Chapter 2. The answer to this problem is found in statistics. One decreases the degree of approximating polynomial so long as there is a statistically significant decrease in the variance, σ^2, which is computed by

$$\sigma^2 = \frac{\sum e_i^2}{N - n - 1}.$$

We shall not pursue this matter further.

3. FITTING DATA WITH A CUBIC SPLINE

The fitting of a polynomial curve to a set of data may be considered from another point of view, that of the draftsman. We suppose that the data are not necessarily subject to error so that the least squares approach is not appropriate, but we wish to fit a "smooth curve" to the points. One may use a French curve on the drafting table, but this a very subjective operation. Fitting a polynomial of high degree to a set of six or eight points, say, does not appeal, since we do not expect that the functional relationship is that complicated.

One technique that seems to be increasingly important is the so-called *spline fitting* of a curve. The name derives from another draftsman's device. A spline is a flexible strip which can be held by weights so that it passes through each of the given points, but goes smoothly from each interval to the next according to the laws of beam flexure. The present mathematical procedure is an adaptation of this idea.

The conditions for a cubic spline fit are that we pass a set of cubics through the points, using a new cubic in each interval. To correspond to the idea of the draftsman's spline, we require that both the slope and the curvature be the same for the pair of cubics that join at each point. We now develop the equations subject to these conditions.

Write the cubic for the ith interval, which lies between the points (x_i, y_i) and (x_{i+1}, y_{i+1}) in the form

$$y = a_i(x - x_i)^3 + b_i(x - x_i)^2 + c_i(x - x_i) + d_i. \tag{3.1}$$

Since it fits at the two end points of the interval,

$$y_i = a_i(x_i - x_i)^3 + b_i(x_i - x_i)^2 + c_i(x_i - x_i) + d_i = d_i, \tag{3.2}$$

$$y_{i+1} = a_i(x_{i+1} - x_i)^3 + b_i(x_{i+1} - x_i)^2 + c_i(x_{i+1} - x_i) + d_i$$
$$= a_i h_i^3 + b_i h_i^2 + c_i h_i + d_i. \tag{3.3}$$

In the last equation, we use h_i for Δx in the ith interval. We need the first and second derivatives to relate the slopes and curvatures of the joining polynomials, so we differentiate Eq. (3.1):

$$y' = 3a_i(x - x_i)^2 + 2b_i(x - x_i) + c_i, \tag{3.4}$$

$$y'' = 6a_i(x - x_i) + 2b_i. \tag{3.5}$$

The mathematical procedure is simplified if we write the equations in terms of the second derivatives of the interpolating cubics. Let S_i represent the second derivative at the point (x_i, y_i) and S_{i+1} at the point (x_{i+1}, y_{i+1}).

From Eq. (3.5) we have

$$S_i = 6a_i(x_i - x_i) + 2b_i$$
$$= 2b_i,$$
$$S_{i+1} = 6a_i(x_{i+1} - x_i) + 2b_i$$
$$= 6a_i h_i + 2b_i.$$

Hence we can write

$$b_i = S_i/2, \qquad\qquad (3.6)$$
$$a_i = (S_{i+1} - S_i)/6h_i. \qquad\qquad (3.7)$$

We substitute the relations for a_i, b_i, d_i given by Eqs. (3.2), (3.6), and (3.7) into Eq. (3.3) and then solve for c_i:

$$y_{i+1} = \left(\frac{S_{i+1} - S_i}{6h_i}\right)h_i^3 + \frac{S_i}{2}h_i^2 + c_i h_i + y_i,$$

$$c_i = \frac{y_{i+1} - y_i}{h_i} - \frac{2h_i S_i + h_i S_{i+1}}{6}.$$

We now invoke the condition that the slopes of the two cubics that join at (x_i, y_i) are the same. For the equation in the ith interval, Eq. (3.4) becomes, with $x = x_i$,

$$y_i' = 3a_i(x_i - x_i)^2 + 2b_i(x_i - x_i) + c_i = c_i.$$

In the previous interval, from x_{i-1} to x_i, the slope at its right end will be

$$y_i' = 3a_{i-1}(x_i - x_{i-1})^2 + 2b_{i-1}(x_i - x_{i-1}) + c_{i-1}$$
$$= 3a_{i-1}h_{i-1}^2 + 2b_{i-1}h_{i-1} + c_{i-1}.$$

Equating these, and substituting for a, b, c, d their relationships in terms of S and y, we get

$$y_i' = \frac{y_{i+1} - y_i}{h_i} - \frac{2h_i S_i + h_i S_{i+1}}{6}$$

$$= 3\left(\frac{S_i - S_{i-1}}{6h_{i-1}}\right)h_{i-1}^2 + 2\left(\frac{S_{i-1}}{2}\right)h_{i-1} + \frac{y_i - y_{i-1}}{h_{i-1}} - \frac{2h_{i-1}S_{i-1} + h_{i-1}S_i}{6}.$$

On simplifying this equation we get

$$h_{i-1}S_{i-1} + (2h_{i-1} + 2h_i)S_i + h_i S_{i+1} = 6\left(\frac{y_{i+1} - y_i}{h_i} - \frac{y_i - y_{i-1}}{h_{i-1}}\right). \qquad (3.8)$$

Equation (3.8) applies at each internal point, from $i = 2$ to $i = n - 1$, n being the total number of points. This gives $n - 2$ equations relating the n values of S_i. We get two additional equations involving S_1 and S_n when we specify conditions pertaining to the end intervals of the whole curve. To some extent these end conditions are arbitrary. If we take that S_1 is a linear extrapolation from S_2 and S_3, with analogous linearity of S_n, S_{n-1}, and S_{n-2}, one finds that, for a set of data which are fitted throughout by a single cubic, the spline curve is this same cubic. Alternate end conditions do not have this property.*

We shall assume linearity, so

$$\frac{S_2 - S_1}{h_1} = \frac{S_3 - S_2}{h_2}, \qquad h_2 S_1 - (h_1 + h_2)S_2 + h_1 S_3 = 0,$$

$$\frac{S_n - S_{n-1}}{h_{n-1}} = \frac{S_{n-1} - S_{n-2}}{h_{n-2}}, \qquad h_{n-1}S_{n-2} - (h_{n-2} + h_{n-1})S_{n-1} + h_{n-2}S_n = 0.$$

It is convenient to write the set of equations in matrix form:

$$
\begin{vmatrix}
h_2 & -(h_1 + h_2) & h_1 & 0 & 0 & \cdots & 0 \\
h_1 & 2(h_1 + h_2) & h_2 & 0 & 0 & & 0 \\
0 & h_2 & 2(h_2 + h_3) & h_3 & 0 & & 0 \\
0 & 0 & h_3 & 2(h_3 + h_4) & h_4 & & 0 \\
\vdots & & & & & & \\
0 & \cdots & & 0 & h_{n-1} & -(h_{n-2} + h_{n-1}) & h_{n-2}
\end{vmatrix}
\begin{vmatrix}
S_1 \\ S_2 \\ S_3 \\ S_4 \\ \vdots \\ S_n
\end{vmatrix}
$$

$$
= 6
\begin{vmatrix}
0 \\
\dfrac{y_3 - y_2}{h_2} - \dfrac{y_2 - y_1}{h_1} \\
\dfrac{y_4 - y_3}{h_3} - \dfrac{y_3 - y_2}{h_2} \\
\dfrac{y_5 - y_4}{h_4} - \dfrac{y_4 - y_3}{h_3} \\
\vdots \\
0
\end{vmatrix}. \quad (3.9)
$$

If the data are equally spaced, so each h_i is the same, we have a particularly simple matrix equation. The vector on the right has components equal to $\Delta^2 y$:

$$
\begin{vmatrix}
1 & -2 & 1 & 0 & 0 & \cdots & & & & & 0 \\
1 & 4 & 1 & 0 & 0 & & & & & & 0 \\
0 & 1 & 4 & 1 & 0 & & & & & & 0 \\
0 & 0 & 1 & 4 & 1 & & & & & & 0 \\
\vdots & & & & & & & & & & \\
0 & & & 0 & 0 & 0 & 1 & 4 & 1 \\
0 & & \cdots & & 0 & 0 & 0 & 1 & -2 & 1
\end{vmatrix}
\quad S = \frac{6}{h^2}
\begin{vmatrix}
0 \\ \Delta^2 y_1 \\ \Delta^2 y_2 \\ \Delta^2 y_3 \\ \vdots \\ \Delta^2 y_{n-2} \\ 0
\end{vmatrix}. \quad (3.10)
$$

After the S_i values are calculated, the values of a_i, b_i, c_i, and d_i are obtained, which gives the y equations. Note that while the coefficient matrices are not tridiagonal, they are diagonally dominant, so that iterative methods may be alternately applied.

---

* Other conditions that might be imposed are $S_1 = S_n = 0$. Taking $S_1 = S_2$ and $S_n = S_{n-1}$ gives parabolas for the end intervals.

Table 3.1

x	y	Δy	$\Delta^2 y$
0	0		
		1	
1	1		2
		3	
2	4		0
		3	
3	7		5
		8	
4	15		0
		8	
5	23		

Example. Fit the data in Table 3.1 by a cubic spline curve.   Since the x-values are uniformly spaced, the simpler set of equations in Eq. (3.10) apply.   Since $h = 1$, we get

$$
\begin{vmatrix}
1 & -2 & 1 & 0 & 0 & 0 \\
1 & 4 & 1 & 0 & 0 & 0 \\
0 & 1 & 4 & 1 & 0 & 0 \\
0 & 0 & 1 & 4 & 1 & 0 \\
0 & 0 & 0 & 1 & 4 & 1 \\
0 & 0 & 0 & 1 & -2 & 1
\end{vmatrix}
\begin{vmatrix}
S_1 \\ S_2 \\ S_3 \\ S_4 \\ S_5 \\ S_6
\end{vmatrix}
= \frac{6}{1}
\begin{vmatrix}
0 \\ 2 \\ 0 \\ 5 \\ 0 \\ 0
\end{vmatrix}.
$$

Solving the set of equations by elimination gives

$$S_1 = 6.53, \qquad S_2 = 2.00, \qquad S_3 = -2.53,$$
$$S_4 = 8.13, \qquad S_5 = 0, \qquad S_6 = -8.13.$$

We now compute the constants of the cubic polynomials for y on each interval. The results are:

x-values:	0 to 2	2 to 3	3 to 5
a_i	−0.755	1.778	−1.355
b_i	3.265	−1.265	4.065
c_i	−1.510	2.487	5.290
d_i	0	4	7

These give the coefficients of the cubics in each interval.   For example, between $x = 2$ and $x = 3$,

$$y = 1.788(x - 2)^3 - 1.265(x - 2)^2 + 2.487(x - 2) + 4.$$

Our end conditions give the same equation for the first and second intervals and for the last and next to last intervals.   This is because linearity of the second derivative is a property of a cubic within its region of applicability.

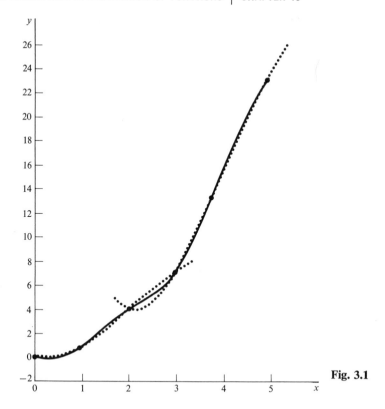

Fig. 3.1

Figure 3.1 shows the resulting curve fitted to the points. Also shown by dotted lines are cubic interpolating polynomials derived by the methods of Chapter 2. The first of these passes through the first four points taken as a group, and the second through the last four. The advantage of the spline fit is readily apparent at the interval where the two interpolating polynomials overlap. Not only are the values of y ambiguous between $x = 2$ and $x = 3$, but the slopes, as given by the two curves, are completely different. The spline curve is not subject to this difficulty.

4. APPROXIMATION OF FUNCTIONS WITH ECONOMIZED POWER SERIES

We turn now to the problem of representing a function with minimum error. This is a central problem in the software development of digital computers because it is more economical to compute the values of the common functions using an efficient approximation than to store a table of values and employ interpolation techniques. Since digital computers are essentially only arithmetic devices, the most elaborate function they can compute is a rational function, a ratio of polynomials. We will hence restrict our discussion to representation of functions by polynomials or rational functions.

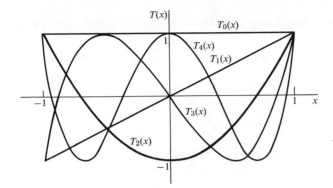

Fig. 4.1

The familiar Taylor series expansion represents the function with very small error near the point of the expansion, but the error increases rapidly (proportional to a power) as we employ it at points farther away. In a digital cc nputer, we have no control over where in an interval the approximation will be used, so the Taylor series is not usually appropriate. We would prefer to trade some of its excessive precision at the center of the interval to reduce the errors at the ends.

We can do this while still expressing functions as polynomials by the use of Chebyshev polynomials. The first few of these are*

$$
\begin{aligned}
T_0(x) &= 1, \\
T_1(x) &= x, \\
T_2(x) &= 2x^2 - 1, \\
T_3(x) &= 4x^3 - 3x, \\
T_4(x) &= 8x^4 - 8x^2 + 1, \\
T_5(x) &= 16x^5 - 20x^3 + 5x, \\
T_6(x) &= 32x^6 - 48x^4 + 18x^2 - 1, \\
T_7(x) &= 64x^7 - 112x^5 + 56x^3 - 7x, \\
T_9(x) &= 256x^9 - 576x^7 + 432x^5 - 120x^3 + 9x.
\end{aligned}
\tag{4.1}
$$

The members of this series of polynomials can be generated from the two-term recursion formula

$$
T_{n+1}(x) = 2xT_n(x) - T_{n-1}(x), \qquad T_0(x) = 1, \quad T_1(x) = x. \tag{4.2}
$$

Note that the coefficient of x^n in $T_n(x)$ is always 2^{n-1}. In Fig. 4.1 we plot the first four polynomials of Eq. (4.1).

These polynomials have some unusual properties. They form an orthogonal set, in that

$$
\int_{-1}^{1} \frac{T_n(x)T_m(x)}{\sqrt{1 - x^2}}\, dx = \begin{cases} 0, & n \neq m, \\ \pi, & n = m = 0, \\ \pi/2, & n = m \neq 0. \end{cases}
$$

The orthogonality of these functions will not be of immediate concern to us.

---

* The commonly accepted symbol $T(x)$ comes from the older spelling, Tschebycheff.

The Chebyshev polynomials are also terms of Fourier series since

$$T_n(x) = \cos n\theta, \tag{4.3}$$

where $\theta = \arccos x$.   Observe that $\cos 0 = 1$, $\cos \theta = \cos(\arccos x) = x$.

In order to demonstrate the equivalence of Eq. (4.3) to Eqs. (4.1) and (4.2), we recall some trigonometric identities, such as

$$\cos 2\theta = 2 \cos^2\theta - 1,$$
$$T_2(x) = 2x^2 - 1;$$
$$\cos 3\theta = 4 \cos^3\theta - 3 \cos\theta,$$
$$T_3(x) = 4x^3 - 3x;$$
$$\cos (n + 1)\theta + \cos (n - 1)\theta = 2 \cos \theta \cos n\theta,$$
$$T_{n+1}(x) + T_{n-1}(x) = 2xT_n(x).$$

Because of the relation $T_n(x) = \cos n\theta$, it is apparent that the Chebyshev polynomials have a succession of maximums and minimums of alternating signs, each of magnitude one.   Further, since $|\cos n\theta| = 1$ for $n\theta = 0$, π, $2\pi, \ldots$, and since θ varies from 0 to π as x varies from 1 to -1, $T_n(x)$ assumes its maximum magnitude of unity $n + 1$ times on the interval $[-1,1]$.

Most important for our present application of these polynomials is the fact that of all polynomials of degree n where the coefficient of x^n is unity, the polynomial

$$\frac{1}{2^{n-1}} T_n(x)$$

has a smaller upper bound to its magnitude in the interval $[-1,1]$ than any other. Because the maximum magnitude of $T_n(x)$ is one, the upper bound referred to is $1/2^{n-1}$.   This is of importance because we shall be able to write power series representations of functions whose maximum errors are given in terms of this upper bound.

We first prove this assertion about bounds on the magnitude of polynomials. The proof is by contradiction.   Let $P_n(x)$ be a polynomial whose leading term is x^n and suppose that its maximum magnitude on $[-1,1]$ is less than that of $T_n(x)/2^{n-1}$.   Write

$$T^n(x)/2^{n-1} - P_n(x) = P_{n-1}(x),$$

where $P_{n-1}(x)$ is a polynomial of degree $n - 1$ or less, since the x^n terms cancel. The polynomial $T_n(x)$ has $n + 1$ extremes (counting end points), each of magnitude one, so $T_n(x)/2^{n-1}$ has $n + 1$ extremes each of magnitude $1/2^{n-1}$, and these successive extremes alternate in sign.   By our supposition about $P_n(x)$, at each of these maximums or minimums, the magnitude of $P_n(x)$ is less than $1/2^{n-1}$; hence $P_{n-1}(x)$ must change its sign at least for every extreme of $T_n(x)$, which is then at least $n + 1$ times. $P_{n-1}(x)$ hence crosses the axis at least n times and would have n zeros.   But this is impossible if $P_{n-1}(x)$ is only of degree $n - 1$, unless it is identically zero.   The premise must then be false and $P_n(x)$ has a larger magnitude than the polynomial we are testing, or alternatively $P_n(x)$ is exactly the same polynomial.

We are now ready to use Chebyshev polynomials to economize a power series. Consider the Maclaurin series for e^x:

$$e^x = 1 + x + \frac{x^2}{2} + \frac{x^3}{6} + \frac{x^4}{24} + \frac{x^5}{120} + \frac{x^6}{720} + \cdots.$$

If we would like to use a truncated series to approximate e^x on the interval $[0,1]$ with a precision of 0.001, we will have to retain terms through that in x^6, since the error after the term in x^5 will be more than $1/720$. Suppose we subtract

$$(1/720)(T_6/32)$$

from the truncated series. We note from Eq. (4.1) that this will exactly cancel the x^6 term and at the same time make adjustments in other coefficients of the Maclaurin series. Since the maximum value of T_6 on the interval $[0,1]$ is unity, this will change the sum of the truncated series by only $\frac{1}{720} \cdot \frac{1}{32} < 0.00005$, which is small with respect to our required precision of 0.001. Performing the calculations, we have

$$e^x \doteq 1 + x + \frac{x^2}{2} + \frac{x^3}{6} + \frac{x^4}{24} + \frac{x^5}{120} + \frac{x^6}{720} - \frac{1}{720}\left(\frac{1}{32}\right)\left(32x^6 - 48x^4 + 18x^2 - 1\right)$$

$$= 1.000043 + x + 0.499219x^2 + \frac{x^3}{6} + 0.0437x^4 + \frac{x^5}{120}. \tag{4.4}$$

This gives a fifth-degree polynomial that approximates e^x on $[0,1]$ almost as well as the sixth-degree one derived from the Maclaurin series. (The actual maximum error of the fifth-degree expression is 0.000270; for the sixth-degree expression it is 0.000226.) We hence have "economized" the power series.

By subtracting $\frac{1}{120}(T_5/16)$ we can economize further, getting a fourth-degree polynomial that is almost as good as the economized fifth-degree one. It is left as an exercise to do this and to show that the maximum error is now 0.000781, so that we have found a fourth-degree power series that meets an error criterion that requires us to use two additional terms of the original Maclaurin series. Because of the relative ease with which they can be developed, such economized power series are frequently used for approximations to functions and are much more efficient than power series of the same degree obtained by merely truncating a Taylor or Maclaurin series.

By rearranging the Chebyshev polynomials, we can express powers of x in terms of them:

$$
\begin{aligned}
1 &= T_0, \\
x &= T_1, \\
x^2 &= \tfrac{1}{2}(T_0 + T_2), \\
x^3 &= \tfrac{1}{4}(3T_1 + T_3), \\
x^4 &= \tfrac{1}{8}(3T_0 + 4T_2 + T_4), \\
x^5 &= \tfrac{1}{16}(10T_1 + 5T_3 + T_5), \\
x^6 &= \tfrac{1}{32}(10T_0 + 15T_2 + 6T_4 + T_6), \\
x^7 &= \tfrac{1}{64}(35T_1 + 21T_3 + 7T_5 + T_7), \\
x^8 &= \tfrac{1}{128}(35T_0 + 56T_2 + 28T_4 + 8T_6 + T_8), \\
x^9 &= \tfrac{1}{256}(126T_1 + 84T_3 + 36T_5 + 9T_7 + T_9).
\end{aligned}
\tag{4.5}
$$

By substituting these identities into a Taylor series and collecting terms in $T_i(x)$, we create a Chebyshev series. For example, we can get the first four terms of a Chebyshev series by starting with the Maclaurin expansion for e^x. Such a series converges more rapidly than does a Taylor series on $[-1,1]$:

$$e^x = 1 + x + \frac{x^2}{2} + \frac{x^3}{6} + \frac{x^4}{24} + \cdots.$$

Replacing terms by Eq. (4.5), but omitting polynomials beyond $T_3(x)$, since we want only four terms, we have

$$e^x = T_0 + T_1 + \tfrac{1}{4}(T_0 + T_2) + \tfrac{1}{24}(3T_1 + T_3) + \tfrac{1}{192}(3T_0 + 4T_2 + \cdots)$$
$$+ \tfrac{1}{1920}(10T_1 + 5T_3 + \cdots) + \tfrac{1}{23,040}(10T_0 + 15T_2 + \cdots) + \cdots$$
$$= 1.2661T_0 + 1.1303T_1 + 0.2715T_2 + 0.0444T_3 + \cdots.$$

In order to compare the Chebyshev expansion with the Maclaurin series, we convert back to powers of x, using Eq. (4.1):

$$e^x = 1.2661 + 1.1303(x) + 0.2715(2x^2 - 1) + 0.0444(4x^3 - 3x) + \cdots$$
$$= 0.9946 + 0.9971x + 0.5430x^2 + 0.1776x^3 + \cdots. \tag{4.6}$$

Table 4.1 and Fig. 4.2 compare the error of the Chebyshev expansion, Eq. (4.6), with the Maclaurin series using terms through x^3 in each case. The figure shows how the Chebyshev expansion attains a smaller maximum error by permitting the error at the origin to increase. The errors can be considered to be distributed more or less uniformly throughout the interval. In contrast to this, the Maclaurin expansion, which gives very small errors near the origin, allows the error to bunch up at the ends of the interval.

Table 4.1. Comparison of Chebyshev Series for e^x with Maclaurin Series

$$e^x = 0.9946 + 0.9971x + 0.5430x^2 + 0.176x^3$$
$$e^x = 1 + x + 0.5x^2 + 0.1667x^3$$

x	e^x	Chebyshev	Error	Maclaurin	Error
-1.0	0.3679	0.3629	0.0050	0.3333	0.0346
-0.8	0.4493	0.4535	-0.0042	0.4347	0.0146
-0.6	0.5488	0.5535	-0.0047	0.5440	0.0048
-0.4	0.6703	0.6713	-0.0010	0.6693	0.0010
-0.2	0.8187	0.8155	0.0032	0.8187	0.0000
0	1.0000	0.9946	0.0054	1.0000	0.0000
0.2	1.2214	1.2172	0.0042	1.2213	0.0001
0.4	1.4918	1.4917	0.0001	1.4907	0.0011
0.6	1.8221	1.8267	-0.0046	1.8160	0.0061
0.8	2.2255	2.2307	-0.0052	2.2053	0.0202
1.0	2.7183	2.7123	0.0060	2.6667	0.0516

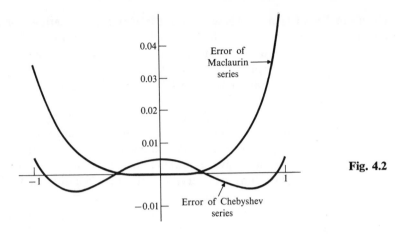

Fig. 4.2

If the function is to be expressed directly as an expansion in Chebyshev polynomials, the coefficients can be obtained by integration. Based on the orthogonality property, the coefficients are computed from

$$a_i = \frac{2}{\pi} \int_{-1}^{1} \frac{f(x)T_i(x)}{\sqrt{1-x^2}} \, dx,$$

and the series is expressed as

$$f(x) = \frac{a_0}{2} + \sum_{i=1}^{\infty} a_i T_i(x).$$

A change of variable will be required if the desired interval is other than $(-1,1)$. In some cases, the definite integral which defines the coefficients can be profitably evaluated by numerical methods as described in Chapter 4.

Since the coefficients of the terms of a Chebyshev expansion usually decrease even more rapidly than the terms of a Maclaurin expansion, one can get an estimate of the magnitude of the error from the next nonzero term after those which were retained. For the truncated Chebyshev series given by Eq. (4.6), the $T_4(x)$ term would be

$$\tfrac{1}{192}(T_4) + \tfrac{1}{23,040}(6T_4) + \cdots = 0.00525T_4.$$

Since the maximum value of $T_4(x)$ on $(-1,1)$ is 1.0, we estimate the maximum errors of Eq. (4.6) to be 0.00525. The maximum error in Table 4.1 is 0.0060. This good agreement is caused by the very rapid decrease in coefficients in this example.

5. APPROXIMATION WITH RATIONAL FUNCTIONS

We have seen that expansion of a function in terms of Chebyshev polynomials gives a power series expansion that is much more efficient on the interval $(-1,1)$ than the Maclaurin expansion, in that it has a smaller maximum error with a given number of terms.   These are not the best approximations for use in most digital computers, however.   In this application, we measure efficiency by the computer time required to evaluate the function, plus some consideration of storage requirements for the constants.   Since the arithmetic operations of a computer can directly evaluate only polynomials, we limit our discussion of more efficient approximations to rational functions, which are the ratios of two polynomials.

Our discussion of methods of finding efficient rational approximations will be elementary and introductory only.   Obtaining truly best approximations is a difficult subject.   In its present stage of development it is as much art as science, and requires one to use successive approximations from a "suitably close" initial approximation.   Our study will serve to introduce the student to some of the ideas and procedures used.   The topic is of great importance, however, since the saving of just 1 millisecond of time in the generation of a frequently used elementary function may be worth hundreds of dollars worth of machine time each year.

We start with a discussion of Padé approximations.   Suppose we wish to represent a function as the quotient of two polynomials:

$$f(x) \doteq R_N(x) = \frac{a_0 + a_1 x + a_2 x^2 + \cdots + a_n x^n}{1 + b_1 x + b_2 x^2 + \cdots + b_m x^m}, \qquad N = n + m.$$

The constant term in the denominator can be taken as unity without loss of generality, since we can always convert to this form by dividing numerator and denominator by b_0.   The constant b_0 will generally not be zero, else the fraction is undefined at $x = 0$.   The most useful of the Padé approximations are with degree of the numerator equal to or one greater than the degree of the denominator. Note that the number of constants in $R_N(x)$ is $N + 1 = n + m + 1$.

The Padé approximations are related to Maclaurin expansions in that the coefficients are determined in a similar fashion to make $f(x)$ and $R_N(x)$ agree at $x = 0$ and also to make the first N derivatives agree at $x = 0$.*

We begin with the Maclaurin series for $f(x)$ (we use only terms through x^N) and write

$$f(x) - R_N(x) = (c_0 + c_1 x + c_2 x^2 + \cdots + c_N x^N) - \frac{a_0 + a_1 x + \cdots + a_n x^n}{1 + b_1 x + \cdots + b_m x^m}$$

$$= \frac{(c_0 + c_1 x + \cdots + c_N x^N)(1 + b_1 x + \cdots + b_m x^m) - (a_0 + a_1 x + \cdots + a_n x^n)}{1 + b_1 x + \cdots + b_m x^m}. \quad (5.1)$$

The coefficients c_i are $f^{(i)}(0)/i!$ of the Maclaurin expansion.   Now if $f(x) = R_N(x)$ at $x = 0$, the numerator of (5.1) must have no constant term.   Hence

$$c_0 - a_0 = 0. \qquad (5.2)$$

For the first N derivatives of $f(x)$ and $R_N(x)$ to be equal at $x = 0$, the coefficients of the powers of x up to and including x^N in the numerator must all be zero also.

---

* A similar development can be derived for the expansion about a nonzero value of x, but the manipulations are not as easy.   By a change of variable we can always make the region of interest contain the origin.

This gives N additional equations for the a's and b's. The first n of these involve a's, the rest only b's and c's:

$$b_1 c_0 + c_1 - a_1 = 0,$$
$$b_2 c_0 + b_1 c_1 + c_2 - a_2 = 0,$$
$$b_3 c_0 + b_2 c_1 + b_1 c_2 + c_3 - a_3 = 0,$$
$$\vdots$$
$$b_m c_{n-m} + b_{m-1} c_{n-m+1} + \cdots + c_n - a_n = 0, \tag{5.3}$$
$$b_m c_{n-m+1} + b_{m-1} c_{n-m+2} + \cdots + c_{n+1} = 0,$$
$$b_m c_{n-m+2} + b_{m-1} c_{n-m+3} + \cdots + c_{n+2} = 0,$$
$$\vdots$$
$$b_m c_{N-m} + b_{m-1} c_{N-m+1} + \cdots + c_N = 0.$$

Note that in each equation, the sum of the subscripts on the factors of each product is the same, and equal to the exponent of the x-term in the numerator. The $N + 1$ equations of Eqs. (5.2) and (5.3) give the required coefficients of the Padé approximation. We illustrate by an example.

Example. Find $\arctan x \doteq R_9(x)$. Use in the numerator a polynomial of degree five.

The Maclaurin series through x^9 is

$$\arctan x \doteq x - \tfrac{1}{3}x^3 + \tfrac{1}{5}x^5 - \tfrac{1}{7}x^7 + \tfrac{1}{9}x^9. \tag{5.4}$$

We form, analogously to Eq. (5.1),

$$f(x) - R_9(x)$$

$$= \frac{(x - \tfrac{1}{3}x^3 + \tfrac{1}{5}x^5 - \tfrac{1}{7}x^7 + \tfrac{1}{9}x^9)(1 + b_1 x + b_2 x^2 + b_3 x^3 + b_4 x^4) - (a_0 + a_1 x + \cdots + a_5 x^5)}{(1 + b_1 x + b_2 x^2 + b_3 x^3 + b_4 x^4)}. \tag{5.5}$$

Making coefficients through that of x^9 in the numerator equal to zero, we get

$$a_0 = 0,$$
$$a_1 = 1,$$
$$a_2 = b_1,$$
$$a_3 = -\tfrac{1}{3} + b_2,$$
$$a_4 = -\tfrac{1}{3}b_1 + b_3,$$
$$a_5 = \tfrac{1}{5} - \tfrac{1}{3}b_2 + b_4,$$

$$\tfrac{1}{5}b_1 - \tfrac{1}{3}b_3 = 0,$$
$$-\tfrac{1}{7} + \tfrac{1}{5}b_2 - \tfrac{1}{3}b_4 = 0,$$
$$-\tfrac{1}{7}b_1 + \tfrac{1}{5}b_3 = 0,$$
$$\tfrac{1}{9} - \tfrac{1}{7}b_2 + \tfrac{1}{5}b_4 = 0.$$

Solving first the last four equations for the b's, and then getting the a's, we have

$$a_0 = 0, \quad a_1 = 1, \quad a_2 = 0, \quad a_3 = \tfrac{7}{9}, \quad a_4 = 0, \quad a_5 = \tfrac{64}{945},$$
$$b_1 = 0, \quad b_2 = \tfrac{10}{9}, \quad b_3 = 0, \quad b_4 = \tfrac{5}{21}.$$

Table 5.1. Comparison of Padé Approximation to Maclaurin Series for Arctan X

x	True Value	Padé Eq. (5.6)	Error	Maclaurin Eq. (5.4)	Error
0.2	0.19740	0.19740	0.00000	0.19740	0.00000
0.4	0.38051	0.38051	0.00000	0.38051	0.00000
0.6	0.54042	0.54042	0.00000	0.54067	−0.00025
0.8	0.67474	0.67477	−0.00003	0.67982	−0.00508
1.0	0.78540	0.78558	−0.00018	0.83492	−0.04952

The rational function which approximates arctan x is then

$$\arctan x \doteq \frac{x + \frac{7}{9}x^3 + \frac{64}{945}x^5}{1 + \frac{10}{9}x^2 + \frac{5}{21}x^4}. \tag{5.6}$$

In Table 5.1 we compare the errors for Padé approximation (Eq. 5.6) to the Maclaurin series expansion (Eq. 5.4). Enough terms are available in the Maclaurin series to give five-decimal precision at $x = 0.2$ and 0.4, but at $x = 1$ (the limit for convergence of the series) the error is sizeable. Even though we used no more information in establishing it, the Padé formula is surprisingly accurate, having an error only 1/275 as large at $x = 1$. It is then particularly astonishing to realize that the Padé approximation is still not the best one of its form, for it violates the minimax principle. If the extreme precision near $x = 0$ is relaxed, we can make the maximum error smaller in the interval.

Before we discuss such better approximations in the form of rational functions, remarks on the amount of effort to compute using Eq. (5.6) are in order. If we implement the equation in a computer as it stands, we would of course use the constants in decimal form, and we would evaluate the polynomials in nested form:

$$\text{Numerator} = ((0.0677x^2 + 0.7778)x^2 + 1)x,$$
$$\text{Denominator} = (0.2381x^2 + 1.1111)x^2 + 1.$$

Since additions and subtractions are generally much faster than multiplications or divisions, we generally neglect them in a count of operations. We have then three multiplications for the numerator, two for the denominator, plus one to get x^2, and one division, for a total of six operations. The Maclaurin series is evaluated with six multiplications using the nested form. If division and multiplication consume about the same time, there is a standoff in effort, but greater precision for Eq. (5.6).*

Since small differences in effort accumulate for a frequently used function, it is of interest to see if we can further decrease the number of operations to evaluate Eq. (5.6). By means of a succession of divisions we can re-express it in continued

---

* On some computers division is slower than multiplication. This will modify the conclusion reached here.

fraction form:

$$\frac{0.0677x^5 + 0.7778x^3 + x}{0.2381x^4 + 1.1111x^2 + 1} = \frac{0.2844x^5 + 3.2667x^3 + 4.2x}{x^4 + 4.6667x^2 + 4.2}$$

$$= \frac{0.2844x(x^4 + 11.4846x^2 + 14.7659)}{x^4 + 4.6667x^2 + 4.2} = \frac{0.2844x}{\dfrac{x^4 + 4.6667x^2 + 4.2}{x^4 + 11.4846x^2 + 14.7659}}$$

$$= \frac{0.2844x}{1 - \dfrac{6.879x^2 + 10.5659}{x^4 + 11.4846 + 14.7659}} = \frac{0.2844x}{1 - \dfrac{6.879x^2 + 1.5497}{x^4 + 11.4846x^2 + 14.7659}}$$

$$= \frac{0.2844x}{1 - \dfrac{6.8179}{\dfrac{x^4 + 11.846x^2 + 14.7659}{x^2 + 1.5497}}} = \frac{0.2844x}{1 - \dfrac{6.8179}{x^2 + 9.9348 - \dfrac{0.6304}{x^2 + 1.5497}}}.$$

In this last form, we see that three divisions and two multiplications are needed (one multiplication by x and one to get x^2), for a total of five operations. We have saved one step. In most cases there is an even greater advantage to the continued fraction form; in this example the missing powers of x favored the evaluation as polynomials.

The error of a Padé approximation can often be roughly estimated by computing the next nonzero term in the numerator of Eq. (5.5). For the above example, the coefficient of x^{10} is zero, and the next term is

$$(-\tfrac{1}{7}b_4 + \tfrac{1}{9}b_2 - \tfrac{1}{11})x^{11} = (-\tfrac{1}{7}(\tfrac{5}{21}) + \tfrac{1}{9}(\tfrac{10}{9}) - \tfrac{1}{11})x^{11}$$
$$= -0.0014x^{11}.$$

Dividing by the denominator, we have

$$\text{Error} \doteq \frac{-0.0014x^{11}}{1 + 1.1111x^2 + 0.2381x^4}.$$

At $x = 1$ this estimate gives -0.00060, which is about three times too large, but still of the correct order of magnitude. It is not unusual that such estimates be rough; analogous estimates of error by using the next term in a Maclaurin series behave similarly. The validity of the rule of thumb that "next term approximates the error" is poor when the coefficients do not decrease rapidly.

The preference for Padé approximations with the degree of the numerator the same or one more than the degree of the denominator rests on the empirical fact that the errors are usually less for these. Ralston (1965) gives examples demonstrating this.

One can get somewhat improved rational function approximations by starting with the Chebyshev expansion and operating analogously to the method for

Padé approximations.   We illustrate with an approximation for e^x.   The Chebyshev series was derived in Section 4, Eq. (4.6):

$$e^x \doteq 1.2661T_0 + 1.1303T_1 + 0.2715T_2 + 0.0444T_3.$$

Using this approximation, we form the difference

$$f(x) - \frac{P_n(x)}{Q_m(x)}$$

$$= \frac{(1.2661 + 1.1303T_1 + 0.2715T_2 + 0.0444T_3)(1 + b_1T_1) - (a_0 + a_1T_1 + a_2T_2)}{1 + b_1T_1}.$$

Here we have chosen the numerator as a second-degree Chebyshev polynomial and the denominator as first degree.   We again make the first $N = n + m$ powers of x in the numerator vanish.   Expanding the numerator, we get

$$\text{Numerator} = 1.2661 + 1.1303T_1 + 0.2715T_2 + 0.044T_3 + 1.2661b_1T_1$$
$$+ 1.1303b_1T_1^2 + 0.2715b_1T_1T_2 + 0.0444b_1T_1T_3 - a_0$$
$$- a_1T_1 - a_2T_2.$$

Before we can equate coefficients to zero, we need to resolve the products of Chebyshev polynomials that occur.   Recalling that $T_n(x) = \cos n\theta$, we can use the trignometric identity

$$\cos n\theta \cos m\theta = \tfrac{1}{2}(\cos(n + m)\theta + \cos(n - m)\theta),$$
$$T_n(x)T_m(x) = \tfrac{1}{2}(T_{n+m}(x) + T_{|n-m|}(x)).$$

The absolute value of the difference $n - m$ occurs because $\cos(z) = \cos(-z)$. Using this relation we can write the equations

$$a_0 = 1.2661 + \frac{1.1303}{2}b_1,$$

$$a_1 = 1.1303 + \left(\frac{0.2715}{2} + 1.2661\right)b_1,$$

$$a_2 = 0.2715 + \left(\frac{1.1303}{2} + \frac{0.0444}{2}\right)b_1,$$

$$0 = 0.0444 + \frac{0.2715}{2}b_1.$$

Solving, we get $b_1 = -0.3266$, $a_0 = 1.0815$, $a_1 = 0.6724$, $a_2 = 0.07966$, and

$$e^x \doteq \frac{1.0815 + 0.6724T_1 + 0.07966T_2}{1 - 0.3266T_1}$$

$$= \frac{1.0018 + 0.6724x + 0.1593x^2}{1 - 0.3266x}. \tag{5.7}$$

The last expression results when the Chebyshev polynomials are written in terms of powers of x.   In Table 5.2 the error of this rational approximation is compared

Table 5.2. Comparison of Rational Approximations (Eq. 5.7) with Chebyshev Series for e^x

x	e^x	Chebyshev	Error	Rational function	Error
−1.0	0.3679	0.3629	0.0050	0.3684	−0.0005
−0.8	0.4493	0.4535	−0.0042	0.4486	0.0007
−0.6	0.5488	0.5535	−0.0047	0.5482	0.0006
−0.4	0.6703	0.6713	−0.0010	0.6707	−0.0004
−0.2	0.8187	0.8155	0.0032	0.8201	−0.0014
0	1.0000	0.9946	0.0054	1.0018	−0.0018
0.2	1.2214	1.2172	0.0042	1.2225	−0.0011
0.4	1.4918	1.4917	0.0001	1.4911	0.0007
0.6	1.8221	1.8267	−0.0046	1.8191	0.0030
0.8	2.2255	2.2307	−0.0052	2.2225	0.0031
1.0	2.7183	2.7123	0.0060	2.7230	−0.0047

to the Chebyshev expansion. We see that the maximum error is reduced by 22%. Note that we do not have a "best approximation" even so. The error should reach equal maximums at five points in the interval—instead the error is large near $x = 1$ and too small elsewhere.

To obtain the optimum rational function which approximates the function with equal magnitude errors distributed through the interval is beyond the scope of this text. The approach that is used is to improve an initial estimate of the function, such as Eq. (5.7), by successive trials, often modifying the constants on the basis of experience until eventually one has a satisfactory formula. Hastings (1955) gives examples of the process. Systematic methods of determining the constants in such minimax rational approximations have also been determined. They are iteration methods beginning from an initial "sufficiently good" approximation.

PROBLEMS

Section 1

1. Find the individual deviations of the data in Fig. 1.1 from those computed from the least squares line, $R = 702.2 + 3.395T$. Compare these deviations with those from the line drawn by eye, $R = 700 + 3.500T$. Find the sum of squares of deviations in each case and compare. Note that even though the maximum deviations from each of the two lines are not too different, the sums of squares differ significantly.

2. Show that the point whose x-coordinate is the mean of all the x-values and whose y-coordinate is the mean of all the y-values satisfies the least squares line. Often a change of variable is made to relocate the origin at this point with a corresponding reduction in the magnitude of the numbers worked with, making them more readily handled by hand or on some desk calculators.

3. Find the least squares line that fits to the following data, assuming the x-values are free of error:

x	y	x	y
1	2.04	4	7.18
2	4.12	5	9.20
3	5.64	6	12.04

(The data are tabulated from $y = 2x$, with perturbations from a table of random numbers.)

4. In the data for Problem 3, consider that the y-values are free of error and that all the errors are in the x-values. By suitable modifications of the normal equations, now determine the least squares line for $x = ay + b$. Note this is not the same line as determined in Problem 3.

Section 2

5. Observe that the following data seem to be fit by a curve $y = ae^{bx}$ by plotting on semi-log paper and noting the points then fall near a straight line. The data are for the solubilities of n-butane in anhydrous hydrofluoric acid at high pressures, and were needed in the design of petroleum refineries.

Temperature, °F	Solubility, weight percent
77	2.4
100	3.4
185	7.0
239	11.1
285	19.6

By plotting on rectilinear graph paper, observe that the relationship is nonlinear.

6. Determine the constants for $y = ae^{bx}$ for the data in Problem 5 by the least squares method by fitting to the relation $\ln y = \ln a + bx$.

7. During the early history of the United States, population growth was nearly exponential. Determine the best curve from the census data for 1790 to 1860.

Year	Population, millions	Year	Population, millions
1790	3.9	1830	12.9
1800	5.3	1840	17.1
1810	7.3	1850	23.2
1820	9.6	1860	31.4

8. It is suspected (from theoretical considerations) that the rate of flow from a fire hose is proportional to some power of the pressure at the nozzle. Determine if the speculation seems to be true, and what the exponent is from these data. (Assume the pressure data are more accurate.)

Flow, gallons per minute	Pressure, psi
94	10
118	16
147	25
180	40
230	60

9. Since the data of Problem 8, when plotted on log-log paper (pressure as a function of flow), seem to have a slope of nearly 2, we should expect that fitting a quadratic to the data would be successful. Do this and compare the deviations with those from the power function relation of Problem 8.

10. The data given in Problem 3, while perturbations from a linear relation, seem to plot better along a curve because of the accidental occurrence of three negative deviations in succession. Fit a quadratic to the data.

11. To compare the results of an exact polynomial fit according to Chapter 2 with the least squares procedure, find y-values at $x = 1.5, 2.5, 3.5, 4.5,$ and 5.5 for the data of Problem 3, utilizing a fifth-degree interpolating polynomial. Sketch the interpolating polynomial and compare to the least-squares line of Problem 3, and the least-squares quadratic of Problem 10. The slope of the "true" function is 2.0. How do the maximum and minimum values of the slope as determined from the three approximations (use a graphical procedure) compare to the "true" value?

Section 3

12. Confirm the statement that for a set of data exactly fitted by a cubic, the values of S at the two ends will be linearly related to the adjacent S-values if the spline curve and the cubic polynomial are the same function. If other conditions are imposed on S in the end intervals ($S_1 = S_n = 0$ or $S_1 = S_2$ and $S_n = S_{n-1}$), how should the set of Eqs. (3.9) be modified? Will this change the portions of the spline curve in intervals other than the first and last?

13. Find the coefficient matrix and the right-hand side vector for fitting a cubic spline to the following data. Use linearity condition on the terminal S-values:

x	y	x	y
0.15	0.3945	1.07	0.2251
0.76	0.2989	1.73	0.0893
0.89	0.2685	2.11	0.0431

14. Solve the set of equations of Problem 13 (you may wish to utilize a computer program for this), and then determine the constants of the various cubics. The data are the ordinates of the normal probability function. Compare a few interpolated values with tabulated values of the function say at $x = 0.30, 0.80, 1.50, 2.00$.

15. Fit a cubic spline to the exponential function, utilizing values of e^x for $x = 0.0 \ (0.5)$ 2.0, and compare the interpolated with the true values for $x = 0.25, 1.25, 1.75$.

Section 4

16. Reduce Eq. (4.4) to a fourth-degree economized polynomial and show that its maximum error is 0.000781.

17. The error curve for a truncated Maclaurin series approximation increases monotonically as x varies from 0 to the ends of the interval. This is not true for an economized power series. Exhibit the form of the error curve by plotting the error of Eq. (4.4) on the interval [0,1].

18. Given

$$\arctan x = x - \frac{x^3}{3} + \frac{x^5}{5} - \frac{x^7}{7} + \frac{x^9}{9} - \cdots .$$

Plot the error over the interval $[-1,1]$ when the above series is truncated after the term in x^7. Economize the ninth-degree truncated power series three times (giving a third-degree expression), and plot its error over the interval $[-1,1]$.

19. Find the first few terms of the Chebyshev series for $\sin x$ by rewriting the Maclaurin series in terms of $T_i(x)$ and collecting terms. Express this as a power series. Compare the errors when both series are truncated to fifth-degree polynomials.

20. The series for $\ln(1 + x)$ converges slowly for $-1 < x < 1$:

$$\ln(1 + x) = x - \tfrac{1}{2}x^2 + \tfrac{1}{3}x^3 - \tfrac{1}{4}x^4 + \tfrac{1}{5}x^5 - \cdots .$$

Since it is an alternating series, its error is less than the first abandoned term. Convert to Chebyshev series, including terms up to $T_5(x)$. What is the maximum error of the truncated Chebyshev series on $[-1,1]$? How many terms of the Maclaurin series are needed to give a smaller maximum error?

21. The Chebyshev series, and economized polynomials as well, require us to approximate the function on the interval $[-1,1]$ only. Show that an appropriate change of variable will change $f(x)$ on $[a,b]$ to $f(y)$ on $[-1,1]$. Find the linear relation between x and y that makes this transformation.

Section 5

22. Find Padé approximations to the following functions, with numerators and denominators each of degree three: (a) $\sin x$, (b) $\cos x$, (c) e^x.

23. Compare the errors on the interval $[-1,1]$ for the Padé approximations of Problem 22 with the errors of the corresponding Maclaurin series.

24. Express these rational functions in continued fraction form:

(a) $\dfrac{x^2 - 2x + 2}{x^2 + 2x - 2}$ (b) $\dfrac{2x^3 + x^2 + x + 3}{x^2 - x - 4}$ (c) $\dfrac{2x^4 + 45x^3 + 381x^2 + 1353x + 1511}{x^3 + 21x^2 + 157x + 409}$.

In each case, compare the number of multiplication and division operations in continued fraction form with initial evaluation of the polynomials by nested multiplication.

25. Express the Padé approximations of Problem 22 as continued fractions.

26. Estimate the errors of the Padé approximations of Problem 22 by computing the coefficient of the next nonzero term in the numerator. Compare these estimates with the actual errors at $x = 1$, and at $x = -1$.

27. The Chebyshev series for $\cos(\pi x/4)$ is

$$1.7033 - 0.1464\, T_2(x) + 0.001921\, T_4(x) - 0.000009965\, T_6(x) + \cdots.$$

Develop a Padé-like rational function from this by the method of Section 5, where the numerator and denominator are both of degree three.

appendix

appendix

Since a number of results and theorems from the calculus are frequently used in the text, we collect here a number of these items for ready reference, and to refresh the student's memory.

Open and Closed Intervals. We use for the open interval $a < x < b$, the notation (a,b), and for the closed interval $a \leq x \leq b$, the notation $[a,b]$.

Continuous Functions. If a real-valued function is defined on the interval (a,b), it is said to be continuous at a point x_0 in that interval if for every $\epsilon > 0$ there exists a positive nonzero number δ such that $|f(x) - f(x_0)| < \epsilon$ whenever $|x - x_0| < \delta$ and $a < x < b$. In simple terms we can meet any criterion of matching the value of $f(x_0)$ (the criterion is the quantity ϵ) by choosing x near enough to x_0, without having to make x equal to x_0, when the function is continuous.

If a function is continuous for all x-values in an interval, it is said to be continuous on the interval. A function that is continuous on a closed interval $[a,b]$ will assume a maximum value and a minimum value at points in the interval (perhaps the end points). It will also assume any value between the maximum and the minimum at some point in the interval.

Similar statements can be made about a function of two or more variables. We then refer to a domain in the space of the several variables instead of an interval.

Sums of Values of Continuous Functions. When x is in $[a,b]$, the value of a continuous function $f(x)$ must be no greater than the maximum and no less than the minimum value of $f(x)$ on $[a,b]$. The sum of n such values must be bounded by $(n)(m)$ and $(n)(M)$, where m and M are the minimum and maximum values. Consequently the sum is n times some intermediate value of the function. Hence

$$\sum_{i=1}^{n} f(\xi_i) = nf(\xi) \qquad \text{if } a \leq \xi_i \leq b, \qquad i = 1,2,\cdots,n, \qquad a \leq \xi \leq b.$$

Similarly, it is obvious that

$$c_1 f(\xi_1) + c_2 f(\xi_2) = (c_1 + c_2) f(\xi), \qquad \xi_1, \xi_2, \xi \text{ in } [a,b],$$

for the continuous function f when c_1 and c_2 are both equal to or greater than one. If the coefficients are positive fractions, dividing by the smaller gives

$$c_1 f(\xi_1) + c_2 f(\xi_2) = c_1 \left[f(\xi_1) + \frac{c_2}{c_1} f(\xi_2) \right] = c_1 \left(1 + \frac{c_2}{c_1} \right) f(\xi) = (c_1 + c_2) f(\xi),$$

so the rule holds for fractions as well. If c_1 and c_2 are of unlike sign, this rule does not hold unless the values of $f(\xi_1)$ and $f(\xi_2)$ are narrowly restricted.

Mean Value Theorem for Derivatives. When $f(x)$ is continuous on the closed interval $[a,b]$, then at some point ξ in the interior of the interval

$$f'(\xi) = \frac{f(b) - f(a)}{b - a}, \qquad a < \xi < b,$$

provided, of course, that $f'(x)$ exists at all interior points. Geometrically this means that the curve has at one or more interior points a tangent parallel to the secant line connecting the ends of the curve (Fig. 1).

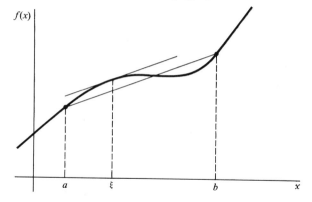

Fig. 1.

Mean Value Theorems for Integrals. If $f(x)$ is continuous and integrable on $[a,b]$, then

$$\int_a^b f(x)\, dx = (b - a) f(\xi), \qquad a < \xi < b.$$

This says, in effect, that the value of the integral is an average value of the function times the length of the interval. Since the average value lies between the maximum and minimum values, there is some point ξ at which $f(x)$ assumes this average value.

If $f(x)$ and $g(x)$ are continuous and integrable on $[a,b]$, and if $g(x)$ does not change sign on $[a,b]$, then

$$\int_a^b f(x)g(x)\, dx = f(\xi) \int_a^b g(x)\, dx, \qquad a < \xi < b.$$

Note that the previous statement is a special case $(g(x) = 1)$ of this last theorem which is called the *second theorem of the mean for integrals*.

Taylor Series. If a function $f(x)$ can be represented by a power series on the interval $(-a,a)$, then the function has derivatives of all orders on that interval and the power series is

$$f(x) = f(0) + f'(0)x + \frac{f''(0)}{2!}x^2 + \frac{f'''(0)}{3!}x^3 + \cdots .$$

The above power series expansion of $f(x)$ about the origin is called a Maclaurin series. Note that if the series exists, it is unique and any method of developing the coefficients gives this same series.

If the expansion is about the point $x = a$, we have the Taylor series

$$f(x) = f(a) + f'(a)(x - a) + \frac{f''(a)}{2!}(x - a)^2 + \frac{f'''(a)}{3!}(x - a)^3 + \cdots .$$

We frequently represent a function by a polynomial approximation, which we can consider as a truncated Taylor series. Usually we cannot represent a function exactly by this means, so we are interested in the error. Taylor's formula with a remainder gives us the error term. The remainder term is usually derived in elementary calculus texts in the form of an integral:

$$f(x) = f(a) + f'(a)(x - a) + \frac{f''(a)}{2!}(x - a)^2 + \cdots + \frac{f^{(n)}(a)}{n!}(x - a)^n$$

$$+ \int_a^x \frac{(x - t)^n}{n!} f^{(n+1)}(t)\, dt.$$

Since $(x - t)$ does not change sign as t varies from a to x, the second theorem of the mean allows us to write the remainder term as

$$\text{Remainder of Taylor series} = \frac{(x - a)^{n+1}}{(n + 1)!} f^{(n+1)}(\xi).$$

The derivative form is the more useful for our purposes. It is occasionally useful to express a Taylor series in a notation that shows how the function behaves at a distance h from a fixed point a. If we call $x = a + h$ in the above, so $x - a = h$, we get

$$f(a + h) = f(a) + f'(a)h + \frac{f''(a)}{2!}h^2 + \cdots + \frac{f^{(n)}(a)}{n!}h^n + \frac{f^{(n+1)}(\xi)}{(n + 1)!}h^{n+1}.$$

Taylor Series for Functions of Two Variables. For a function of two variables, $f(x,y)$, the rate of change of the function can be due to changes in either x or y. The derivatives of f can be expressed in terms of the partial derivatives. For the expansion in the neighborhood of the point (a,b),

$$f(x,y) = f(a,b) + f_x(a,b)(x - a) + f_y(a,b)(y - b)$$

$$+ \frac{1}{2!}[f_{xx}(a,b)(x - a)^2 + 2f_{xy}(a,b)(x - a)(y - b) + f_{yy}(a,b)(y - b)^2] + \cdots .$$

bibliography

bibliography

Allen, D. N. (1954): *Relaxation Methods*, McGraw-Hill, New York.

Davis, Philip J., and Philip Rabinowitz (1967): *Numerical Integration*, Blaisdell, Waltham, Mass.

Fox, L. (1962): *Numerical Solution of Ordinary and Partial Differential Equations*, Addison-Wesley, Reading, Mass.

Hamming, R. W. (1962): *Numerical Methods for Scientists and Engineers*, McGraw-Hill, New York.

Hastings, C. (1955): *Approximations for Digital Computers*, Princeton University Press, Princeton, N. J.

Henrici, Peter (1964): *Elements of Numerical Analysis*, Wiley, New York.

Kopal, Zdenek (1955): *Numerical Analysis*, Wiley, New York.

Kunz, Kaiser S. (1957): *Numerical Analysis*, McGraw-Hill, New York.

Love, Carl H. (1966): *Abscissas and Weights for Gaussian Quadrature*, National Bureau of Standards, Nomograph 98.

Muller, D. E. (1956): A Method of Solving Algebraic Equations Using an Automatic Computer, *Math Tables and other Aids to Comp.*, 10:208-215.

Peaceman, D. W., and H. H. Rachford (1955): The Numerical Solution of Parabolic and Elliptic Differential Equations, *J. Soc. Ind. Appl. Math.*, 3:28-41.

Ralston, Anthony (1965): *A First Course in Numerical Analysis*, McGraw-Hill, New York.

Scarborough, J. B. (1950): *Numerical Mathematical Analysis*, Johns Hopkins Press, Baltimore.

Smith, G. D. (1965): *Numerical Solution of Partial Differential Equations*, Oxford, London.

Varga, Richard (1959): *p*-Cyclic Matrices: A Generalization of the Young-Frankel Successive Overrelaxation Scheme, *Pacific J. Math*, 9:617-628.

Warten, R. M. (1963): Automatic Step-Size Control for Runge-Kutta Integration, *IBM Journal*, October, p. 340.

**answers to
selected
problems**

answers to
selected
problems

Chapter 1

1. Thirteen iterations give 1.4141844, which rounds correctly to four decimals. The maximum error at any step is the difference between the last two iterates, which is $1/2^n$. When $n = 13$, $1/2^{13} = 0.0001222$.

4. The root is at $x = 0.6191$. Substitute in either equation to get y-value at intersection, $y = 2.0572$.

5. b) 0.4450   c) 0.9206

8. x-values of intersections are -0.80194, 0.55496, 2.24698.

10. -0.4590

11. Same result as in Problem 10, of course.

14. Let $f(x) = x^2 - N = 0$, substitute into $x_{i+1} = x_i - f(x_i)/f'(x_i)$, then rearrange.

16. b) $x_0 = 0.2$ error = 0.05

$x_1 = 0.24$	0.01	1-place accuracy
$x_2 = 0.2496$	0.0004	3-place accuracy
$x_3 = 0.24999936$	0.00000064	6-place accuracy
$x_4 = 0.2499999999$	0.0000000001?	10^+-place accuracy.

17. b) ± 1.732, 0.618, -1.618.

19. Roots are -0.45896, 0.91001, 3.73308. Beginning at $x_0 = 3.73$ or $x_0 = 3.74$ gives divergence with given form, but $x = \ln (3x^2)$ converges to the largest root.

22. Root is 1.07816. Three convergent forms are

$$x = \sqrt{\frac{2x+5}{2x+4}}, \quad x = \tfrac{1}{2}\sqrt{5 + 2x - 2x^3}, \quad x = \left(\frac{5 + 2x - 4x^2}{2}\right)^{\frac{1}{3}}.$$

The last form converges very slowly.

25. Write $P(x)$ in the form $(x - r)^2 Q(x)$ and differentiate. Since both terms of the derivative contain the factor $(x - r)$, $P'(r) = 0$.

26. Use induction on Problem 25.

28. Convergence is only linear. See Problem 29 for a method of increasing the rate of convergence.

31. The four roots are 1.61803399, 0.79128785, −0.61803399, −3.79128785.

32. The quadratic factors are $(x^2 − 4.1x + 5.2)$, $(x^2 + x + 1)$. Find the roots from these by the quadratic formula.

36. a) Roots are 1.8019, 0.4450, −1.2470.
    c) Roots are 1, 1, 1, 2.

38. Roots are 1, 0.6180, −1.6180.

47. Solve the equation $4/3\pi r^3(0.6) = 1/3\pi(3r\alpha^2r^2 − \alpha^3r^3)$ for $\alpha = h/r$. The only root with physical significance lies between 0 and 2, $\alpha = 1.1341$.

49. $(s + 0.311108)(s + 1.99762)(s − 3.30635)$.

52. Zeros are ±0.23861, ±0.66120, ±0.93246.

Chapter 2

2. For an exact fit to four decimal places, a sixth-degree polynomial is required. Since the average fourth-order difference is nearly zero and its signs oscillate, the third differences could be considered constant. Hence a third-degree polynomial will nearly fit the data.

3. $f(1.15) = 0.1399$, ln $1.15 = 0.1398$,
   $f(1.55) = 0.4383$, ln $1.55 = 0.4383$

6. y-values are: 0.0000, 0.0875, 0.1763, 0.2679, 0.3640, 0.4663.

9. The fourth differences show alternating signs with magnitudes proportional to 1, 4, 6, 4, 1. If the datum written as 8.83 is changed to 8.90, the fourth differences all become zero. Observe that the error, $8.83 − 8.90 = −0.07$, is one-fourth the largest fourth difference of −0.42, which occurs on the same line as the datum in error.

10. $f(0.158) = 0.787935$, $f(0.636) = 0.651814$.

13. $y(0.58) = 0.1855$.

15. Since the fifth differences are zero, the fourth differences are constant. Hence a fourth-degree polynomial will fit exactly.

17. Identical results to those of Problem 10 are secured if round-off during the computation is prevented by carrying enough decimal places.

19. $f(0.385) = 0.74090976$ from each calculation. (Excess figures must be carried to avoid round-off errors. Accuracy of original data would suggest using 0.74091). Results are identical because each polynomial is a variant of the same polynomial.

21. b) $f(x) = 0.518 + s(0.148) + \dfrac{(s + 1)(s)}{2}(0.026) + \dfrac{(s + 2)(s + 1)(s)}{6}(0.004)$,

    $s = (x − 1.1)/0.2$.

    c) $f(x) = 0.248 + s(0.122) + \dfrac{(s)(s − 1)}{2}(0.022) + \dfrac{(s + 1)(s)(s − 1)}{6}(0.004)$,

    $s = (x − 0.7)/0.2$.

    e) $f(x) = \dfrac{0.248 + 0.370}{2} + \dfrac{s + (s − 1)}{2}(0.122) + \dfrac{(s)(s − 1)}{2}\dfrac{0.022 + 0.026}{2}$

    $+ \dfrac{(s + 1)(s)(s − 1) + (s)(s − 1)(s − 2)}{(2)(6)}(0.004)$, $s = (x − 0.7)/0.2$.

22. Since Stirling polynomials use a horizontal set of differences beginning at y_0, they cannot be written to terminate on a given odd-order difference. They can terminate on even-order differences because these are on the same horizontal line as the y-entries. Because the Bessel polynomials go horizontally through the odd-order differences, this type cannot be included in the set that terminates on the second difference, whose value is 0.022.

25. Since all the polynomials of Problem 21 terminate at the same point in the difference table, they are variations of the same polynomial and must have the same error terms. Evaluate the coefficient of any one of them, using the corresponding value of s.

$$\text{Error} = \frac{-0.2079}{24} h^4 f^{iv}(\xi), \qquad 0.3 < \xi < 0.9$$

28. $\text{Error} = \dfrac{(s)(s-1)}{2} h^2 f''(\xi) = \dfrac{(x - x_0)(x - x_1)}{2} f''(\xi), \qquad x_0 < \xi < x_1,$

$|(x - x_0)(x - x_1)/2|$ is a maximum at $x = (x_0 + x_1)/2$ by simple calculus. Hence, maximum magnitude of coefficient is $(x_1 - x_0)^2/8$.

29. a) Maximum error for $3.0 < x < 3.1$ is 0.0138. Maximum error for $4.9 < x < 5.0$ is 0.0927. The second decimal place in values obtained by linear interpolation may be in error. A higher-degree interpolating polynomial would give greater accuracy.

b) If $x \le 0.0033$, maximum error < 0.0001 (required for $4.9 < x < 5.0$).

34. a) $\Delta(f(x)g(x)) = f(x + h)g(x + h) - f(x)g(x)$. Show by expanding that right-hand side is the same as this.

35. $\Delta^3 E^{-2} \nabla^2 y_4 = \Delta^3 (E^2 \nabla^2) E^{-4} y_4 = \Delta^5 y_0 = y_5 - 5y_4 + 10y_3 - 10y_2 + 5y_1 - y_0.$

36. $\Delta^n y_s = E^n \nabla^n y_s = \nabla^n y_{s+n}.$ If this equals $\nabla^n y_r$, $r = s + n$.

39. Because the odd-order central differences require $y_{n/2}$ values which are not available.

45. $f(x) = \frac{1}{2}(-x^3 + 8x^2 + x - 8).$

48. $\text{Error} = \dfrac{-0.002}{6} e^\xi, \ 0 < \xi < 0.3.$ Maximum value -0.00045, minimum value -0.00033.

51. Interpolated value is $x = 0.6$.

$$\text{Error} = \frac{(25 - 1)(25 - 4)(25 - 9)}{6}\left(\frac{3}{8} \xi^{-5/2}\right), 1 < \xi < 25.$$

Maximum error is 504, minimum error is 0.16. Such wide error bounds are of little value.

55. With values for $-1(1)2$ for x, interpolated value is 0.6913 versus the true value of 0.76292. Using the closer-spaced points, the interpolated value is 0.76293.

57. Interpolated value $= 0.916381698$. Value in tables is 0.90863873.

Chapter 3

1. 1.7683 with $h = 0.1$, 1.7728 with $h = 0.2$, 1.7904 with $h = 0.4$.

2. Errors are -0.0013, -0.0058, -0.0234. Error/h^2: -0.13, -0.145, -0.146. There are additional errors due to round-off of original data.

4. Maximum error -0.160, minimum error -0.032.

6. Integral $= 0.6933$ with $h = 0.05$; analytical value $= \ln 2 = 0.6932$.

8. 0.87556.

9. Using $h = 0.1$ up to $v = 2.0$, $h = 0.5$ up to $v = 5$, and $h = 1$ up to $v = 10$ gives estimate of 0.565. Smaller intervals would give improved value.

11. Extrapolated answers are 1.7668 and 1.7669 with errors $O(h^4)$. Second-order extrapolation is 1.7668 with error $O(h^6)$. The true value is 1.7670. The extrapolated values are subject to the round-off errors in the original data.

13. Beginning with $h = 1$, successive values and their extrapolations are:

    0.750000
                 0.694444
    0.708333   0.693175
                 0.693254   0.693148
    0.697024   0.693148
                 0.693155
    0.694122

15. Integral $= h(f_0 + \frac{1}{2}\Delta f_{-1} + \frac{5}{12}\Delta^2 f_{-2} + \frac{3}{8}\Delta^3 f_{-3} + \cdots)$.

18. There are many formulas, of course; even though they all have $O(h^6)$ errors, they will not give identical results, since they may end on different difference table entries. One answer is 1.7663.

19. Identical answers will not be obtained in all cases since the set of fourth-degree polynomials which are integrated will be different unless they end on the same set of fourth differences (or unless the fourth differences are constant).

21. 1.7668, 1.7669.

23. Since $\frac{1}{90} h^4 e^{\xi} \leq 0.000005$, h must be < 0.113. With $h = 0.1$, the integral is estimated as 1.718283. The exact value is 1.718282.

25. With $h = 0.5$, integral is 0.94615. With $h = 0.25$, integral is 0.94608. Extrapolated value is 0.94608, error is $O(h^6)$. Tabulated value is 0.94608.

28. Choosing that part of the data for which the fourth differences are smallest (fourth derivative is then near zero) will minimize the magnitude of the error. But this approach is not certain to give the most precise answer because there could be a cancellation of errors.

30. The line of coefficients is given by

$$I = h(E^2 + E + 1)\frac{(E - 1)}{\ln E}f_0 = h(\Delta^2 + 3\Delta + 3)\frac{(E - 1)}{\ln E}f_0$$

$$= h(3f_0 + \tfrac{9}{2}\Delta f_0 + \tfrac{9}{4}\Delta^2 f_0 - \tfrac{5}{8}\Delta^3 f_0 + \cdots)$$

34. 0.94606. Tabulated value is 0.94608.

37. Error terms are not necessarily the same. Formulas from the lozenge diagram assume a series of polynomials of degree $2n - 1$ which fit the data. The single $(2n - 1)$-degree polynomial assumed by the Gaussian method will usually be different.

39. a) 0.8427007929

    c) $1.57079633 = \pi/2$

    d) 1.35064388

Chapter 4

2. At $x = 0.15$, $f'(x) = 2.715, 2.862, 2.877$.   Analytical value is 2.8953.   At $x = 0.23$, $f'(x) = 1.810, 1.878, 1.881$.   Analytical value is 1.8883.

3. At $x = 0.15$, error estimates are (a) 0.1503 to 0.1930, (b) 0.0168 to 0.0343, (c) 0.0027 to 0.0103.

4. $f'(x) = \dfrac{1}{h}(\Delta f_{-1} + \frac{1}{2}\Delta^2 f_{-2} + \frac{1}{3}\Delta^3 f_{-3} + \frac{1}{4}\Delta^4 f_{-4} + \cdots)$, $f'(0.29) \doteq 1.491$.

5. Error $= (1/n)h^n f^{(n)}(\xi)$, $x_{-n} < \xi < x_0$, for a formula whose last difference is of order $n - 1$.

8. Use $\Delta^n f$ as an estimate of $h^n f^{(n)}(\xi)$.   These give, at $x = 0.15$, estimates of 0.150, 0.015, 0.010.

13. One term, 1.8925; two terms same because next coefficient is zero; three terms, 1.8881. Analytical value is 1.8882.   Errors are $-0.0043, 0.0001$.

15. We prefer to use a polynomial centered at $\theta = 22$.   The difference table shows that round-off errors in the original data are already affecting fourth differences, so use a polynomial that fits at $\theta = 20$ to $\theta = 24$.   Estimate of second derivative is -0.3994 versus -0.37461 analytically.   (Did you remember to use $\Delta\theta$ in radians?)

19. Error $= -\frac{1}{2}h(1 - \alpha)f''(x_0) - \frac{1}{6}h\dfrac{(1 + \alpha^3)}{(1 + \alpha)}f''(\xi)$.

20. 1.97062

    1.88646

1.9075   1.8876   (1.8882 analytically)

    1.8875

1.8925

25. b) Extrapolated value $=$ more accurate $+ \dfrac{1}{r^n - 1}$ (more $-$ less accurate), where

    $O(h^n)$ is order of error of the two estimates.

Chapter 5

2. $f_0' \doteq \dfrac{1}{6h}(-11y_0 + 18y_1 - 9y_2 + 2y_3)$.

5. $f_0'' \doteq \dfrac{1}{h^2}\dfrac{2}{\alpha(\alpha + 1)}[\alpha f_L - (1 + \alpha)f_0 + f_R]$.

8. a) 0.497, error 0.0433.

   c) 0.5395, error 0.0008.

9. a) 0.85, error -0.00085.

18. For part (a), actual error is 0.0433.   Upper bound 0.0493, lower bound 0.0322.   For part (c), actual error is 0.0008.   Upper bound 0.00127, lower bound 0.00028.

20. $I \doteq (3h/8)(f_0 + 3f_1 + 3f_2 + f_3)$.

22. Error is 0.0014 with $h = 0.1$ versus 0.0089 with $h = 0.25$.

23. Error is 0.0000 with Simpson's $1/3$ rule, versus 0.0089.

24. $I \doteq \dfrac{3(b - a)}{8(3m)}[f_0 + 3f_2 + 3f_3 + 2f_4 + 3f_5 + \cdots + 3f_{3n-1} + f_{3n}]$.

25. a) 65.7, b) 65.0, c) 64.725.

27. 1.5980 versus exact value of 1.7183.

30. Error 0.0005 versus 0.0000 in Problem 23.

Chapter 6

1. 1.1159.

3. $y(x) = (x-1) + \frac{2}{2}(x-1)^2 + \frac{4}{6}(x-1)^3 + \frac{8}{24}(x-1)^4 + \frac{40}{120}(x-1)^5 + \frac{176}{720}(x-1)^6 + \cdots$.

4. $y(0.2) = 1.2015$, $y(0.4) = 1.4128$, $y(0.6) = 1.6471$.

7. 1.1142, error is 0.0017. For four-decimal accuracy, reduce h about 17-fold.

9. Analytical accuracy of 1.1159 is obtained. Each step is about twice the effort, but there are 68-fold fewer steps. Combining these, we find that the modified Euler method is perhaps 34-fold less work.

11. $y(0.1) = 2.2158$, $y(0.2) = 2.4647$, $y(0.3) = 2.7502$, $y(0.4) = 3.0757$, $y(0.5) = 3.4452$.

13. 1.11588, which is accurate to five decimals. Somewhat less effort than modified Euler method, much less than would be required with simple Euler.

15. $y(0.2) = 2.0933$, $y(0.4) = 2.1755$, $y(0.6) = 2.2493$.

19. $y(1.2) = 2.5200$. Analytical solution, $y = -e^x + x^2 + 2x + 2$, gives 2.51988 at $x = 1.2$.

22. $y(0.8) = 2.3162$ predicted   2.3164 corrected
    $y(1.0) = 2.3779$ 2.3780
    $y(1.2) = 2.4350$ 2.4350
    $y(1.6) = 2.5365$ 2.5381
    $y(2.0) = 2.6288$ 2.6294

26. $y(0.8) = 2.0146$ predicted   2.0145 corrected
    $y(1.0) = 2.2820$ 2.2818
    $y(1.2) = 2.5202$ 2.5200

28. $y(0.2) = 0.0004$, $y(0.4) = 0.0064$, $y(0.6) = 0.0324$ by Runge-Kutta. Using Adams-Moulton, $y(0.8) = 0.1033$. At $x = 1.0$, accuracy criterion not met and interpolation is required: $y(0.7) = 0.0611$, $y(0.5) = 0.0149$. Using $h = 0.1$, we have $y(0.9) = 0.1667$, $y(1.0) = 0.2571$, $y(1.1) = 0.3833$, $y(1.2) = 0.5577$.

30. a) $3/\sin x$,   b) 10 in the Nth place,   c) there is no simple relation between the maximum in part (a) and the accuracy criterion.

35.

x:	0	0.02	0.04	0.06	0.08	0.10
Predicted errors	0	0.0004	0.0008	0.0012	0.0016	0.0021
Actual errors	0	0.0004	0.0008	0.0013	0.0018	0.0022

37. $Q' = I$, $Q(0) = 0$, $I' = 48 \sin 10t - 12I - 100Q$, $I(0) = 0$.

40.

t	0.2	0.4	0.6
x	0.022	0.093	0.221
y	0.982	0.933	0.864.

43.

x	y	y'
0	1.0000	-1.0000
0.1	0.8950	-1.0995
0.2	0.7802	-1.1956
0.3	0.6561	-1.2848
0.4	0.5237	-1.3627
0.5	0.3841	-1.4261
0.6	0.2390	-1.4716

Chapter 7

1. a) $A + B$ is not defined.

$$A + C = \begin{bmatrix} 1 & 3 & 3 & 5 \\ -1 & 2 & 1 & 2 \\ 3 & 0 & 3 & 0 \end{bmatrix}, \qquad A - 2C = \begin{bmatrix} 1 & 0 & -9 & -1 \\ 2 & 8 & -8 & -1 \\ 3 & -3 & -3 & 3 \end{bmatrix}$$

b)
$$AB = \begin{bmatrix} 7 & 0 & -2 \\ 4 & 8 & -9 \\ 2 & -2 & 5 \end{bmatrix}, \qquad BA = \begin{bmatrix} 0 & 4 & -2 & 1 \\ -2 & 11 & -6 & 4 \\ 10 & -5 & 4 & 5 \\ 5 & -5 & 3 & 5 \end{bmatrix}$$

AC is not defined.

2. a)
$$M^2 = \begin{bmatrix} -1 & 2 & 1 \\ 0 & -2 & 0 \\ 1 & 2 & -1 \end{bmatrix}, \qquad M^3 = \begin{bmatrix} -2 & -2 & 2 \\ 2 & 0 & -2 \\ -2 & 2 & 2 \end{bmatrix}$$

5. $x_1 = 2, x_2 = 1, x_3 = -1$.

6. $x_1 = 2.5555, x_2 = 1.7222, x_3 = -1.0555$.

11. $x_1 = 2.5555, x_2 = 1.7222, x_3 = -1.0555$.

12. $x_1 = 2.5550, x_2 = 1.7225, x_3 = -1.0555$.

13. a) $x_1 = 1.4497, x_2 = -1.5836, x_3 = -0.2749$.

   b) $x_1 = 1.4528, x_2 = -1.5889, x_3 = -0.2749$.

14. $x_1 = 2.5556, x_2 = 1.7222, x_3 = -1.0556$.

17. $x = (46.1549, 84.6195, 92.3117, 84.6143, 46.1536)$.   The elements of x should be symmetrical about the center.   Round-off errors cause some distortion.

18. a) -12

20. Inverse $= \dfrac{1}{115} \begin{bmatrix} 24 & 5 & 1 \\ 5 & 25 & 5 \\ 1 & 5 & 24 \end{bmatrix}$

24. $x_1 = 2, x_2 = 1, x_3 = -1$.

29. $x = (46.1538, 84.6154, 92.3077, 84.6154, 46.1538)$.

30. $x_1 = 1, x_2 = -1, x_3 = 2$.

35. One solution is $(0.5176, 1.9319)$.

36. $(0.7259, 0.5028), (-1.6700, 0.3451)$.

39. $(0.5716, 1.9319), (1.9319, 0.5716), (-0.5716, -1.9319), (-1.9319, -0.5716)$.

Chapter 8

1. Initial trials with $y'(0) = 1.0$ and $y'(0) = 0.90$ were used and linearly interpolated to adjust $y(1.0)$ to a value of 1.9. Notice that errors in the modified Euler method cause disagreement of the $y(1.0)$ value at the correct initial slope of 0.9.

	y-values			
x	$y'(0) = 1.0$	$y'(0) = 0.9$	$y'(0) = 0.865$	Analytical values
0	0	0	0	0
0.25	0.275	0.249	0.240	0.2406
0.50	0.648	0.593	0.574	0.5750
0.75	1.218	1.128	1.096	1.0969
1.0	2.083	1.948	1.901	1.9

5. a) $x = \dfrac{2}{\pi} \theta$ will normalize, giving

$$\frac{d^2y}{dx^2} + \frac{\pi^2}{4} y = 0, \qquad y(0) = 0, \quad y(1) = 1.$$

   b) $y(0.25) = 0.385$, $y(0.50) = 0.710$, $y(0.75) = 0.926$.

   c) $y(\pi/8) = 0.385$, $y(\pi/4) = 0.710$, $y(3\pi/8) = 0.926$

   d) $y(\pi/8) = 0.3826$, $y(\pi/4) = 0.7071$, $y(3\pi/8) = 0.9238$.

   e) $y(0.5) = 0.723$. Extrapolation gives 0.706.

7. $y(0.25) = 0.245$, $y(0.50) = 0.583$, $y(0.75) = 1.108$.

10. $y(0) = 0.505$, $y(\pi/8) = 0.660$, $y(\pi/4) = 0.714$, $y(3\pi/8) = 0.658$, $y(\pi/2) = 0.500$. Compare to analytical solution: $y = \frac{1}{2} \sin\theta + \frac{1}{2} \cos\theta$.

14. $h = 0.2$ was used. $y(0.2) = 0.189$, $y(0.4) = 0.642$, $y(0.6) = 1.217$, $y(0.8) = 1.740$.

17. Eigenvalues and extrapolations are

$$h = \tfrac{1}{2}, \quad \pm 2.8284$$
$$\pm 3.289$$
$$h = \tfrac{1}{8}, \quad \pm 3.0846$$
$$\pm 3.295$$
$$h = \tfrac{1}{4}, \quad \pm 3.1766$$

21. a) $\lambda = 1$, eigenvector $= (0, 1)$.

22. $\lambda = 3.261$, eigenvector $= (0.8944, 1.3416, 1)$.

Chapter 9

4. $\nabla^2 u = \dfrac{1}{12h^2} \left\{ \begin{array}{ccccc} & & -1 & & \\ & & 16 & & \\ -1 & 16 & -60 & 16 & -1 \\ & & 16 & & \\ & & -1 & & \end{array} \right\}$

6. $u_{11} = 50$, $u_{12} = 75$, $u_{21} = 25$, $u_{22} = 50$.

8. The array of temperatures is

0	60	120	180	240	300
0	45	90	135	180	225
0	30	60	90	120	150
0	15	30	45	60	75
0	0	0	0	0	0

9. $u_1 = 74.46$, $u_2 = 27.69$, $u_3 = 47.86$, $u_4 = 65.38$, $u_5 = 78.85$, $u_6 = 12.90$, $u_7 = 23.89$, $u_8 = 34.82$.

10. a) $\phi = 1$ at each point.

   b) One quarter of the cross-section:

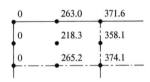

13. $u(1/3,\ 1/3) = -0.00977$, $u(2/3,\ 1/3) = u(1/3,\ 2/3) = -0.01337$, $u(2/3,\ 2/3) = -0.01903$.

14. One quarter of the plate:

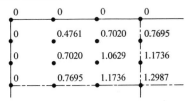

18. Because of symmetry, no heat is transferred across lines that connect the reentrant corners. Likewise, if the insulation in Problem 17 is perfect, no heat is transferred across it. Hence the temperatures will be the same since the same boundary conditions apply.

19. At A, operator is $\left\{ \begin{array}{ccc} & 9/5 & \\ 4/3 & -7 & 8/3 \\ & 6/5 & \end{array} \right\}$, potential = 35.17.

   At B, operator is $\left\{ \begin{array}{ccc} & 1 & \\ 1 & -4 & 1 \\ & 1 & \end{array} \right\}$, potential is 19.07.

   At C, operator is $\left\{ \begin{array}{ccc} & 9/2 & \\ 8/5 & -14 & 32/5 \\ & 3/2 & \end{array} \right\}$, potential is 41.11.

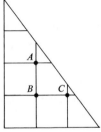

22.

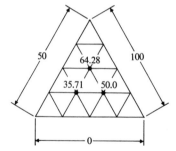

26. a) 0.3026, b) 0.4049, c) 0.4884.

29. The eight internal temperatures are all 33.33°.

30. 53.33°, 33.33°, 33.33°, 13.33°, repeated across the plane of symmetry.

Chapter 10

2. c in Btu/lb,°F; ρ in lb/in³; k in Btu/sec,in²,°F/in.

4. $u(0.25) = u(1.75) = 13.67°$, $u(0.50) = u(1.50) = 27.34°$, $u(0.75) = u(1.25) = 33.00°$, $u(1.00) = 38.67°$.

8. a) After 15 time steps ($t = 252$ sec):

x	0	2	4	6	8	10	12	14	16	18	20
C	0	0.51	1.02	1.65	2.28	3.20	4.12	5.43	6.75	8.38	10.0

Throughout the calculation, the results are much closer to the analytical curve of Fig. 1.2.

9. $u(0.25) = u(1.75) = 14.62°$, $u(0.50) = u(1.50) = 27.01°$, $u(0.75) = u(1.25) = 35.32°$, $u(1.00) = 38.22°$.

12. $u(0) = 633.5°$, $u(1) = 808.3°$, $u(2) = 916.8°$, $u(3) = 969.7°$, $u(4) = 984.6°$.

22. r must be less than $\frac{1}{6}$ for stability, so $\Delta t < 1.26$ sec per time step; more than 12 steps are required. Since there are 125 points with unknown temperatures in the one-inch grid, there are 125 equations at each time step.

Chapter 11

1. With $\Delta x = 10$ cm, $\Delta t = 0.0001908$ sec. Fourteen steps are required to duplicate original displacements, so frequency is 374 cycles per second. The frequency formula gives the same result.

4. $Y = y/L$, $X = x/L$, $\theta = \frac{1}{L}\sqrt{\frac{Tg}{w}}\,t$.

Note that the new variables are all dimensionless.

6. Since $\sin \pi x \cos \pi t = \frac{1}{2}(\sin(\pi x + \pi t) + \sin(\pi x - \pi t))$,

$$F(x + ct) = \tfrac{1}{2}\sin \pi(x + t), \qquad G(x - ct) = \tfrac{1}{2}\sin \pi(x - t).$$

8.

x	0	0.2	0.4	0.6	0.8	1.0
At $t = \Delta t$, y	0	0.56	0.90	0.90	0.56	0
At $t = 10\Delta t$, y	0	0.40	0.80	0.80	0.40	0

11. a) A single characteristic (the differential equation is parabolic) which is a 45° line with equation $t = x - 0.5$.

   c) No characteristics, since $b^2 - 4ac < 0$. The differential equation is elliptic.

   e) Two characteristics through the point, with equations $t = \pm \sqrt{2x^2 - 0.5}$, $t = \pm \sqrt{0.25 - x^2}$.

12. $u(0.4272, 0.044720) = 0.009956$, $u(0.5272, 0.04472) = 0.009955$, $u(0.4553, 0.08944) = 0.01791$.

Chapter 12

2. 1.8329.   Region of fit is $1.0 \le x \le 2.5$, $0.2 \le y \le 0.4$.

4. $u(0.6, 1.0) = 54.96°$.

6. $f(2.3, 0.31) = 1.8403$.

10. Integral $= \frac{1}{24}$.

12. Integral $= \frac{1}{24}$.

13. b) Integral $\doteq 0.14059$.

16. a) 0.1007   b) 0.1068

Chapter 13

1. Sums of squares 584 and 794.   Maximum deviations 12.78 and -14.20.

3. $y = 1.908x + 0.0254$.

6. $y = 1.202 \exp(0.0096x)$.

8. Flow $= 30.2\, P^{0.491}$.

10. $y = 0.1036x^2 + 1.1831x + 0.9919$.   The quadratic fits the points more closely: maximum deviation -0.348 versus -0.566 in Problem 3, sums of squares of deviations 0.3829 versus 0.7834.

13.

$$\begin{bmatrix} 0.13 & -0.74 & 0.61 & 0 & 0 & 0 \\ 0.61 & 1.48 & 0.13 & 0 & 0 & 0 \\ 0 & 0.13 & 0.62 & 0.18 & 0 & 0 \\ 0 & 0 & 0.18 & 1.68 & 0.66 & 0 \\ 0 & 0 & 0 & 0.66 & 2.08 & 0.38 \\ 0 & 0 & 0 & 0.38 & -1.04 & 0.66 \end{bmatrix} S = 6 \begin{bmatrix} 0 \\ -0.07712 \\ -1.5178 \\ 1.5468 \\ 0.08169 \\ 0 \end{bmatrix}$$

16. $e^x = 1.000043 + 0.997396x + 0.499219x^2 + 0.177083x^3 + 0.04375x^4$.   At $x = 1$, error $= 0.0007908$.

18. Arctan $x \doteq \frac{121}{128}x - \frac{1}{8}x^3$, maximum error 0.035.   Maximum error of seventh-degree Maclaurin series is 0.061.

21. $y = \dfrac{2x - a - b}{b - a}$.

22. a) $\sin x \doteq (x - 7/60x^3)/(1 + 1/20x^2)$.   Maximum error on $[-1,1]$ is 0.00020; occurs at $x = \pm 1$.

   b) $\cos x$ cannot be expressed as a Padé approximation with third-degree numerator and denominator, since it is an even function.

24. a)
$$\cfrac{1}{1 + \cfrac{4}{x - 1 + \cfrac{1}{x - 1}}}$$

b) $2x + \cfrac{3}{1 - \cfrac{4}{x + 7/4 - \cfrac{19/16}{x + 5/4}}}$

26. For part (a), maximum error is 0.00020. Estimate of error is $1/2400 - 1/7! = 0.00022$.

index

index